Ceramics: Materials, Processing and Properties

Ceramics: Materials, Processing and Properties

Editor: Frank Shiner

NY RESEARCH
P R E S S

New York

Published by NY Research Press
118-35 Queens Blvd., Suite 400,
Forest Hills, NY 11375, USA
www.nyresearchpress.com

Ceramics: Materials, Processing and Properties
Edited by Frank Shiner

Cataloging-in-Publication Data

Ceramics : materials, processing and properties / edited by Frank Shiner.
    p. cm.
Includes bibliographical references and index.
ISBN 978-1-63238-866-7
1. Ceramics. 2. Building materials. 3. Chemistry, Technical. 4. Materials science. I. Shiner, Frank.
TP807 .C47 2022
620.14--dc23

Contents

Preface

A ceramic is a solid material which comprises of an inorganic compound. Such an inorganic compound can be made up of metal, non-metal or metalloid atoms primarily held in ionic and covalent bonds. Earthenware and brick are some common examples of ceramics. In terms of crystallinity, ceramic materials can range from completely amorphous to semi-crystalline, and highly oriented. Earthenware, stoneware and porcelain are examples of fired ceramics which are either vitrified or semi-vitrified. Most ceramic substances exhibit some common properties such as high melting temperature, low ductility, high moduli of elasticity, chemical resistance and poor conductivity. There are various options for the composition and structure of a ceramic due to the broad range of available combinations of elements, bonding and levels of crystallinity. Accordingly, the hardness and toughness of different ceramics can vary considerably. Different approaches, evaluations, methodologies and advanced studies on ceramics have been included in this book. Those with an interest in the field of ceramics would find this book helpful.

This book unites the global concepts and researches in an organized manner for a comprehensive understanding of the subject. It is a ripe text for all researchers, students, scientists or anyone else who is interested in acquiring a better knowledge of this dynamic field.

I extend my sincere thanks to the contributors for such eloquent research chapters. Finally, I thank my family for being a source of support and help.

Editor

Study of Direct Bonding Ceramics with Metal using Sn2La Solder

Roman Koleňák and Igor Kostolný

Faculty of Materials Science and Technology in Trnava, Slovak University of Technology in Bratislava, Paulínska 16, 917 24 Trnava, Slovakia

Correspondence should be addressed to Igor Kostolný; igor.kostolny@stuba.sk

Academic Editor: Wei Zhou

The aim of this research was to study the direct bonding of ceramic materials, mainly Al_2O_3 and selected metals, with primary attention given to Cu substrate. Soldering was performed with Sn-based solder alloyed with 2% La. We found that the bond formation between Sn2La solder and Al_2O_3 occurs at the activation of lanthanum phases in solder by ultrasound. Lanthanum in the solder becomes oxidised in air during the soldering process. However, due to ultrasonic activation, the lanthanum particles are distributed to the boundary with ceramic material. A uniformly thin layer containing La, 1.5 μm in thickness, is formed on the boundary with Al_2O_3 material, ensuring both wetting and joint formation. The shear strength with Al_2O_3 ceramics is 7.5 MPa. Increased strength to 13.5 MPa was observed with SiC ceramics.

1. Introduction

Sn-based solders are the most used solders in the electronic industry for surface mount technology and similar applications [1–3]. However, industrial production often requires also joining parts made of different material combinations such as ceramics/metal [4–6]. In soldering such joints, it is inevitable that the ceramic material is wetted with metallic solder.

Research is at present oriented toward direct bonding of ceramic materials by application of active solders [7–10]. This approach reduces the time required for joint fabrication, the hygiene of the working environment is improved, and the economy of production is also enhanced.

The direct bonding of copper with Al_2O_3 ceramics was established [11]. The joint was fabricated by ultrasonic soldering with application of Zn14Al solder. High quality joints without visible defects were achieved at optimised soldering parameters. The highest shear strength of Al_2O_3/Zn14Al/Cu joint, 80 MPa, was achieved at a soldering temperature of 480°C and soldering time of 30 s.

In another study [12], the direct soldering of SiC ceramics was performed with ultrasound assistance. The SiC ceramic substrates were soldered in air with Zn8.5Al1Mg solder at a temperature of 420°C. The highest strength (148.1 MPa) was achieved with ultrasound acting for 8 s. A new amorphous layer 2 to 6 nm in thickness was formed in the interface between the solder and substrate. The strong bond between SiC substrate and Zn-Al-Mg solder is attributed to transfer of SiO_2 mass to Zn-Al-Mg solder due to cavitation erosion.

The subject of study [13] was oriented to direct soldering of sapphire by ultrasound with application of Sn10Zn2Al solder. It was found that ultrasound supported the oxidation reaction between Al from the solder and sapphire substrate. A nanocrystalline layer of α-Al_2O_3 (2 nm in thickness) was formed in the Sn-Zn-Al/sapphire interface at soldering in air. This layer allowed the bond formation. The shear strength of joints achieved 43 to 48 MPa, which is a relatively high value when compared to other joints of Al_2O_3 ceramics fabricated with active Sn solders with addition of Ti and/or lanthanides.

The work [14] deals with direct soldering of Al_2O_3 by use of an active Sn solder type Sn3.5Ag4Ti(Ce,Ga). Soldering was performed in air at a temperature of 280°C. Ultrasound with a frequency of 40 kHz was employed for solder activation. The shear strength of Al_2O_3/Sn-AgTi/Al_2O_3 joint was 24 MPa. A reaction layer, 4–7 μm in thickness, was formed on the interface of Al_2O_3/Sn-AgTi joint, formed with Ti oxides, mainly TiO. This layer ensured the wettability of Al_2O_3 ceramics.

The direct soldering of ITO (indium tin oxide) ceramics with a copper substrate, performed with the same Sn3.5Ag4Ti(Ce,Ga) solder in air at temperature of 250°C, was solved in work [15]. The molten solder was agitated by mechanical activation for 30 s. Line scanning has shown that Ti was segregated in the boundary between the ceramics and solder. This new layer is responsible for bond formation. The shear strength of ITO/Cu joints was 3.4 MPa. The Cu/Cu and ITO/ITO joints attained shear strengths of 14.3 MPa and 6.8 MPa, respectively.

In work [16], the authors have dealt with soldering of Al_2O_3/Al_2O_3 and Al_2O_3/Cu at a temperature of 250°C in air. The molten Sn3.5Ag4Ti(Ce,Ga) solder was agitated for 30 s for wetting on bond surfaces and then a copper or alumina specimen was placed on the molten solder to be joined with an alumina specimen by rubbing together for 30 s. The affinity of Ce to oxygen prevents Ti from oxidising and thus Ti can react with Al_2O_3 at a low temperature. The shear strength of alumina/alumina, copper/copper, and alumina/copper joints was 13.5 MPa, 14.3 MPa, and 10.2 MPa, respectively.

In spite of a great number of positive properties, which Sn solder attains by the addition of rare earth elements to alloy, some negative phenomena also occurred, namely, the formation of Sn whiskers. A wide group of authors [19–23] dealt with this issue. The growth of whiskers is, by the opinion of many authors, caused by the oxidation on solder surfaces and due also to internal stress. Author [24] stated that the stress induced by formation of intermetallic Cu_6Sn_5 compound forces the Sn atoms out of the outer surface oxide of Sn layer. By author [25], the growth of Sn whiskers is very slow in most cases. Regarding the fact that the rare earth elements have a high chemical potential and that they react much easier with oxygen, they thus enhance the growth of Sn whiskers in solders containing these elements.

As is obvious from the studies mentioned, the active element is an essential component of the solder, since it ensures wettability and bond formation between the metallic solder and ceramic material.

Many research works dealt with solder where titanium was used as an active element [26–29]. Active solders containing Ti (up to 4 wt.%) and small trace amounts of lanthanides such as Ce [18, 30–32] were also studied. Cerium and other lanthanides in those solders contained around 0.1 to 0.2 wt.% in trace amounts. The authors in the mentioned works did not study whether the presence of lanthanides without the presence of Ti can ensure the wettability of ceramic substrates such as Al_2O_3 and thus create a direct bonding with the ceramic material.

The aim of our study was oriented toward the direct bonding of Al_2O_3 ceramics with copper substrate with the application of Sn2La solder. We studied whether Sn-based solder, alloyed with La, can wet Al_2O_3 ceramics and form a strong bond. We also studied whether La can substitute Ti in the active Sn solders.

For this purpose, the analyses were performed to reveal the mechanism of bond formation and the shear strength

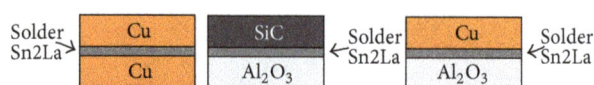

FIGURE 1: Analysed combinations of Cu/Cu, SiC/Al_2O_3, and Cu/Al_2O_3 materials.

of fabricated joints was also measured. Lanthanum, which exerts a high affinity to oxygen, was applied as an active element in this case. The amount of 2 wt.% La was selected for comparison, since in the previous study [26] we dealt with interactions of Sn2Ti solder with the surface of Al_2O_3 ceramics. Soldering was performed at a low temperature in air with application of high-power ultrasound.

2. Experimental

Sn solder with 2 wt.% La was used in the experiments. The solder was manufactured by casting in form of an ingot. Weighing single solder components was done after setting the weight ratio of prepared alloys. Components with high purity from 3 N to 5 N were used for solder fabrication. The manufacture was performed in horizontal tube vacuum furnace with resistance heating. The working temperature used during manufacture was 900°C and vacuum of 10^{-4} Pa was also employed. At this temperature, held up within the time of 20 min, homogenization of soldering alloy took place. Cooling down in vacuum furnace was slow. The cooling rate was 14°C/min.

The substrates of the following materials were used in experiments:

(i) Metallic Cu substrate of 4 N purity in the shape of rings in dimensions Φ 15 × 1.5 mm.

(ii) Ceramic Al_2O_3 substrate of 5N purity in the form of Φ 15 × 2 mm rings (manufacturer Glynwed, GmbH, designation Degussit Al23).

(iii) Ceramic SiC substrate in form of Φ 15 × 3 mm rings (manufacturer CeramTec, GmbH, designation Rocar SiC).

For a more detailed analysis, the material combinations given in Figure 1 were selected.

Soldering was performed on Hanuz UT2 ultrasonic equipment with parameters given in Table 1. Solder activation was realised by using an encapsulated ultrasonic transducer consisting of a piezoelectric oscillation system and titanium sonotrode with an end diameter of Φ 3 mm. The scheme of ultrasonic soldering through the layer of molten solder is shown in Figure 2. The soldering temperature was 20°C above the liquid temperature of the solder. Soldering temperature was controlled by a continuous temperature measurement on hot NiCr/NiSi plate by a thermocouple.

Soldering proceeded as follows: solder layer was deposited on the substrate heated at a soldering temperature. The liquid solder was then subjected to ultrasound activation in air for 5 s. After ultrasonic activation, the excessive layer of molten solder and formed oxides were removed from

TABLE 1: Soldering parameters.

Ultrasound power	[W]	400
Working frequency	[kHz]	40
Amplitude	[μm]	2
Soldering temperature	[°C]	290
Time of ultrasound activation	[s]	5

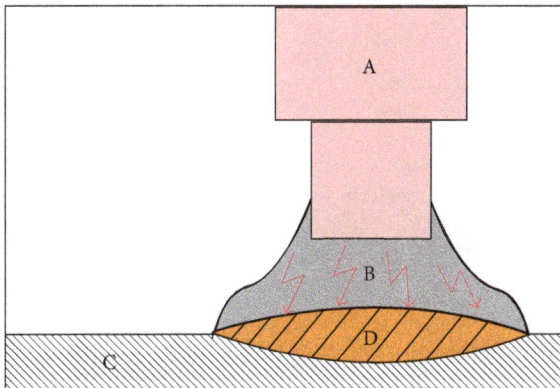

(A) Ultrasonic tool   (D) Reaction layer
(B) Solder   ⚡ Ultrasonic wave
(C) Substrate

FIGURE 2: Ultrasonic soldering through the layer of molten solder.

the substrate surface. Both joined surfaces were prepared in this way. The substrates with deposited layer of molten solder were laid on each other to obtain contact with the liquid phase. Then, they were centred and the joint formed by their slight compression. Graphical representation of this procedure is shown in Figure 3.

Metallographic preparation of specimens from soldered joints was realised by the standard metallographic procedures. Grinding was performed by use of SiC emery papers with granularity of 240, 320, and 1200 grains/cm². Polishing was performed by use of diamond suspensions with grain sizes: 9 μm, 6 μm, and 3 μm. Final polishing was made by use of polishing emulsion OP-S (Struers) with 0.2 μm grain size.

Solder microstructure was studied using

(i) light optical microscope type Neophot 32 with the application of image analyser NIS-Elements, type E,

(ii) electron scanning microscopy (SEM) on FEI Quanta 200 FEG microscope,

(iii) qualitative and semiquantitative chemical analysis of Sn2La solder performed on JEOL 7600 F equipment with microanalyser type Microspec WDX-3PC.

X-ray diffraction analysis was used for identification of phase composition of solder. It was realised with a solder specimen in dimensions of 10×10 mm on XRD diffractometer type PANalytical X'Pert PRO.

DSC analysis of Sn2La solder was performed on equipment type Netzsch STA 409 C/CD in shielding with Ar gas of 6 N purity.

Shear test was carried out for determination of shear strength of soldered joints. Measurements were realised on two ceramic (Al_2O_3 and SiC) and five metallic (Al, Ni, Ti, Cr-Ni steel, and Cu) materials with Sn2La solder. The shear strength was measured on versatile tearing equipment type LabTest 5.250SP1-VM. An especially developed shearing jig was used for changing the direction of tensile loading forces acting on the specimen. This shear jig assures a uniform loading of the specimen in shear in the plane boundary between the solder and substrate (Figure 4). The dwell time at a soldering temperature during specimen fabrication was 30 s and the ultrasound period was 5 s.

3. Experimental Results

3.1. Analysis of Sn2La Solder. Figure 5 shows the macrostructure of Sn2La solder on ingot cross section. Clearly visible La phases, uniformly distributed in tin matrix, may be seen. The uniform distribution of La phases in tin matrix can be seen also on solder microstructure shown in Figure 6. A detailed view of La phase may be seen in Figure 7. No La was observed in the matrix of solder studied. Solder matrix is formed of pure tin. This was proved by EDS analysis.

The results of XRD analysis in Figure 8 prove that the solder matrix is formed of pure tin, where lanthanum phases, $LaSn_3$, also occur. Similarly, the presence of $LaSn_3$ phase was proved by EDS analysis. The point analysis of $LaSn_3$ phase in the solder is documented in Figure 9.

The authors' opinion in [17] and the binary La-Sn system also proves the presence of $LaSn_3$ phase at lanthanum content 2 wt.% La (Figure 10).

DSC analysis was performed to identify the melting point of Sn2La solder. Measurements were taken in three subsequent cycles called RUN1, RUN2, and RUN3. DSC record of Sn2La solder is shown in Figure 11. From the curve course, it is obvious that the start of solder melting is at temperature 232.1°C, which approximately corresponds to the melting point of pure tin. The temperature peak is at 237.2°C. According to the binary La-Sn diagram in Figure 10, the eutectic phase transformation at temperature 235 ± 2°C is concerned. The solder is already fully molten at temperature 243.6°C. The presence of 2 wt.% La in the tin solder matrix resulted in a shift of melting temperature of Sn2La by approximately 5°C, compared to the melting point of pure tin.

3.2. Analysis of Boundary of Al_2O_3/Sn2La Soldered Joint. Solder microstructure obtained by SEM after soldering with power ultrasound on Al_2O_3/Sn2La solder boundary is shown in Figure 12. Since the soldering was performed in air, the La containing phases oxidised. Lanthanum is concentrated in great, considerably oxidised phases in size from 5 to 60 μm, as is clearly visible in Figure 12. Some phases are with sharp edges (Figure 14). However, this phenomenon was not observed in case of soldering Cu/Sn2La solder/Cu

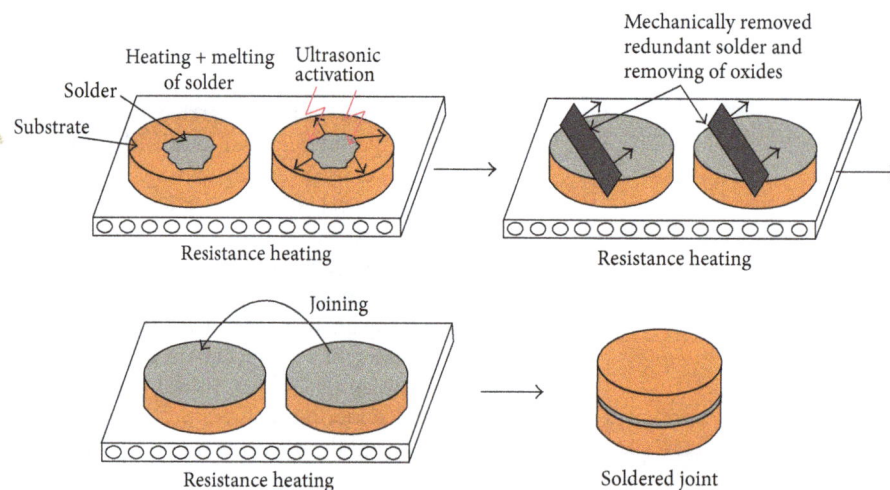

FIGURE 3: Procedure of joint fabrication by ultrasonic soldering.

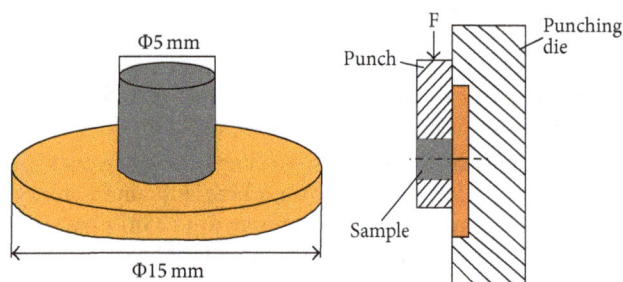

FIGURE 4: Test specimen for shear test and scheme of specimen in a jig during shear strength testing.

FIGURE 5: Macrostructure of Sn2La solder ingot.

joint. In this case, La remains uniformly distributed in the solder matrix structure. Great oxidised La phases scarcely occur. These facts suggest that formation of great oxidised La phases is related to La interaction with the surface of Al_2O_3. Comparison of microstructures of soldered Al_2O_3/SiC and Cu/Cu joints from optical microscopy is documented in Figure 13.

During the course of study with bounced electrons (BSE), lanthanum should glow, but due to the fact that it is enriched by oxygen, the La containing particle is dark (Figure 14).

Oxidised lanthanum particles from Figure 14 contain 9.56 wt.% O_2, 18.20 wt.% Sn, and 72.24 wt.% La. Point analysis of sharp-edged lanthanum particle is documented in Figure 15.

Solder matrix in $Al_2O_3/Sn2La$ soldered joint is composed of almost pure tin (the accuracy of EDS analysis should be taken into account). The point analysis of matrix is documented in Figure 16. The matrix is composed of fine-grained tin structure.

As was shown, considerable amount of lanthanum is oxidised during soldering in the air. The lanthanum particles are distributed to the boundary with ceramic material owing to the effect of ultrasonic activation (Figures 17 and 18). The concentration line of La in Figure 17 proves increased La concentration on the boundary with Al_2O_3 ceramics. In Figure 18, one can see a uniform continuous layer of lanthanum particles on the boundary with ceramic material, which ensures the joint formation. This layer is around 1.5 μm thick. In spite of this layer, the solder is more or less adhered to the ceramic substrate. The bond with ceramic material is of adhesive character. No formation of intermetallic phases was observed, which also causes lower shear strength of joints with ceramic materials.

3.3. Boundary Analysis of Cu/Sn2La/Cu Soldered Joint. When soldering copper substrate with Sn2La solder, it was found that the solder matrix is also formed of pure tin, similarly as in cases of joints with Al_2O_3 ceramics. Redistribution of greater constituents containing lanthanum (Figure 19) was caused by ultrasound activation. The solder structure in Cu/Sn2La/Cu joints differs from the $Al_2O_3/Sn2La/Al_2O_3$ joints. Lanthanum in the case of the Cu/Sn2La/Cu joint was uniformly distributed on the grain boundaries of tin in the form of very fine $LaSn_3$ phases (Figures 19 and 20). The particles

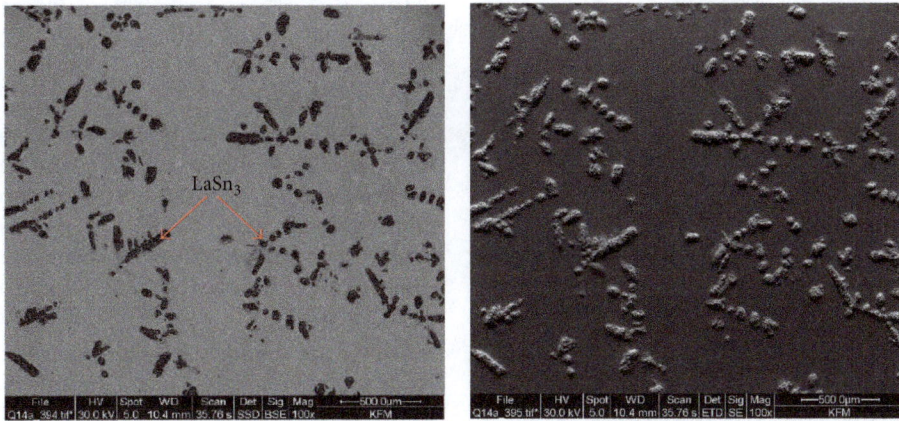

FIGURE 6: Microstructure of Sn2La solder.

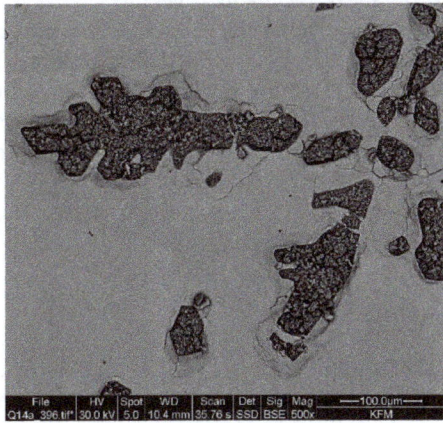

FIGURE 7: Detail of La phase in Sn2La solder matrix.

FIGURE 8: Diffraction records of Sn2La solder.

kV: 20.0 Tilt: 0.0 Take-off: 34.9 Det type: SDD Apollo XP Res: 134
Amp. T: 25.60 FS: 2216 Lsec: 30

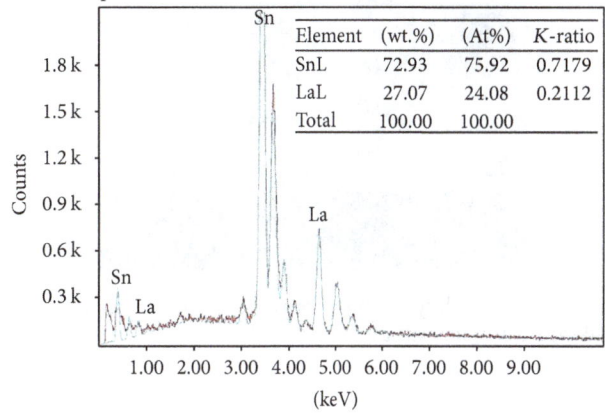

Element	(wt.%)	(At%)	K-ratio
SnL	72.93	75.92	0.7179
LaL	27.07	24.08	0.2112
Total	100.00	100.00	

FIGURE 9: Point analysis of $LaSn_3$ composition in Sn2La solder.

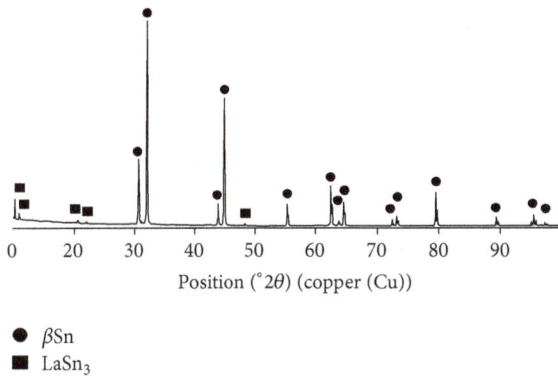

FIGURE 10: Binary La-Sn diagram [18].

of this phase are very fine with their size ranging from 0.25 to 0.35 μm. It is surprising that due to La presence in solder matrix the copper is not dissolved in tin matrix. Normally, the Cu concentration in Sn should be almost 1 wt.%.

The intermetallic Cu_6Sn_5 and Cu_3Sn phases were formed on the boundary of copper substrate and tin solder during ultrasonic soldering. Fine arches of those intermetallic phases

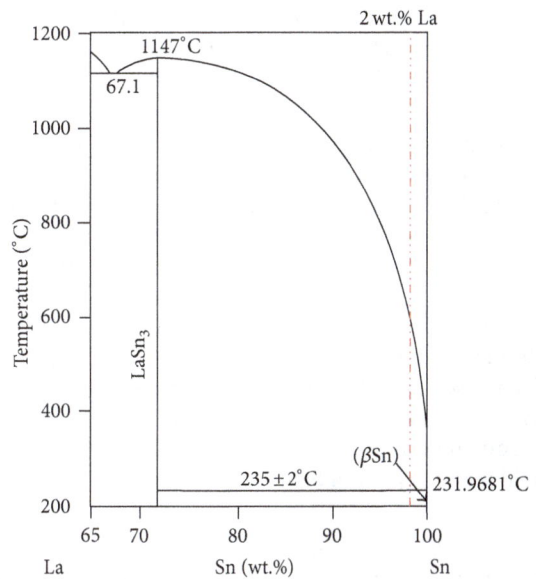

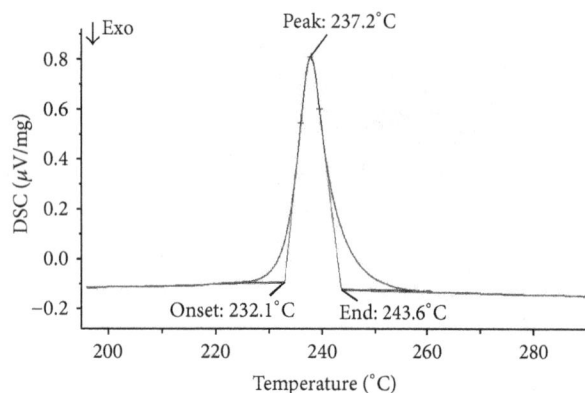

FIGURE 11: DSC analysis of Sn2La solder.

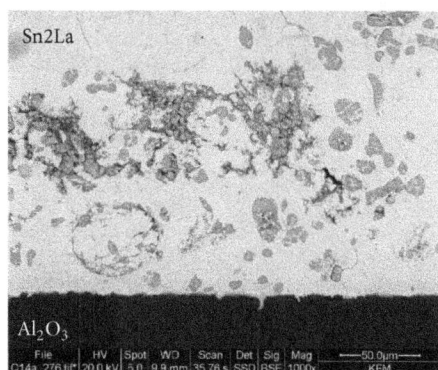

FIGURE 12: Boundary microstructure of Al$_2$O$_3$/Sn2La joint after soldering with active ultrasound.

with maximum thickness of $4\,\mu m$ are seen in Figure 20. The Cu$_3$Sn phase is very thin and therefore less visible. It was identified by EDS analysis. From these results, we may conclude that La in the solder prevents dissolution of Cu in tin solder and hinders the growth of intermetallic Cu$_6$Sn$_5$ and Cu$_3$Sn phases, which is advantageous from the viewpoint of the life of joints subjected to thermal cycling.

Contrary to Al$_2$O$_3$/Sn2La joint, also greater constituent of oxidised lanthanum phases seldom occurs in cases of Cu/Sn2La joints in solder matrix. The incidental oxidised La phase may be observed in Figures 20 and 21. The concentration profiles of La, Cu, Sn, and O elements in Figure 21 show that the greatest oxygen amount, thus oxides, occurs in the lanthanum particle. In Figures 19, 20, and 21, purely visible uniformly distributed fine particles of LaSn$_3$ phase are seen.

Beside the formation of transient metallic Cu$_6$Sn$_5$ and Cu$_3$Sn phases, we also observe planar distribution of La, O, and Sn elements in sharp-edged oxidised particle with high lanthanum contents on the planar analysis of soldered Cu/solder Sn2La joint in Figure 22.

3.4. The Results of Shear Strength in Soldered Joints. Research in this study was primarily oriented to soldering ceramic Al$_2$O$_3$ substrate and copper substrate. These experiments

determining shear strength of soldered joints were also extended to other metallic materials (Al, Ni, Ti, and CrNi steel) and SiC ceramics in order to show the wider applicability of Sn2La solder.

Measurements were performed on 4 specimens of each material. The resulting average shear strengths are documented in Figure 23. The lowest shear strength was observed on Al$_2$O$_3$ ceramics (7.5 MPa). Somewhat higher strength (13.5 MPa) was observed on SiC ceramics and the highest was achieved on Al and Ni. The shear strength on copper substrate was 26.0 MPa.

The Sn2La solder showed relatively wide variables in shear strength on metallic and ceramic materials. We found the shear strength of joints in metallic materials to be nearly 3 times higher than that in case of ceramic materials. For more correct identification, the fractured surfaces of Cu/Sn2La and Al$_2$O$_3$/Sn2La joints (Figures 24 and 25) were also analysed.

The Cu/Sn2La joint showed formation of a typical ductile failure by shear mechanism (Figure 24). The fracture morphology clearly shows the motion of the shearing tool. Figure 25 shows the documented fractured surfaces of Al$_2$O$_3$/Sn2La joint. Fracture surface is without visible tracks of shear tool, so the failure occurred by tearing without a shearing mechanism. 100% Sn2La solder covering remained on the Cu substrate after the test. On the opposite, Sn2La solder was torn from the ceramic substrate. The Al$_2$O$_3$ substrate remained only partially covered by Sn2La.

4. Discussion

Lanthanum present in form of LaSn$_3$ in the volume of Sn2La solder is inefficient at soldering temperature 290°C and therefore it does not wet the ceramic material. Powerful ultrasound is thus necessary for its activation. At direct ultrasound activation through the layer of molten solder (Figure 2), the solder is extremely vibrated by titanium sonotrode. This allows significant diffusion of active La element from the solder matrix to boundary with ceramic and/or metallic material. This depends on mechanism of joint formation. In the zone of ultrasound impact, the temperature may be locally increased even up to 1000°C and more [33]. This enhances the physicochemical processes at joint formation.

The experiments have also shown that pure tin without addition of active element (La) does not wet Al$_2$O$_3$ material. The solder composed of pure tin is after melting first shrunk into globular form. Only fine particles of tin and tin oxides are formed by long-time acting of ultrasound, which are then sputtered over the surface of ceramic material without bond formation.

The results of direct soldering of ceramic and metallic materials have proven that Sn solder containing La ensures their wettability at application of ultrasonic activation, meaning that Sn2La solder is suitable for practical applications.

We suppose from the results of analysis performed in transition zone of soldered joints that the joint with a metallic material is of metallurgical-diffusion character. However, the joint with ceramic material, namely, Al$_2$O$_3$, is of adhesive character, when soldered with solder containing La.

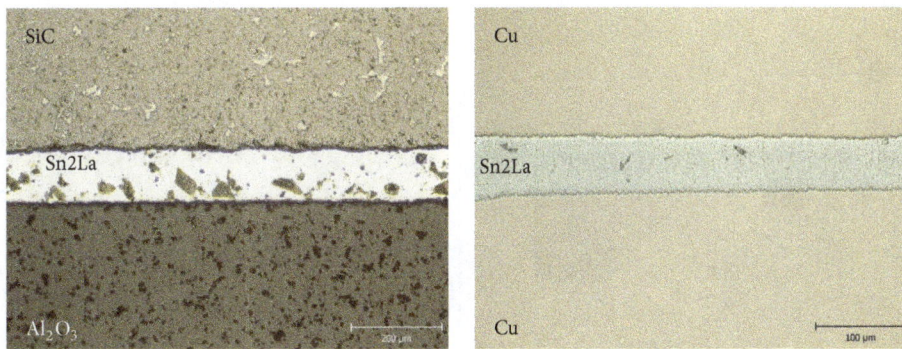

FIGURE 13: Comparison of microstructure of Sn2La solder after UT soldering at the same parameters and conditions of soldering.

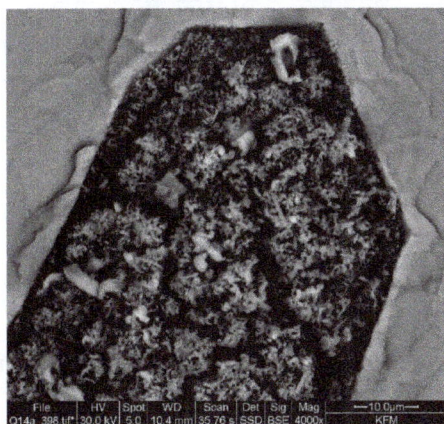

FIGURE 14: Sharp-edged oxidised lanthanum phase in Al_2O_3/Sn2La/ Al_2O_3 joint.

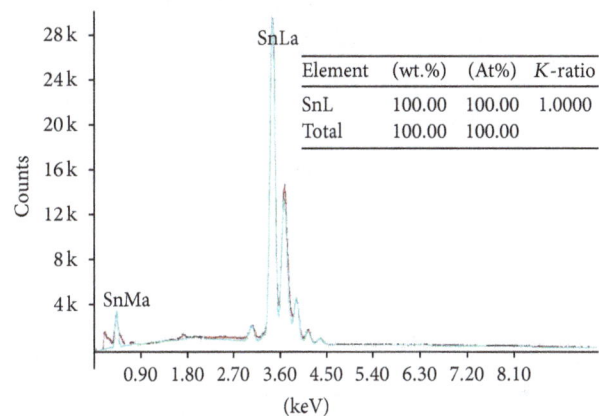

FIGURE 16: Point analysis of solder matrix in Al_2O_3/Sn2La solder joint.

kV: 30.0 Tilt: 0.3 Take-off: 35.6 Det type: SDD Apollo XP RES: 157 Amp. T: 6.40 FS: 4121 Lsec: 39

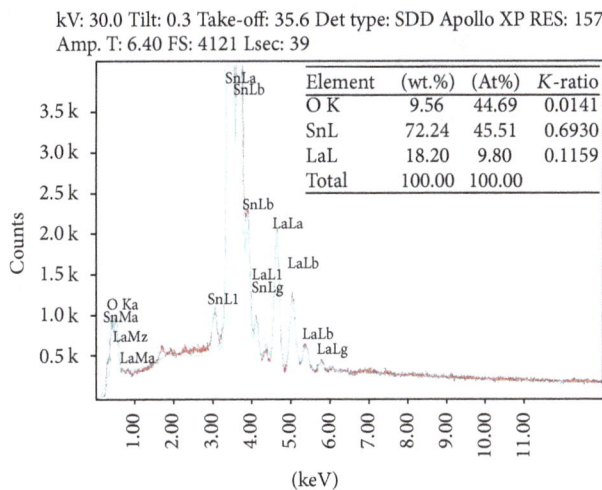

Element	(wt.%)	(At%)	K-ratio
O K	9.56	44.69	0.0141
SnL	72.24	45.51	0.6930
LaL	18.20	9.80	0.1159
Total	100.00	100.00	

FIGURE 15: Point analysis of sharp-edged lanthanum particle.

A new transition layer with thickness around 1.5 μm was observed on the boundary with the ceramic material. For example, in study [10, 11], a transition layer 2 to 6 nm was formed at application of Zn-Al (Zn-Al-Mg, Sn-Zn-Al) based solders. At application of Sn-Ag-Ti based solders, described in

work [12], the transition layer in thickness from 4 to 7 μm was formed. The studies [12–14] proved that this transition layer ensures the joint formation at direct soldering of ceramic materials.

In case of application of Sn2La solder for soldering Al_2O_3 ceramics, we observed no formation of soldered joint of diffusion character with creation of new products. The shear strength of Al_2O_3/Sn2La joint was 7.5 MPa, shear strength of SiC/Sn2La joint was 13.5 MPa, and shear strength of Cu/Sn2La joint was 26.0 MPa.

For comparison of results of shear strength, we also give the results from similar studies, while it must be taken into account that different researchers are using different test methods, shape of specimens, and loading rate during testing. They are also using different compositions of soldering alloys and different soldering parameters. For example, at application of solders type Zn-Al (Zn14Al) in work [11] on Al_2O_3/Zn14Al/Cu joint, the achieved shear strength was 80 MPa at ultrasound power of 200 W. In study [12], the joint of ceramic SiC substrates soldered with Zn8.5Al1Mg solder achieved shear strength of 148.1 MPa at ultrasound acting for 8 s.

In study [13], sapphire was soldered with ultrasound assistance at application of Sn10Zn2Al solder. The shear strength

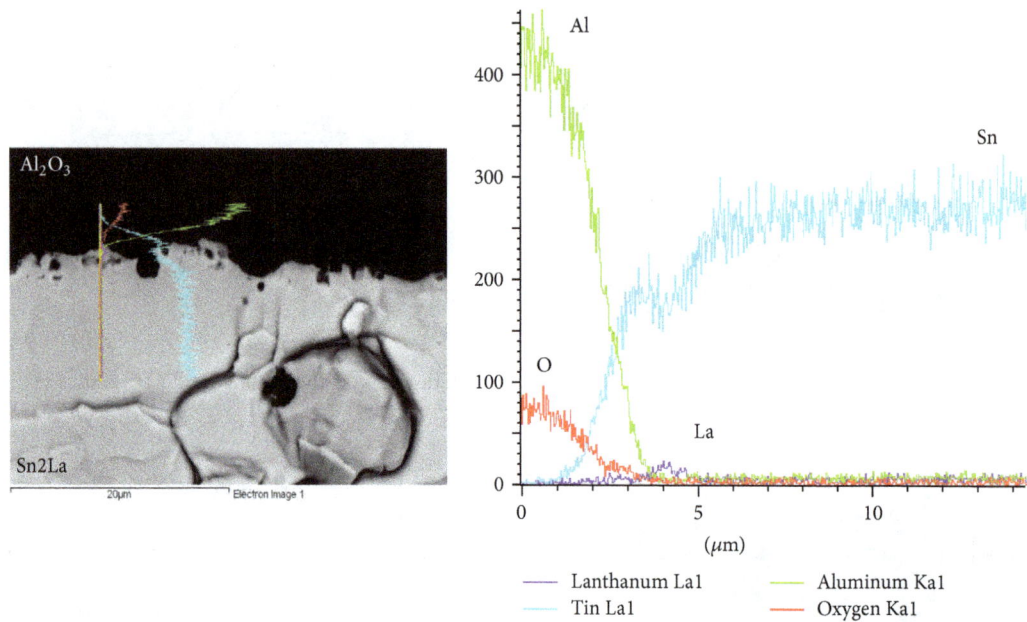

FIGURE 17: Concentration profiles of Al, Sn, La, and O_2 on Al_2O_3/Sn2La solder boundary.

TABLE 2: Comparison of achieved shear strength of Al_2O_3/solder joints.

Substrate	Solder alloy	Shear strength [MPa]
	Sn2La	7.5
	Zn14Al [11]	80
Al_2O_3	Sn10Zn2Al [13]	43–48
	Sn3,5Ag4Ti(Ce,Ga) [14]	13.5
	Sn3,5Ag4Ti(Ce,Ga) [16]	24

of joints in that case attained 43 to 48 MPa. Several studies dealt with type Sn-Ag-Ti solders. The new metallic, ceramic, and nonmetallic materials were also tested. For example, in study [16], the Al_2O_3/Sn-Ag-Ti/Al_2O_3 joint exerted the shear strength of 24 MPa. In [15], the following strength values were achieved: Cu/Cu (14.3 MPa), ITO/ITO (6.8 MPa), and ITO/Cu (3.4) MPa. Similar examples are also mentioned in study [14], where shear strength of alumina/alumina joint was 13.5 MPa, copper/copper 14.3 MPa, and alumina/copper 10.2 MPa.

A brief overview of shear strength of Al_2O_3/solder joints achieved by mentioned authors is documented in Table 2.

5. Conclusions

The aim of this work was to study direct bonding of Al_2O_3 ceramics with copper substrate. We studied whether Sn-based solder alloyed with La can wet the Al_2O_3 ceramics and other ceramic materials and thus form strong joints. For this reason, analyses were performed to reveal the mechanism

of joint formation and also measure the shear strength of fabricated joints. The following results were achieved:

(i) DSC analysis has shown that Sn2La solder has a melting point of 273.2°C, which is approximately 5°C more than the melting temperature of pure tin.

(ii) The matrix of Sn2La solder after soldering Al_2O_3 ceramics as well as the copper substrate is formed of almost pure tin, where fine particles of $SnLa_3$ phase are segregated along the grain boundaries.

(iii) The bond of Sn2La solder with Cu substrate at ultrasound application occurs due to formation of fine bridges of Cu_6Sn_5 and Cu_3Sn phases on the Cu substrate/Sn2La solder boundary. Thickness of these transition phases is extremely low (max. 4 μm). Comparison has shown that the growth of layer of transition intermetallic phases is prevented by fine La phases precipitated along the grain boundaries.

(iv) Bond formation between Sn2La solder and Al_2O_3 substrate occurs at activation of lanthanum phases by ultrasound. Lanthanum contained in the solder volume is oxidised on the air during soldering. The lanthanum particles are distributed to boundary of ceramic material owing to ultrasonic activation. Lanthanum particles are then bound with surface oxides on ceramic material. A thin uniform layer containing La is formed on the boundary with ceramics.

(v) The lowest shear strength was observed on Al_2O_3 ceramics (7.5 MPa). Somewhat higher strength was achieved on SiC ceramics (13.5 MPa). The highest strength within the metallic materials was achieved

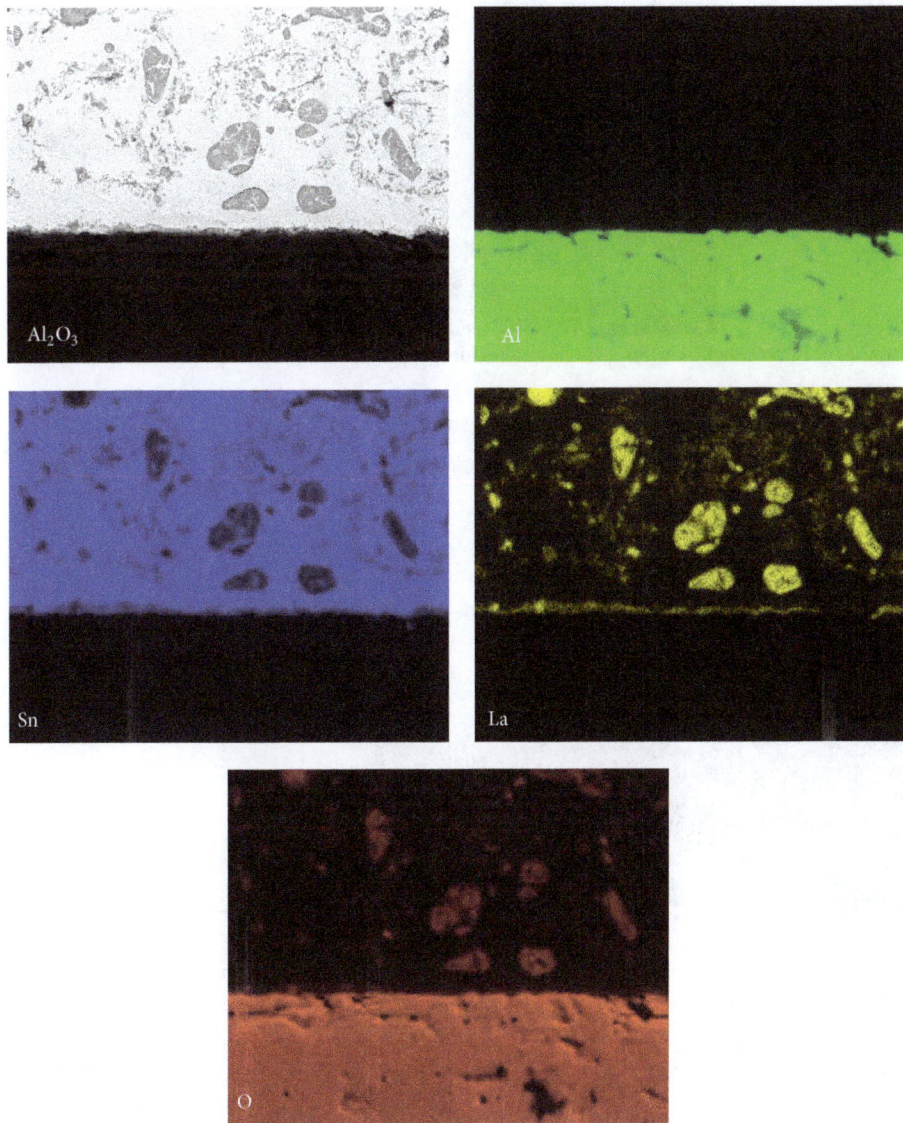

FIGURE 18: Planar EDX analysis of soldered Sn2La/Al$_2$O$_3$ boundary.

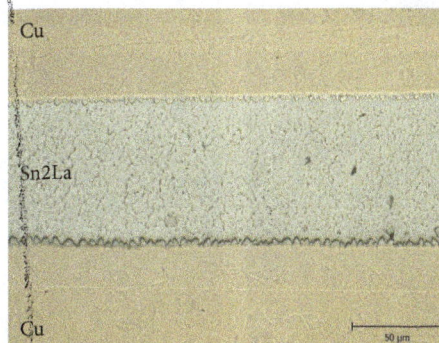

FIGURE 19: Microstructure of Cu/Sn2La/Cu joint from the light optical analysis.

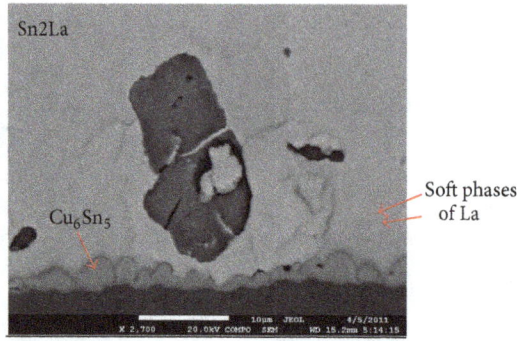

FIGURE 20: Fine arches of intermetallic Cu_6Sn_5 phases and incidental greater oxidised La phase in tin matrix of Cu/Sn2La joint.

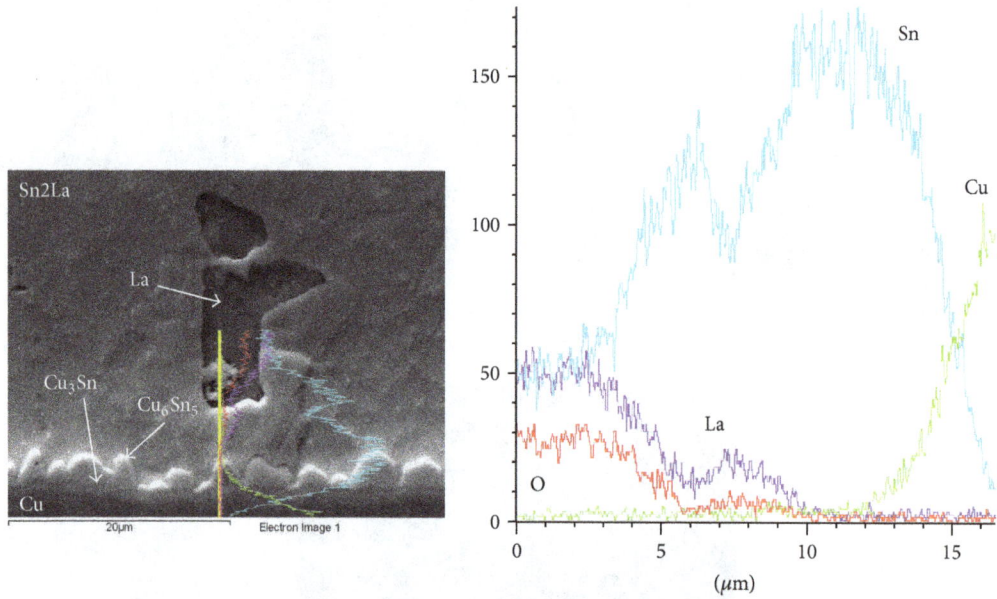

FIGURE 21: Microstructure of Cu/Sn2La boundary (SEM) with concentration profiles of La, Cu, Sn, and O elements.

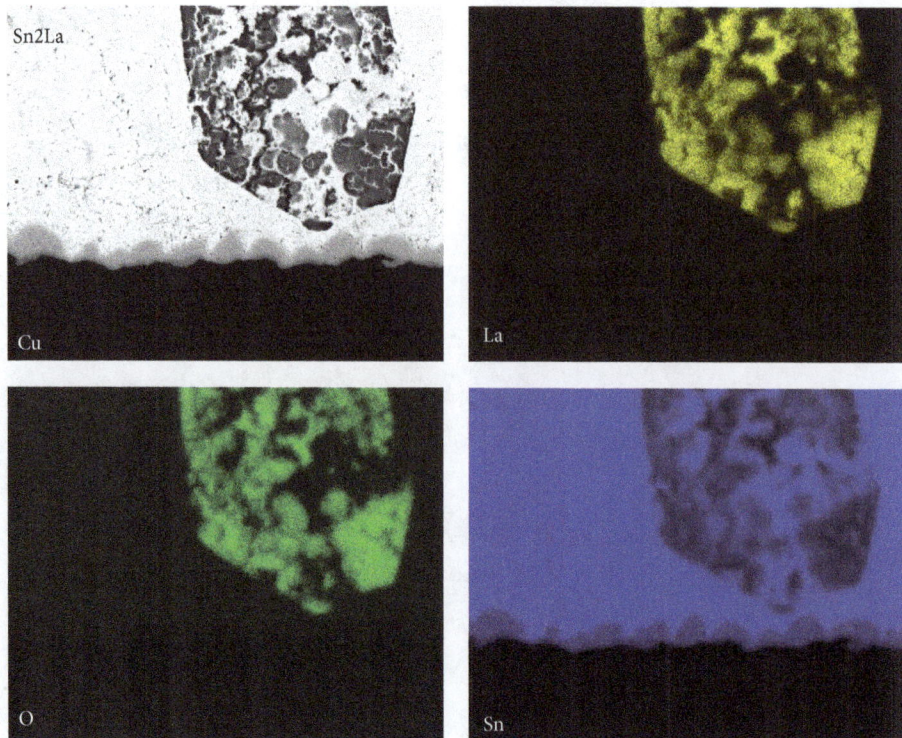

FIGURE 22: Planar analysis of Cu/Sn2La joint boundary with a sharp-edged lanthanum particle.

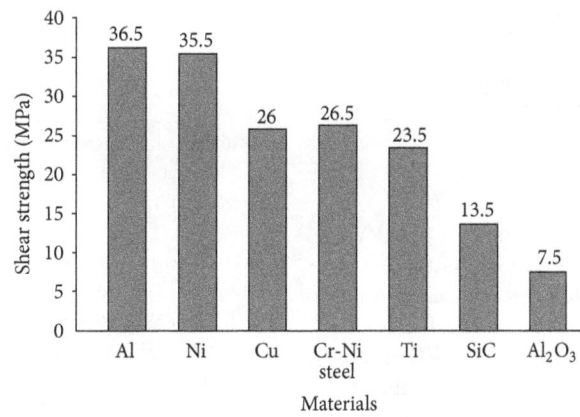

FIGURE 23: Measured results of shear strength of joints fabricated with Sn2La solder.

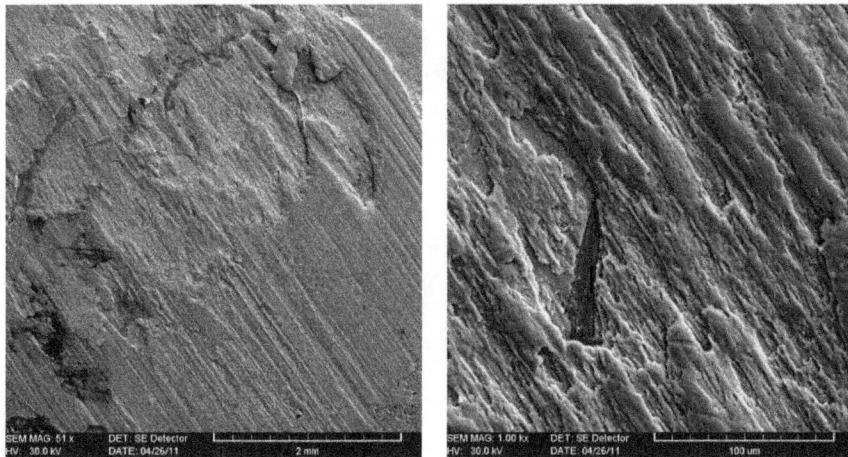

FIGURE 24: Fractured surface of Sn2La/Cu joint.

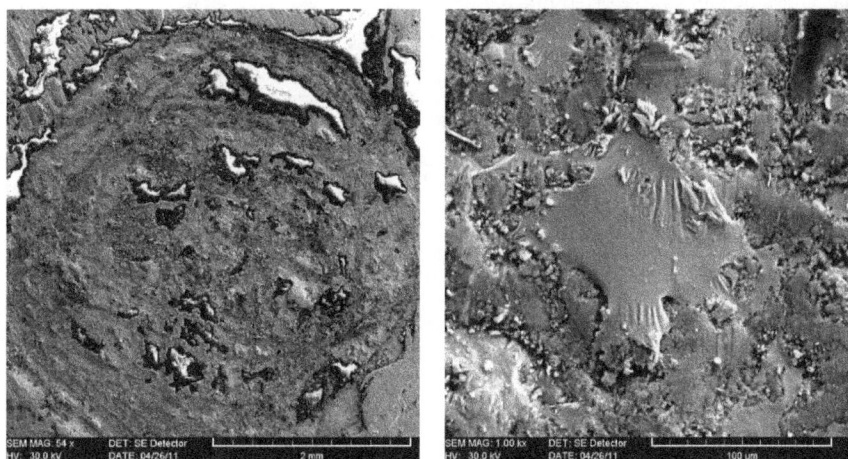

FIGURE 25: Fractured surface of Sn2La/Al$_2$O$_3$ joint.

on Al and Ni (around 35 MPa). The strength attained on copper substrate was 26.0 MPa.

(vi) From the results of analysis of transition zones in soldered joints, measurements of shear strength, and fractographic analysis of fractured surfaces, we suppose that the bond with metallic material is of metallurgical-diffusion character. The bond with ceramic material, namely, Al$_2$O$_3$, soldered with La containing solder is of an adhesive character.

Conflict of Interests

The authors declare that there is no conflict of interests regarding the publication of this paper.

Acknowledgments

The contribution was prepared with the support of APVV-0023-12, research of new soldering alloys for fluxless soldering with application of beam technologies and ultrasound, and VEGA 1/0455/14, research of modified solders for fluxless soldering of metallic and ceramic materials. The authors thank Ing. Marián Drienovský, Ph.D., for DSC analysis; Associate Professor Ing. Maroš Martinkovič, Ph.D., for shear strength measurements, Ing. Jiří Faltus, CSc., for EDS analysis, and Ing. Ivona Černičková, Ph.D., for EDX analysis.

References

[1] C. H. Liu, Y. J. Kim, D. W. Chun et al., "Universal solders for direct bonding and packaging of optical devices," *Materials Letters*, vol. 152, pp. 232–236, 2015.

[2] A. A. El-Daly, A. M. El-Taher, and S. Gouda, "Development of new multicomponent Sn-Ag-Cu-Bi lead-free solders for low-cost commercial electronic assembly," *Journal of Alloys and Compounds*, vol. 627, pp. 268–275, 2015.

[3] E. H. Amalu, W. K. Lau, N. N. Ekere et al., "A study of SnAgCu solder paste transfer efficiency and effects of optimal reflow profile on solder deposits," *Microelectronic Engineering*, vol. 88, no. 7, pp. 1610–1617, 2011.

[4] J. Lemus-Ruiz, A. O. Guevara-Laureano, J. Zarate-Medina, A. Arellano-Lara, and L. Ceja-Cárdenas, "Interface behavior of Al_2O_3/Ti joints produced by liquid state bonding," *Applied Radiation and Isotopes*, vol. 98, pp. 1–6, 2015.

[5] W.-Y. Yu, S.-H. Liu, X.-Y. Liu, M.-P. Liu, and W.-G. Shi, "Interface reaction in ultrasonic vibration-assisted brazing of aluminum to graphite using Sn–Ag–Ti solder foil," *Journal of Materials Processing Technology*, vol. 221, pp. 285–290, 2015.

[6] T. Zaharinie, R. Moshwan, F. Yusof, M. Hamdi, and T. Ariga, "Vacuum brazing of sapphire with inconel 600 using Cu/Ni porous composite interlayer for gas pressure sensor application," *Materials and Design*, vol. 54, pp. 375–381, 2014.

[7] A. P. Xian, "Wetting of Si-Al-O-N ceramic by Sn-5at.%Ti-X ternary active solder," *Materials Science and Engineering: B*, vol. 25, pp. 39–46, 1994.

[8] Y. H. Liu, J. D. Hu, P. Shen, X. H. Han, and J. C. Li, "Microstructural and mechanical properties of jointed ZrO_2/Ti-6Al-4V alloy using $Ti_{33}Zr_{17}Cu_{50}$ amorphous brazing filler," *Materials and Design*, vol. 47, pp. 281–286, 2013.

[9] A. V. Durov, B. D. Kostjuk, A. V. Shevchenko, and Y. V. Naidich, "Joining of zirconia to metal with Cu-Ga-Ti and Cu-Sn-Pb-Ti fillers," *Materials Science and Engineering A*, vol. 290, no. 1, pp. 186–189, 2000.

[10] M. Wu, C.-Z. Cao, R. Ud-Din, X.-B. He, and X.-H. Qu, "Brazing diamond/Cu composite to alumina using reactive Ag-Cu-Ti alloy," *Transactions of Nonferrous Metals Society of China*, vol. 23, no. 6, pp. 1701–1708, 2013.

[11] H. Ji, H. Chen, and M. Li, "Microstructures and properties of alumina/copper joints fabricated by ultrasonic-assisted brazing for replacing DBC in power electronics packaging," in *Proceedings of the 15th International Conference on Electronic Packaging Technology (ICEPT '14)*, pp. 1291–1294, Chengdu, China, August 2014.

[12] X. Chen, J. Yan, S. Ren, Q. Wang, J. Wei, and G. Fan, "Microstructure, mechanical properties, and bonding mechanism of ultrasonic-assisted brazed joints of SiC ceramics with ZnAlMg filler metals in air," *Ceramics International*, vol. 40, no. 1, pp. 683–689, 2014.

[13] W. Cui, J. Yan, Y. Dai, and D. Li, "Building a nano-crystalline α-alumina layer at a liquid metal/sapphire interface by ultrasound," *Ultrasonics Sonochemistry*, vol. 22, pp. 108–112, 2015.

[14] R. Koleňák, P. Šebo, M. Provazník, M. Koleňáková, and K. Ulrich, "Shear strength and wettability of active Sn3.5Ag4Ti(Ce,Ga) solder on Al_2O_3 ceramics," *Materials and Design*, vol. 32, no. 7, pp. 3997–4003, 2011.

[15] S. Y. Chang, L. C. Tsao, M. J. Chiang, C. N. Tung, G. H. Pan, and T. H. Chuang, "Active soldering of indium tin oxide (ITO) with Cu in air using an Sn3.5Ag4Ti(Ce, Ga) filler," *Journal of Materials Engineering and Performance*, vol. 12, no. 4, pp. 383–389, 2003.

[16] S. Y. Chang, T. H. Chuang, and C. L. Yang, "Low temperature bonding of alumina/alumina and alumina/copper in air using Sn3.5Ag4Ti(Ce,Ga) filler," *Journal of Electronic Materials*, vol. 36, no. 9, pp. 1193–1198, 2007.

[17] S. Y. Chang, T. H. Chuang, L. C. Tsao, C. L. Yang, and Z. S. Yang, "Active soldering of ZnS-SiO_2 sputtering targets to copper backing plates using an Sn3.5Ag4Ti(Ce, Ga) filler metal," *Journal of Materials Processing Technology*, vol. 202, no. 1–3, pp. 22–26, 2008.

[18] L. C. Tsao, "Direct active soldering of micro-arc oxidized Ti/Ti joints in air using Sn3.5Ag0.5Cu4Ti(RE) filler," *Materials Science and Engineering A*, vol. 565, pp. 63–71, 2013.

[19] T. H. Chuang, H. J. Lin, and C. C. Chi, "Rapid growth of tin whiskers on the surface of Sn-6.6Lu alloy," *Scripta Materialia*, vol. 56, no. 1, pp. 45–48, 2007.

[20] H. Hao, H. He, and Y. Lu, "Study of tin whisker growth accelerated by rare earth phase and the mechanism of tin whisker growth," in *Proceedings of the15th International Conference on Electronic Packaging Technology (ICEPT '14)*, pp. 1120–1126, Chengdu, China, August 2014.

[21] T.-H. Chuang, "Temperature effects on the whiskers in rare-earth doped Sn-3Ag-0.5Cu-0.5Ce solder joints," *Metallurgical and Materials Transactions A: Physical Metallurgy and Materials Science*, vol. 38, no. 5, pp. 1048–1055, 2007.

[22] M. Liu and A.-P. Xian, "Tin whisker growth on the surface of sn-0.7Cu lead-free solder with a rare earth (Nd) addition," *Journal of Electronic Materials*, vol. 38, no. 11, pp. 2353–2361, 2009.

[23] T.-H. Chuang and H.-J. Lin, "Inhibition of whisker growth on the surface of Sn-3Ag-0.5Cu-0.5Ce solder alloyed with Zn," *Journal of Electronic Materials*, vol. 38, no. 3, pp. 420–424, 2009.

[24] K. N. Tu, "Interdiffusion and reaction in bimetallic Cu-Sn thin films," *Acta Metallurgica*, vol. 21, no. 4, pp. 347–354, 1973.

[25] T.-H. Chuang, "Rapid whisker growth on the surface of Sn-3Ag-0.5Cu-1.0Ce solder joints," *Scripta Materialia*, vol. 55, no. 11, pp. 983–986, 2006.

[26] R. Kolenak, M. Chachula, P. Sebo, and M. Kolenakova, "Wettability and shear strength of active Sn2Ti solder on Al_2O_3 ceramics," *Soldering and Surface Mount Technology*, vol. 23, no. 4, pp. 224–228, 2011.

[27] C. Peng, M. Chen, and S. Liu, "Die bonding of silicon and other materials with active solder," in *Proceedings of the 11th International Conference on Electronic Packaging Technology &*

High Density Packaging (ICEPT-HDP '10), pp. 275–278, IEEE, Xi'an, China, August 2010.

[28] R. Koleňák and M. Prach, "Research of joining graphite by use of active solder," *Advanced Materials Research*, vol. 875–877, pp. 1270–1274, 2014.

[29] L. Zhang, S.-B. Xue, L.-L. Gao et al., "Microstructure characterization of SnAgCu solder bearing Ce for electronic packaging," *Microelectronic Engineering*, vol. 88, no. 9, pp. 2848–2851, 2011.

[30] M. A. Dudek and N. Chawla, "Nanoindentation of rare earth-Sn intermetallics in Pb-free solders," *Intermetallics*, vol. 18, no. 5, pp. 1016–1020, 2010.

[31] M. Pei and J. Qu, "Effect of lanthanum doping on the microstructure of tin-silver solder alloys," *Journal of Electronic Materials*, vol. 37, no. 3, pp. 331–338, 2008.

[32] M. Sadiq, R. Pesci, and M. Cherkaoui, "Impact of thermal aging on the microstructure evolution and mechanical properties of lanthanum-doped tin-silver-copper lead-free solders," *Journal of Electronic Materials*, vol. 42, no. 3, pp. 492–501, 2013.

[33] V. L. Lanin, "Physical effects of ultrasound in the liquid media and its applications in the technics," in *Technology in the Electrotechnical Industry*, vol. 2, pp. 10–15, 2013.

Preparation and Coloration of Colored Ceramics Derived from the Vanadium-Titanium Slags

Yi Deng,[1,2,3] Yang Zhou,[1] Yuanyi Yang,[4] Xiuyuan Shi,[1] Kewei Zhang,[1] Ping Zhang ⓘ,[1] and Weizhong Yang[1]

[1]School of Materials Science and Engineering, Sichuan University, Chengdu 610065, China
[2]School of Chemical Engineering, Sichuan University, Chengdu 610065, China
[3]Department of Mechanical Engineering, The University of Hong Kong, Hong Kong 999077, China
[4]Department of Materials Engineering, Sichuan College of Architectural Technology, Deyang 618000, China

Correspondence should be addressed to Ping Zhang; zhp@scu.edu.cn

Academic Editor: Davide Palumbo

Vanadium-titanium slag is a type of industrial solid waste from vanadium-titanium magnetite, and the resource utilization of vanadium-titanium slag is of great significance. The chemical composition of vanadium-titanium slag is similar to ceramic raw materials. Besides, the Fe, Ti, and Mn elements in vanadium-titanium slags possess strong color rendering ability. In the present study, we explored the influence of Fe, Ti, and Mn in vanadium-titanium slags on the coloration of ceramics. Besides, the crystal structures and chemical constituents of the sintered ceramics were studied by XRD, SEM, EDS, XPS, and ultraviolet-visible near-infrared spectrophotometer, and the color-rendering ability ($L^*a^*b^*$ values) of the ceramics was investigated. The results showed that the main crystal phases of the ceramics were quartz and anorthite. In addition, the color of the ceramics with the addition of vanadium-titanium slags was significantly affected by Fe and Mn ions, which decreased their L^* and a^* values. Meanwhile, Ti ion alone had no apparent coloring effect but strengthened the coloring effects of Fe and Mn ions. This work could provide the feasibility for preparation of colored ceramics based on vanadium-titanium slag, which could be used in the fields of high-value added building decoration, infrared-emission ceramic, and artwork.

1. Introduction

With the development of industrial technology and population growth, industrial wastes have become a serious social and environmental problem. These wastes cause many serious problems associated with transportation, storage, and air and environmental pollutions. Thus, recycling waste materials is urgently necessary. Some waste materials are adopted to make concrete, cement, and other building materials, and this approach reduces environmental pollution, sustainable consumption and production, and energy efficiency. Previously, numerous researchers have used industrial waste materials to produce new products. For instance, Lu et al. mixed fly ash and blast furnace slag and fluxed with borax and potash feldspar to prepare glass-ceramic glazes with low water absorption, proper stain resistance, and alkali and acid resistance [1]. Jonker and Potgieter reported that

Fe-rich wastes could be used as a good flux for ceramic production and increase the strength of ceramics [2]. Sarkar et al. investigated the possibility of fabrication of vitreous ceramics in a steel-melting electric arc furnace slag [3]. They found that a certain amount of slag with other conventional raw materials could be used for sintering at a range of 1100–1150°C. Furthermore, other researchers used industrial waste materials, such as sewage sludge, fly ash, and steel slags, to produce ceramics, cement, and various building materials [4–8]. Simultaneously, vanadium-titanium slags are industrial wastes mainly concentrated in the Panxi area in south-western China. High titanium slag emissions amount to 3.6 million tons annually in the Panzhihua area and have reached nearly 70 million tons nationwide, thereby causing severe pollution. Vanadium-titanium slags are mainly composed of Fe_2O_3, TiO_2, CaO, SiO_2, and trace amounts of Al_2O_3, K_2O, and Na_2O. Chinese scholars have started studying the

TABLE 1: Main chemical composition of vanadium and titanium slags.

Constituent	Na_2O	MgO	Al_2O_3	SiO_2	TiO_2	V_2O_5	MnO_2	Fe_2O_3	CaO
wt. %	8.18	2.10	4.04	12.96	11.67	0.89	7.72	51.29	1.15

comprehensive use of vanadium-titanium slags since the early 1960s and have made some promising achievements. For example, Wang used titanium slags to fabricate concretes with other raw materials [9], and Sun et al. produced building bricks with titanium slags from Pangang factory and clays [10]. Besides, for the purpose of vanadium-titanium slag recycling, building bricks were prepared using vanadium-titanium slags or by extracting titanium from vanadium-titanium slags [11, 12].

Ceramic bodies are heterogeneous materials, and they mainly consist of a mixture of natural raw materials with various compositions. Currently, many previous studies mostly focus on the harmless treatment of wastes, preparation of the basic formula, and characterization of sample performances. There are few reports regarding the mechanism of ceramic coloring. Because the main ions in vanadium-titanium industrial waste slags are Fe, Mn, and Ti, vanadium-titanium industrial waste slags can replace a certain amount of clays and quartz, which are widely used to prepare ceramics. Herein, the aims of the present work are to explore the impact of Fe_2O_3, MnO_2, and TiO_2 on ceramic color performance, to study the color mechanism of ceramics after adding vanadium-titanium, and to assess the feasibility for preparation of colored ceramics, which could be used in the high-value-added building decoration field, infrared-emission field [13], and artwork.

2. Materials and Method

The basic materials were vanadium-titanium slags collected from PanGang Iron and Steel Group Co. The chemical composition of these slags is listed in Table 1. Other raw materials used to formulate the ceramic body were provided by Baita Ceramic Group Co. The mixtures were subjected to wet ball milling for 20 min and sieves using 140 mesh with a sieve margin of less than 3%.

The mixture was then filtrated, dried at 80°C for 2 h, ground, and humidified with 7.0 wt.% water. The humidified mixtures were then mold-shaped to the wafer samples with a diameter of 50 mm under 20 MPa pressure. Finally, the samples were dried at 100°C for 30 min and sintered in muffle furnace at 1100°C for 10 min.

The ceramic material colorant was characterized through the CIE1976 $L^*a^*b^*$ color system, which was established by the International Commission on Illumination in 1976. In the $L^*a^*b^*$ color system, the color is represented by three parameters, namely, L^* (lightness), a^* (color), and b^* (color). L^* shows lightness, and the L^* value ranges from 0 to 100 (black-white). a^* indicates the range from red to green, and its value ranges from +127 to −128 (red-green). b^* indicates the range from yellow to blue, and its value ranges from +127 to −128 (yellow-blue).

The phase composition of sintered ceramics was studied by X-ray diffraction (XRD, DX-1000X, Japan). The samples

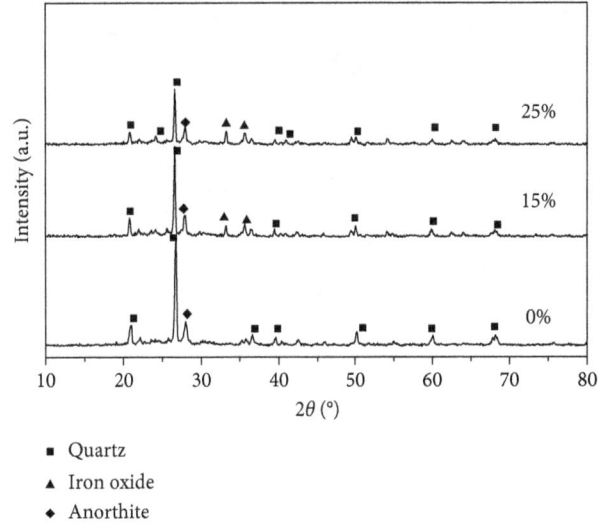

- ■ Quartz
- ▲ Iron oxide
- ◆ Anorthite

FIGURE 1: XRD spectra of ceramics with different amounts of vanadium and titanium industrial slags at sintering temperature of 1100°C.

were scanned from 10° to 80° at a scanning speed of 2°/min. The microstructure of the samples was observed by the scanning electron microscopy (SEM, S-3400N, Japan) equipped with an energy dispersive X-ray spectrometer for elemental analysis. All samples were coated by gold for 1 min before SEM observation. The $L^*a^*b^*$ value was measured by WR-10 colorimeter. The reflectance spectrum was observed by UV-visible near-infrared spectrophotometer (UV-3600, USA) at a wavelength range of 250–1500 nm.

3. Results and Discussion

3.1. Phase Composition. From previous researches, we can find that vanadium and titanium industrial waste slags were employed to partly replace ceramic raw materials for the preparation of ceramic bodies. Vanadium and titanium industrial waste slags exhibited the following favorable conditions: the addition amount was 15%, and the sintering temperature was 1100°C. Thus, the flexural strength of the ceramics was 31.28 MPa, and the water absorption was 7.07%. The ceramics were prepared with different amounts of vanadium-titanium slags and then sintered at 1100°C. The XRD analysis results are shown in Figure 1.

Figure 1 shows that the main crystalline phases of the sintered ceramics were quartz, anorthite, and iron oxide. Anorthite has been used in ceramics because of its high mechanical strength, dilatability, and low sintering temperature [14, 15]. The crystallization of anorthite increases the chemical stability and strength of the material and improves its physical properties including high thermal shock resistance and low thermal expansion coefficient [16–19]. The increased

(a)

(b)

(c)

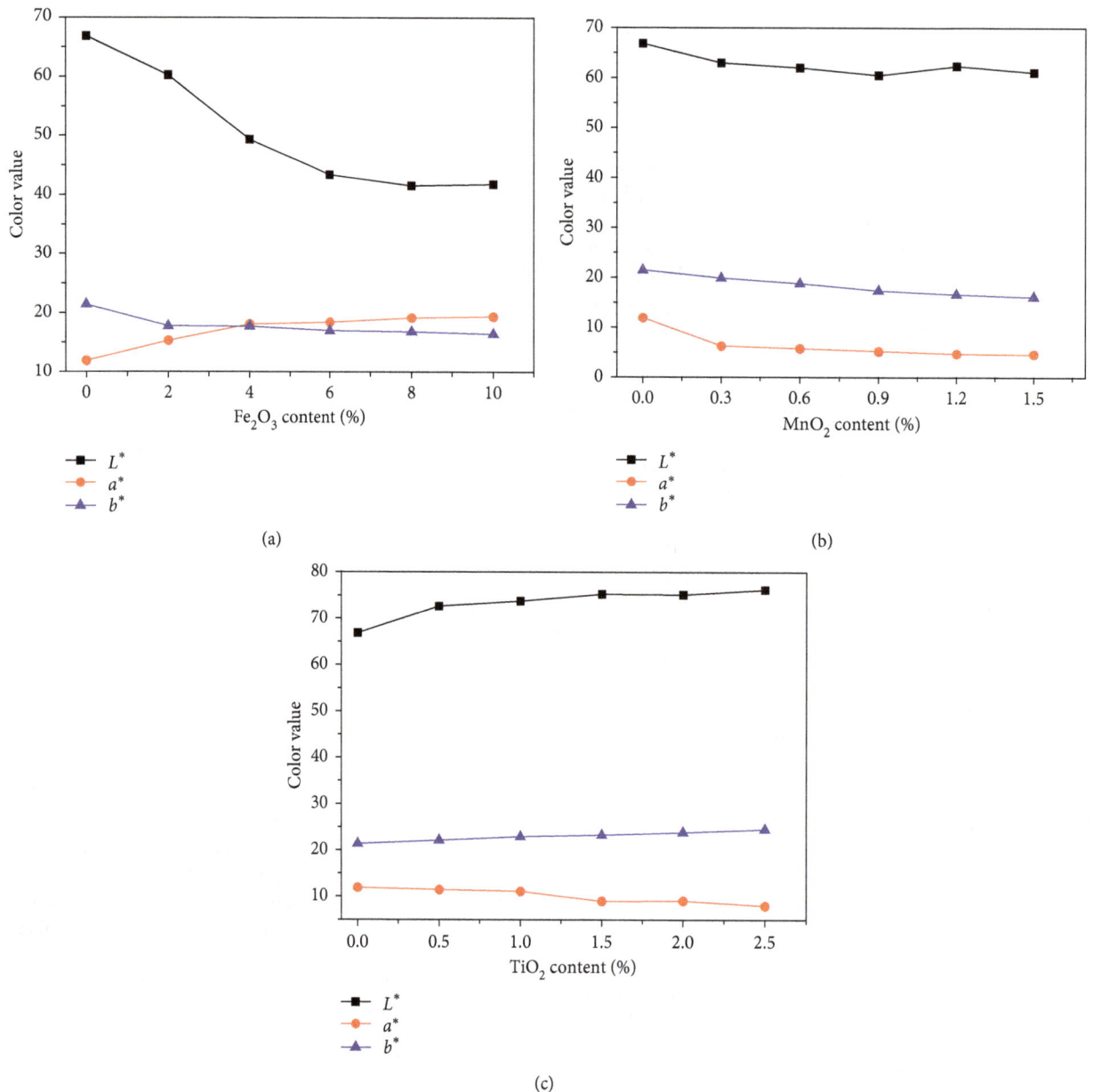

FIGURE 2: Effect of different colorants on the color of ceramics: (a) Fe_2O_3/addition, (b) MnO_2/addition, and (c) TiO_2/addition.

amount of the vanadium-titanium industrial waste slag resulted in the decrease of SiO_2 content and the increase of Fe_2O_3 content. The calculated area of the XRD spectra peaks indicated that when the amount of vanadium-titanium industrial waste slag increased to 15% and 25%, the iron oxide crystal phase ratios were 5.76% and 9.41%, respectively, and the composition of the raw materials decreased by 7.69% and 12.82%, respectively. The iron oxide in the ceramics was not entirely crystal and was partly in the glass phase. Similar to the network effect, the Fe^{3+} with eight ligands can damage the ceramic glass network structure by reducing the viscosity of the ceramic glass phase [20]. This effect contributes to the diffusion of atoms and ions during the sintering process and can reduce the sintering temperature, densify the ceramics, and increase the flexural strength.

3.2. Effect of Colorant on Color of Ceramics. The colorants were Fe_2O_3, MnO_2, and TiO_2 with contents of 0%–10% (wt.%), 0%–1.5% (wt.%), and 0%–2.5% (wt.%), respectively. Each colorant at various contents was added to different ceramics, and the products were sintered at 1100°C. The $L^*a^*b^*$ analysis of the ceramics is shown in Figure 2.

As shown in Figure 2(a), the lightness of the ceramics gradually decreased with the increase of the Fe_2O_3 content. When the amount of Fe_2O_3 reached 8%, the L^* value decreased gradually and then remained generally stable, whereas the a^* value increased gradually. The value of b^* decreased initially and then remained unchanged at the end. These results indicated that Fe_2O_3 could effectively reduce the lightness of materials and provide red coloring to the materials, thereby producing red ceramics. As shown in

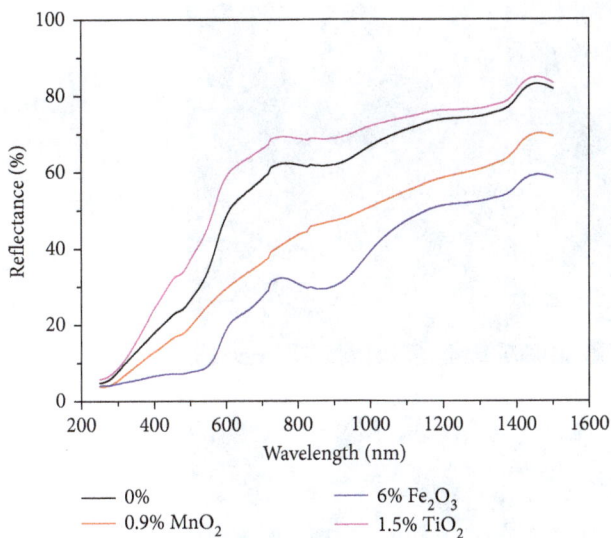

FIGURE 3: Reflectance spectra of ceramics with different colored oxides.

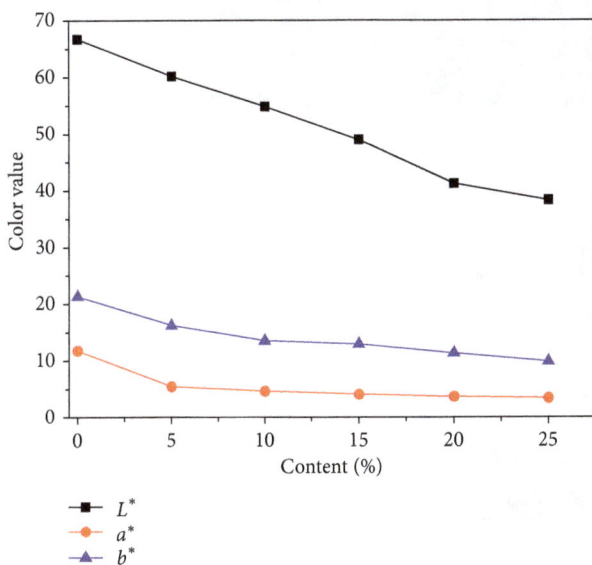

FIGURE 4: Effect of different vanadium-titanium slags on the color of ceramics.

TABLE 2: Color parameters of ceramics with different components.

Components	L^*	a^*	b^*
15% vanadium-titanium slags	49.08	4.15	13.03
6% Fe_2O_3	43.43	18.44	17.02
0.9% MnO_2	60.52	5.28	17.39
1.5% TiO_2	75.29	9.07	23.36

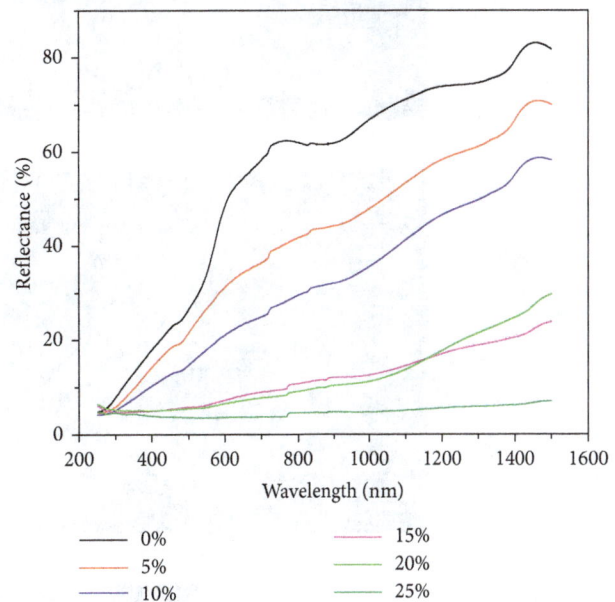

FIGURE 5: Reflectance spectra of ceramics with different vanadium-titanium slags.

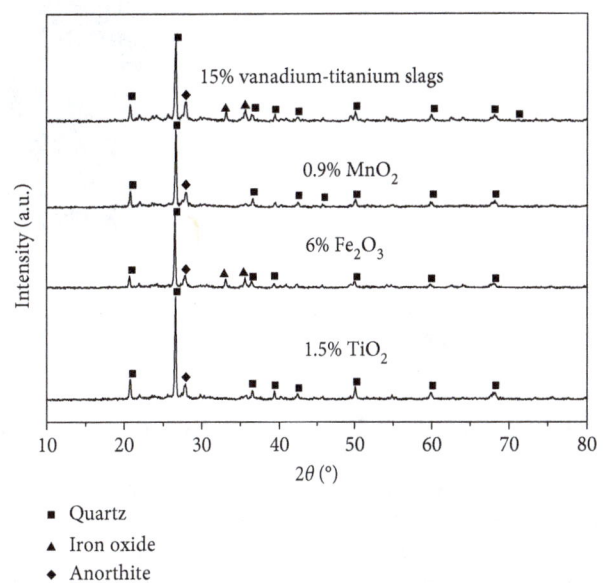

FIGURE 6: XRD spectra of ceramics with different additions at sintering temperature of 1100°C.

Figure 2(b), when the addition of MnO_2 was minimal, it had little effect on the lightness of the ceramics. The a^* value decreased initially and kept unchanged at the end. The b^* value also decreased gradually, but the change was subtle. MnO_2 minimally reduced the redness and yellowness of the materials. Compared with Fe_2O_3 and MnO_2, TiO_2 improved the lightness, and it increased the L^* value of the materials in Figure 2(c). The a^* value increased, whereas the b^* value decreased gradually. TiO_2 improved the yellowness and reduced the redness of the materials. Therefore, the relationship among the contents of Fe_2O_3, MnO_2, TiO_2, and color parameters of the ceramics showed a monotone linear change, and single color oxide could affect the color properties of the materials.

The first transition metal elements (fourth cycles) (Co, Cr, Ti, V, Mn, Fe, Ni, Cu, etc.) present generally in the form of ions in the silicate structure and have a common feature in terms of atomic structure, particularly maximum capacity of

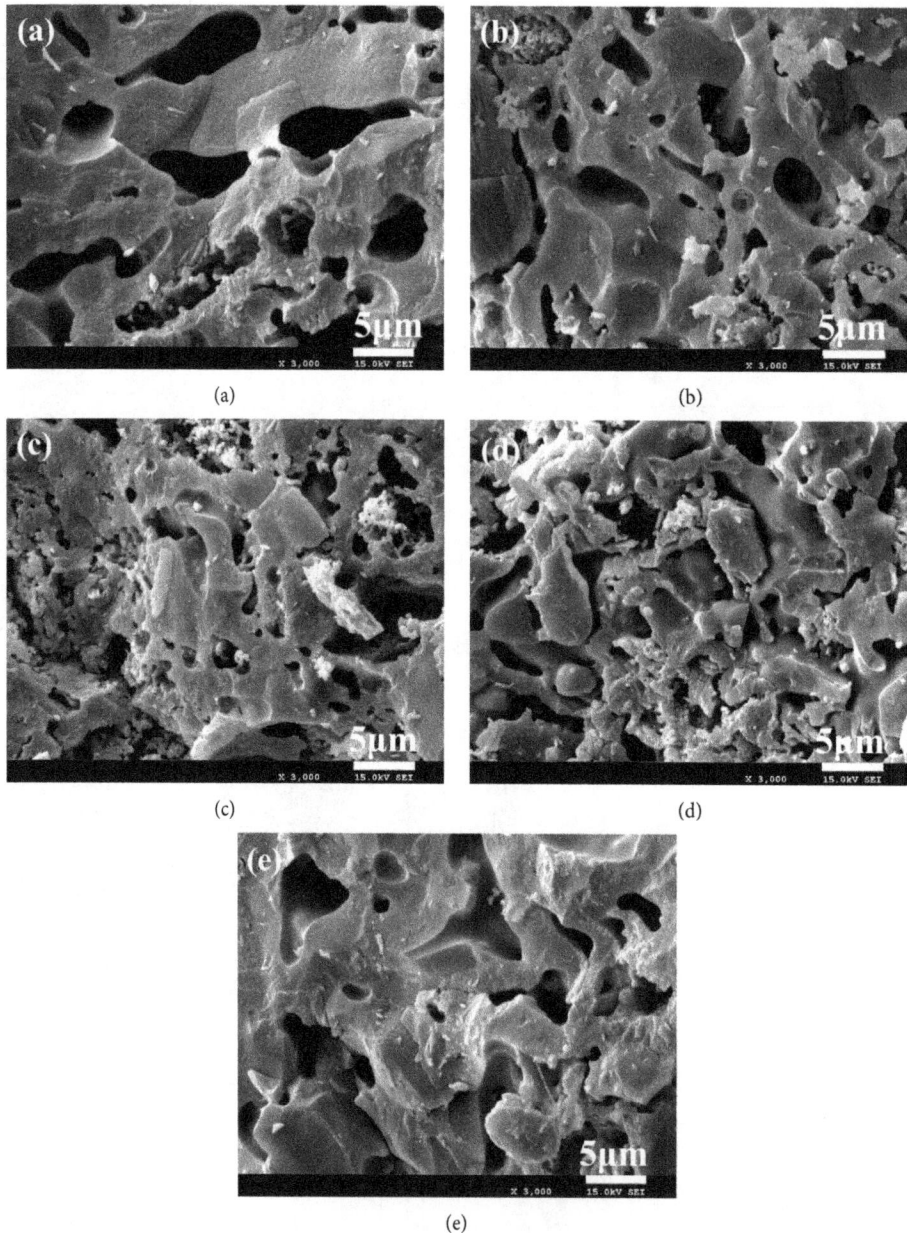

FIGURE 7: Microstructures of ceramics with different additions at the sintering temperature of 1100°C: (a) no addition, (b) 15% vanadium-titanium slags, (c) 6% Fe_2O_3, (d), 1.5% TiO_2, and (e) 0.9% MnO_2.

10 electrons in 3d orbit, 1–9 electron transitions, and valence electrons jump between different energy levels (called d-d transition). Thus, the ions exhibit selective absorption in the visible region and thus render the ceramics colorful. The Fe ion has two valence states, namely, Fe^{3+} and Fe^{2+}. Fe^{2+} is easily oxidized and rare, whereas Fe^{3+} is extremely unstable at high temperatures and easily decomposes into Fe^{2+}. The coexistence of two valence states of iron oxides produces ferrous ferrite (Fe^{3+}-O-Fe^{2+}). Ferrous ferrite is a kind of Fe_3O_4 structure that renders ceramics colorful. At sintering temperature of 1100°C, the ceramics were slight red because the trivalent iron dominated.

In silicate ceramics, Ti is generally in the form of Ti^{4+}. The valence state of Ti^{4+} means that the $3d_24s_2$ of the outermost electrons of Ti is lost completely, and $3d$ orbital is completely empty. The "d-d" transition between the electrons in the d orbitals cannot occur. Thus, the valence state of Ti^{4+} should be colorless. However, Ti^{4+} ions strongly absorb ultraviolet light, and the absorption band can cross into the purple and blue part of visible light, subsequently rendering the materials pale brown. Ti^{4+} alone does not cause a darker color, but it can affect other transition elements and thus may strengthen and intensify the effect of the other ion colorants. In ceramic materials, Mn usually exists in the form of MnO_2, which reduces the lightness of materials.

The ceramics were prepared with colorants at varying contents and then sintered at 1100°C. The reflectance spectra of the ceramics are shown in Figure 3. The ceramics rarely

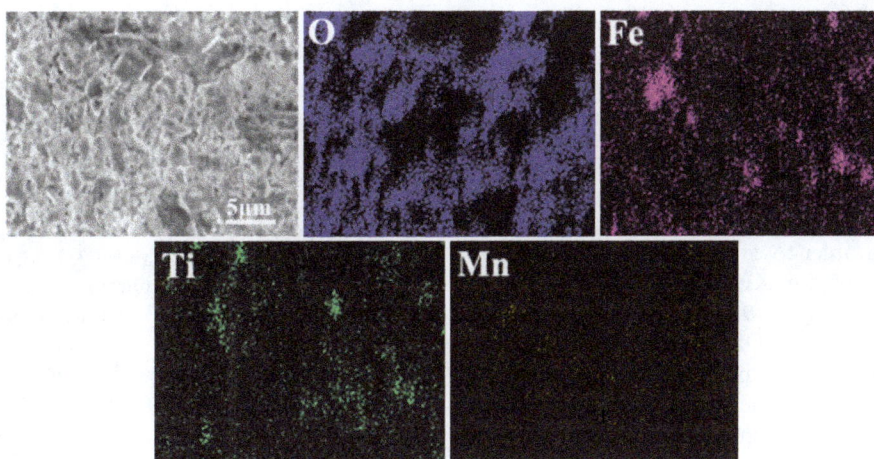

FIGURE 8: Element distribution of ceramics with 15% vanadium-titanium slags at the sintering temperature of 1100°C.

absorb light without colorants (Figure 3), and the ceramics were slightly yellow. No evident absorption peak was observed in the wavelength range of a single colorant, and the reflectivity increased gradually with increased wavelength. The reflectivity of ceramics with TiO_2 was higher than those of ceramics without colorants. However, the reflectivity of ceramics with Fe_2O_3 or MnO_2 decreased.

3.3. Effect of Vanadium-Titanium Slags on Color of Ceramics. The ceramics were prepared with different additions of vanadium-titanium slags and then sintered at 1100°C. The results of $L^*a^*b^*$ analysis on the ceramics are shown in Figure 4. It was found that the L^*, a^*, and b^* values all decreased with the addition of vanadium-titanium slags, in contrast to those of ceramics with different colorants, where the L^* value considerably decreased. The ceramics were purple black, which darkened gradually after the addition of vanadium-titanium slags. The color parameters of the ceramics with titanium-vanadium slags or single colorant are listed in Table 2. The color parameters of the ceramics with vanadium-titanium slags were different from those of the ceramics with a single colorant, and the L^*, a^*, and b^* values decreased. The alteration of the ceramics color with vanadium-titanium slags was attributed to the combined colors of various ions. Fe ions mainly reduced the L^* value of the material, and Mn ions mainly reduced the a^* value of the material. Ti ion alone had no evident coloring effect but strengthened the coloring effects of the Mn and Fe ions. The ion content, valence state, and sintering atmosphere cocontributed to the final color formation.

Figure 5 presents the reflectance spectra of ceramics with different amount of vanadium-titanium slags. The reflectivity of ceramics gradually decreased with the increase of vanadium-titanium slags. Moreover, the reflectivity of ceramics gradually increased with the increase of the wavelength of light, and there was no obvious absorption peak in all wavelength area, which corresponded to the color of ceramics.

3.4. Phase Composition and Microstructure. The XRD analysis of ceramics is shown in Figure 6. The main crystal phases of ceramics with different additions were quartz and

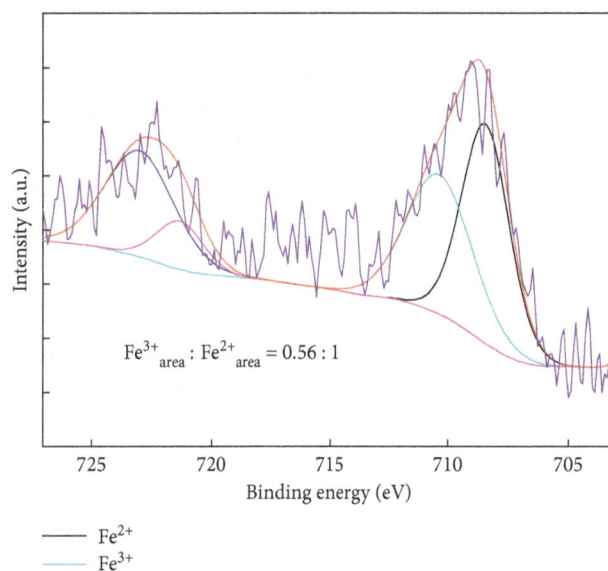

FIGURE 9: The XPS spectra of ceramics with 15% vanadium-titanium slags at the sintering temperature of 1100°C.

anorthite and that of ceramics with vanadium-titanium slags or Fe_2O_3 was iron oxide. The microstructures of ceramics with different additions at sintering temperature of 1100°C are also depicted in Figure 7. These ceramics all displayed porous structures, and the numbers and sizes of the pores of the ceramics varied at different additions.

The content of SiO_2 in vanadium and titanium industrial waste slags was small, and the contents of Fe_2O_3 and Na_2O were relatively higher when vanadium-titanium slags or single oxide (such as MnO_2, Fe_2O_3, and TiO_2) were added to the ceramics. The addition reduced the content of SiO_2 in the body. Similarly, the content of SiO_2/R_2O decreased. In the sintering process, the skeleton effect of SiO_2 was partially destroyed. The aluminosilicates in clay and calcium oxide derived from the decomposition of raw materials to form anorthite. Simultaneously, iron oxide, sodium oxide, calcium oxide, and silicon dioxide formed a eutectic mixture, which produced a large number of liquid phases that promoted the sintering process.

3.5. Element Distribution. Figure 8 shows the element distribution of ceramics with 15% vanadium-titanium slags. From Figure 8, it could be seen that the elements of Fe, Ti, and Mn were distributed in the ceramics evenly, and the content of Fe was more than that of Ti and Mn. Therefore, this distribution can well describe the color of ceramics.

3.6. XPS Analysis. In order to confirm the content of Fe^{3+} and Fe^{2+}, the high-resolution XPS spectra of Fe 2p for ceramics are presented in Figure 9. And the calculated peak area ratio (Fe^{3+} area/Fe^{2+} area) of the sample has been marked in the spectra. It could be found that the content of Fe^{3+}/Fe^{2+} was 0.56. Under this condition, the valence of Mn was +4, and the valence of Ti was also +4, which was in accordance with the color of ceramics.

4. Conclusions

The vanadium-titanium industrial waste slags can partly replace ceramic raw materials for the preparation of ceramic bodies. The XRD results showed that the main crystalline phases of the ceramics with vanadium-titanium slags or single colorant were quartz and anorthite. The color of the ceramics with the addition of vanadium-titanium slags was significantly affected by various ions. The Fe and Mn ions mainly decreased the L^* and a^* values of the materials; however, Ti ions had a minimal coloring effect but affected the coloring effects of the Fe and Mn ions. The experimental results indicated that the ceramics with 15% vanadium-titanium slags could be fabricated to colored ceramics, and the color parameters were as follows: $L^* = 49.08$, $a^* = 4.15$, and $b^* = 13.03$. Our work could provide the feasibility for the preparation of colored ceramics.

Conflicts of Interest

The authors declare that they have no conflicts of interest.

Acknowledgments

This work was jointly supported by the financial support of Sichuan Science and Technology Program (2015GZ0175 and 2017FZ0046), Full-Time Postdoctoral Research Fund of Sichuan University (2017SCU12016), Research Program of Star of Chemical Engineering (School of Chemical Engineering, Sichuan University), and Hong Kong Scholars Program. The authors would like to thank Wang Hui (Analytical & Testing Center, Sichuan University) for her help in SEM observation.

References

[1] H. X. Lu, M. He, Y. Y. Liu et al., "A preparation and performance study of glass-ceramic glazes derived from blast furnace slag and fly ash," *Journal of Ceramic Processing Research*, vol. 12, p. 588, 2011.

[2] A. Jonker and J. H. Potgieter, "An evaluation of selected waste resources for utilization in ceramic materials applications," *Journal of the European Ceramic Society*, vol. 25, no. 13, pp. 3145–3149, 2005.

[3] R. Sarkar, N. Singh, and S. K. Das, "Utilization of steel melting electric arc furnace slag for development of vitreous ceramic tiles," *Bulletin of Materials Science*, vol. 33, no. 3, pp. 293–298, 2010.

[4] A. Zimmer and C. P. Bergmann, "Fly ash of mineral coal as ceramic tiles raw material," *Waste Management*, vol. 27, no. 1, pp. 59–68, 2007.

[5] S. N. Monteiro, J. Alexandre, J. I. Margem, R. Sánchez, and C. M. F. Vieira, "Incorporation of sludge waste from water treatment plant into red ceramic," *Construction and Building Materials*, vol. 22, no. 6, pp. 1281–1287, 2008.

[6] S. Ghosh, M. Das, S. Chakrabarti, and S. Ghatak, "Development of ceramic tiles from common clay and blast furnace slag," *Ceramics International*, vol. 28, no. 4, pp. 393–400, 2002.

[7] E. Karamanova, G. Avdeev, and A. Karamanov, "Ceramics from blast furnace slag, kaolin and quartz," *Journal of the European Ceramic Society*, vol. 31, no. 6, pp. 989–998, 2011.

[8] F. He, Y. Fang, J. L. Xie, and J. Xie, "Fabrication and characterization of glass-ceramics materials developed from steel slag waste," *Materials & Design*, vol. 42, pp. 198–203, 2012.

[9] H. B. Wang, "The research on preparation of high-performance concrete using Pangang high titaniferous blast-furnace slag," *Panzhihua Sci-Tech and Information*, vol. 4, pp. 13–18, 2006.

[10] X. W. Sun, J. T. Zhang, Z. Y. Yang, T C. You, and Y. Liu, "The development of high titaniferous building slag brick," *New Building Materials*, vol. 3, pp. 5–7, 2003.

[11] M. G. Zhou, X. Y. Li, and J. Wang, "Influence of property changes of panzhihua titanium magnetite on iron concentrate grade," *Multipurpose Utilization of Mineral Resources*, vol. 5, p. 29, 2012.

[12] S. J. Gao and W. L. Zhang, "The present state and perspective of the titaniferous blast-furnace's utilization," *Shandong Ceramics*, vol. 5, pp. 29–31, 2006.

[13] K. Zhang, Y. Deng, Y. Yang et al., "Effect of lanthanum doping on the far-infrared emission property of vanadium–titanium slag ceramic," *RSC Advances*, vol. 7, no. 22, pp. 13509–13516, 2017.

[14] C. W. Wu, C. J. Sun, S. H. Gao et al., "Mechanochemically induced synthesis of anorthite in MSWI fly ash with kaolin," *Journal of Hazardous Materials*, vol. 244-245, pp. 412–420, 2013.

[15] S. Kurama and E. Ozel, "The influence of different CaO source in the production of anorthite ceramics," *Ceramic International*, vol. 35, no. 2, pp. 827–830, 2009.

[16] X. Cheng, S. Ke, Q. Wang, H. Wang, A. Shui, and P. Liu, "Fabrication and characterization of anorthite-based ceramic using mineral raw materials," *Ceramic International*, vol. 38, no. 4, pp. 3227–3235, 2012.

[17] V. M. F. Marques, D. U. Tulyaganov, S. Agathopoulos, V. K. Gataullin, G. P. Kothiyal, and J. M. F. Ferreira, "Low temperature synthesis of anorthite based glass-ceramics via sintering and crystallization of glass-powder compacts," *Journal of the European Ceramic Society*, vol. 26, no. 13, pp. 2503–2510, 2006.

[18] A. A. Capoglu, "A novel low-clay translucent whiteware based on anorthite," *Journal of the European Ceramic Society*, vol. 31, no. 3, pp. 321–329, 2011.

[19] W. P. Tai, K. Kimura, and K. Jinnai, "A new approach to anorthite porcelain bodies using nonplastic raw materials," *Journal of the European Ceramic Society*, vol. 22, no. 4, pp. 463–470, 2002.

3

Evaluating the Effects of Chemical Composition on Induction Heating Ability of Fe$_2$O$_3$-CaO-SiO$_2$ Glass Ceramics

Y. Y. Wang, B. Li, Y. L. Yu, and P. S. Tang

Department of Material Chemistry, Huzhou University, Huzhou 313000, China

Correspondence should be addressed to Y. Y. Wang; arince@hutc.zj.cn

Academic Editor: Michele Iafisco

In order to investigate the relationship between induction heating ability of Fe$_2$O$_3$-CaO-SiO$_2$ glass ceramics and chemical composition, a series of glass ceramic samples with different chemical compositions were prepared by the sol-gel method. The structural, textural, and magnetic properties of the samples were analyzed and correlated with the Fe$_2$O$_3$ content. This is the first time work of its kind that evaluates the relationships between induction heating ability and chemical composition of Fe$_2$O$_3$-CaO-SiO$_2$ glass ceramics. The results showed that induction heating ability of Fe$_2$O$_3$-CaO-SiO$_2$ glass ceramics increased gradually with increasing magnetite content. Also, the induction heating ability became considerably better when a small amount of phosphorus was introduced. This study thus reveals a methodology to control the induction heating ability of Fe$_2$O$_3$-CaO-SiO$_2$ glass ceramics through modifying the chemical composition.

1. Introduction

Magnetic induction hyperthermia has been found to be a useful modality for cancer therapy recently and the thermoseed materials used in hyperthermia therapy have become a topic of increasing research focus [1–3]. The thermoseed material must be biocompatible and contain a magnetic phase to generate and dissipate the proper amount of heat under alternating magnetic fields. Ferromagnetic and bioactive glass ceramics are considered to be effective thermoseed materials for cancer therapy. In particular, Fe$_2$O$_3$-CaO-SiO$_2$ glass ceramics are of great interest to researchers, due to their excellent magnetic properties and good biological activity. Several Fe$_2$O$_3$-CaO-SiO$_2$ glass ceramics having a considerable amount of magnetic phases have been developed for this application [4–7]. In fact, hyperthermic cancer therapy using Fe$_2$O$_3$-CaO-SiO$_2$ ferromagnetic glass ceramics has already been reported to show excellent results in animal experiment [8–10].

Fe$_2$O$_3$-CaO-SiO$_2$ glass ceramics absorb magnetic energy and generate heat under the alternating magnetic field, as a result of eddy currents, Néel losses, and hysteresis losses [11–13]. There are several factors that influence the heat production rate [14], including the internal factors such as permeability, magnetic energy product, material size, and microstructure, as well as external factors such as magnetic field frequency and amplitude. The magnetic property was found to be one of the most significant factors affecting the induction heating ability of glass ceramics. Lee et al. explored the effect of the iron state on crystallization in Fe$_2$O$_3$-CaO-SiO$_2$ glasses [15], while Bretcanu et al. investigated the influence of crystallized Fe$_3$O$_4$ on magnetic properties of ferrimagnetic glass ceramics [16]. Magnetic properties of CaO-P$_2$O$_5$-Na$_2$O-Fe$_2$O$_3$-SiO$_2$ glass upon heat treatment were also analyzed by Shankhwar et al. [17]. However, to date, the effects of chemical compositions on the induction heating ability of Fe$_2$O$_3$-CaO-SiO$_2$ glass ceramics have not been reported.

The Fe$_2$O$_3$-CaO-SiO$_2$ glass ceramics with different compositions were prepared by the sol-gel method in this work. A small amount of phosphorus was introduced to increase the biological activity at the same time. Magnetic properties of the glass ceramic samples were evaluated and correlated mainly with the iron oxide content. The effect of Fe$_2$O$_3$ content on the induction heating ability of Fe$_2$O$_3$-CaO-SiO$_2$ glass ceramics has been investigated by analyzing the crystal phase and magnetic properties of the samples. This study also reveals a methodology to control the magnetic properties

TABLE 1: Chemical compositions of different glass ceramic samples (wt%).

Sample	Fe_2O_3	CaO	SiO_2	P_2O_5
A1S1	15	42	43	0
A1S3	20	39	41	0
A1S5	25	36	39	0
A1S7	30	34	36	0
A1S9	35	32	33	0
A2P1	30	39	30	1
A2P7	28	37	28	7

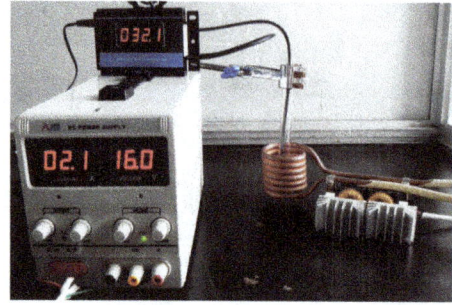

FIGURE 1: Homemade device for measuring induction heating ability.

of Fe_2O_3-CaO-SiO_2 glass ceramics through the chemical composition and its potential application in hyperthermia treatment of cancer.

2. Materials and Methods

2.1. Preparation of Fe_2O_3-CaO-SiO_2 Glass Ceramics. Several different Fe_2O_3-CaO-SiO_2 glass ceramics samples were prepared by the sol-gel method and classified as groups A1 and A2 (without phosphorous and with phosphorous, resp.).

The chemical compositions of the samples are shown in Table 1. The dosages of reagents calculated on the base of the chemical compositions were listed in Table 2. Briefly, the procedure involved preparing a solution of tetraethyl orthosilicate (TEOS) in ethanol. Distilled water was added to the solution such that the molar ratio of TEOS and water was 1 : 15. Then hydrochloric acid was added to the solution and stirred with a magnetic stirrer for 10 min, and then calcium nitrate and iron nitrate were added to the solution. Triethyl phosphate was also added to the solution in group A2 at this time. The resulting solution was placed in a water bath at 45°C for 150 min. Then the temperature was raised up to 65°C until the solution became semisolid. The samples were aged for 7 days at room temperature, and the gels obtained were baked in the drying oven at 120°C for 12 h. Finally, the dried samples were heated to 650°C for an hour, followed by further heating at 950°C for 1 hour in a heat treatment furnace in a reducing atmosphere. The glass ceramic samples thus obtained were cooled to room temperature and characterized as described in the following sections.

2.2. Characterization. X-ray diffraction (XRD) analysis was performed with a XD-6 X-ray diffractometer at room temperature, with CuKα radiation, 36 kV voltage, and 20 mA current. X-ray data were collected in the $5 < 2\theta < 80$ range. Prior to the XRD test, the samples were ground into powder and sieved through a 300-mesh gauze.

Microarea elemental compositions of the samples were determined using a scanning electron microscopy (SEM) system (S-3400N) equipped with energy dispersive spectrometer (EDS). The samples were coated with a layer of gold approximately 10 nm thick, and the samples were analyzed for a live time of 30 s. The total amounts of oxide in the samples were normalized to 100%, and the Fe^{2+} and Fe^{3+} contents were converted to the equivalent Fe_2O_3 content.

Magnetization measurements were performed using a vibrating sample magnetometer (VSM, Lakeshore model 7407) at room temperature.

A homemade device was built for measuring the induction heating ability (Figure 1). The glass ceramics samples were introduced into a copper coil, which was part of a resonant RLC circuit producing an AC magnetic field in the frequency range 80 kHz–100 kHz and with amplitudes up to 0.15 kA/m. The copper coil was cooled by circulating water, and the temperature was monitored using a fiber thermometer placed in the center of the sample. The induction heating ability was determined by plotting the initial linear rise in temperature versus time, normalized to the mass of the sample.

3. Results and Discussion

3.1. Crystal Phase Analysis. XRD patterns of the glass ceramic samples of group A1 are shown in Figure 2(a), demonstrating the presence of a mixture of crystalline phases for all the materials. Three major crystalline phases, wollastonite (JCPDS No. 84-0654), magnetite (JCPDS No. 85-1436), and hematite (JCPDS No. 85-0599) were observed in group A1 samples. The hematite level of sample A1S9 was much higher than those of other samples. XRD patterns of the glass ceramic samples of group A2 with phosphorus were characterized by the major hydroxyapatite phase (JCPDS No. 09-0432) in addition to the above three major crystalline phases (Figure 2(b)). Hydroxyapatite is the main constituent of bones, and the presence of this bone mineral phase along with the magnetic phase suggests both the biocompatible nature and the induction heating ability of the samples.

The relative mass of each phase, which was also calculated according to the XRD patterns, was listed in Table 3. The amounts of both magnetite and hematite phase of group A2 were more than that of group A1, and the increase of magnetite content indicates that the induction heating ability would get stronger.

3.2. Chemical Composition Analysis. The SEM-EDS spectra for the samples are shown in Figure 3. The results confirmed the presence of silicon, calcium, and iron in samples A1S5 and A1S7 while silicon, phosphorus, calcium, and iron were present in all the samples of group A2. Samples A1S5 and A1S7

TABLE 2: The amount of the reagent.

Sample	Iron nitrate (g)	Calcium nitrate (g)	TEOS (mL)	Triethyl phosphate (mL)	Ethanol (mL)
A1S1	3.62	8.43	7.60	0	22.80
A1S3	4.82	7.83	7.25	0	21.74
A1S5	6.03	7.22	6.89	0	20.68
A1S7	7.23	6.82	6.36	0	19.09
A1S9	8.44	6.42	5.83	0	17.50
A2P1	7.23	8.03	5.30	0.12	15.91
A2P7	6.75	7.59	4.82	0.76	14.46

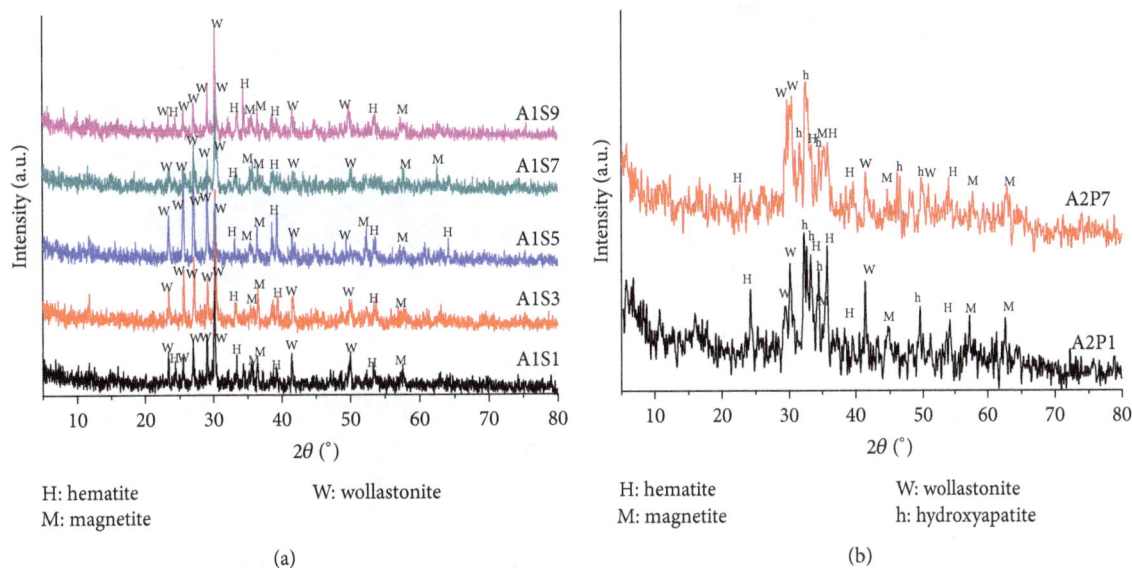

(a)

H: hematite   W: wollastonite
M: magnetite

(b)

H: hematite   W: wollastonite
M: magnetite   h: hydroxyapatite

FIGURE 2: Room temperature XRD patterns of glass ceramic samples.

TABLE 3: The relative mass of the phases of the samples (wt%).

Sample	Hematite	Magnetite	Wollastonite	Hydroxyapatite
A1S1	4.7	2.8	92.5	0
A1S3	2.5	2.6	94.8	0
A1S5	2.4	1.6	96.1	0
A1S7	3.8	4.8	91.4	0
A1S9	5.9	4.2	89.9	0
A2P1	15	6	45.1	33.9
A2P7	7.5	4.5	50	38

TABLE 4: Chemical composition of the samples determined by SEM-EDS (wt%).

	Fe_2O_3	CaO	SiO_2	P_2O_5
A1S5	28.71	37.75	33.54	0
A1S7	22.73	32.79	45.01	0
A2P1	32.57	34.26	30.40	2.76
A2P7	32.05	30.95	25.31	11.70

showed the presence of lathlike particles while the mineral particles in group A2 were granular. Moreover, the particle sizes of A2P1 were about $3\,\mu$m, which was much smaller than that of A1S7 which was about $16\,\mu$m. The chemical compositions of the samples calculated from EDS are given in Table 4. No other elements were identified, which ruled out the influence of impurities on the induction heating ability of the samples.

3.3. Magnetic Properties. Figure 4 shows the magnetization (*M-H*) curves obtained for the samples as a function of applied magnetic field. The samples exhibit hysteresis with narrow hysteresis loop and low coercivity, suggesting the soft ferromagnetic nature of the samples. This is due to the formation of the magnetic phase upon heat treatment, as already confirmed by the XRD results.

The different magnetic parameters of the samples are listed in Table 5. Sample A1S7 shows higher saturation magnetization M_s (19 emu/g) and remanent magnetization M_r (2.3 emu/g) than the other samples of group A1. This may be due to the enhancement in the magnetic phase in A1S7. However, M_s and M_r values began to decrease when the Fe_2O_3 content was over 30% (Figure 5), which can be attributed to the higher hematite content and the lower magnetite content. A2P1 has a lower saturation magnetization value but a larger coercive force compared to A1S7. The lower saturation magnetization may be attributed to some possible reactions

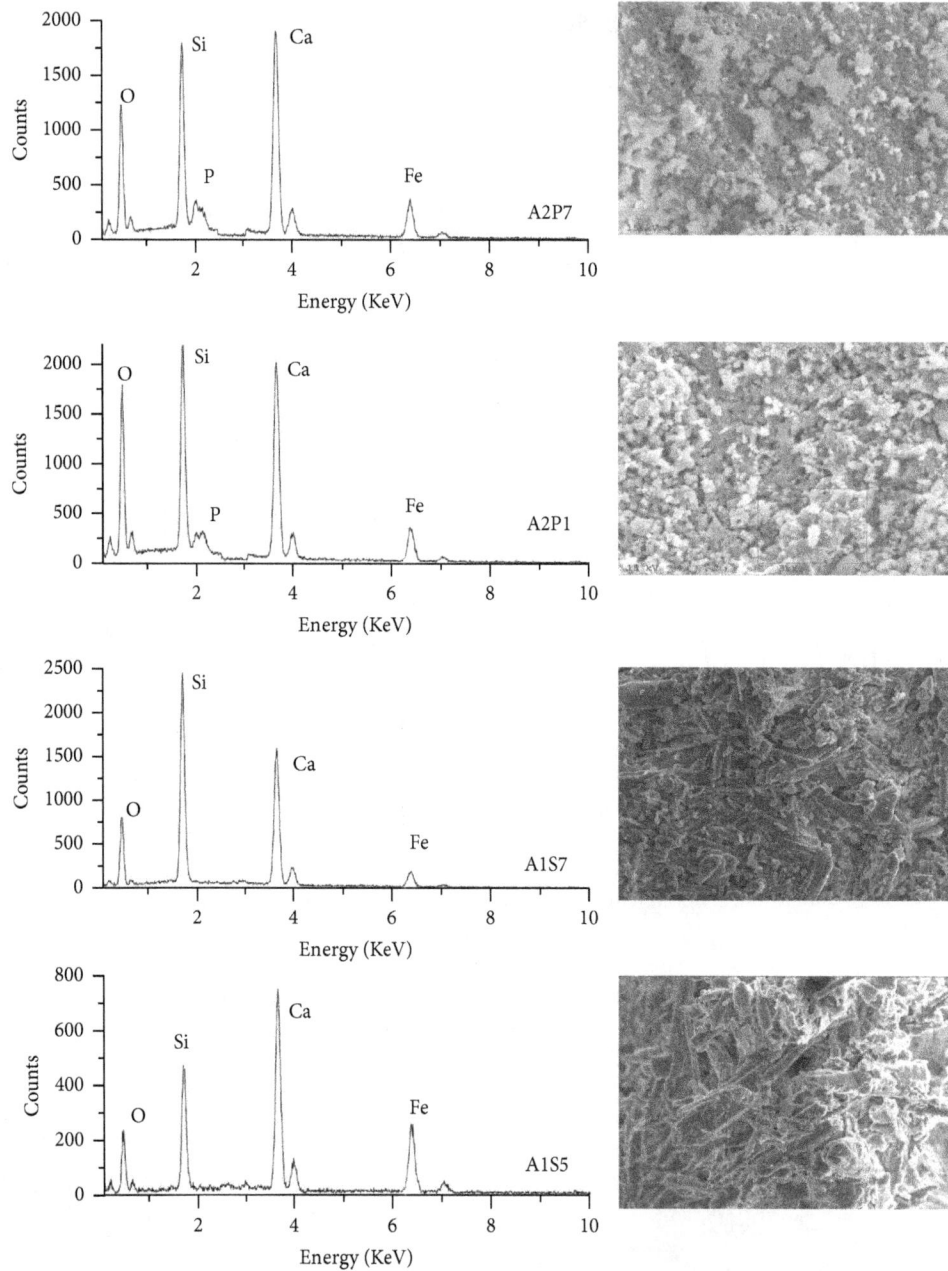

FIGURE 3: SEM-EDS spectra and images of the samples.

TABLE 5: Magnetic parameters of the samples.

	A1S1	A1S3	A1S5	A1S7	A1S9	A2P1	A2P7
H_c (G)	147.76	124.82	206.83	106.24	82.45	113.85	87.87
M_s (emu/g)	1.99	2.06	3.32	19	17.09	10.99	10.65
M_s^* (emu/g)	13.27	10.30	13.28	63.33	48.83	36.63	38.04
M_r (emu/g)	0.33	0.41	0.60	2.30	0.78	2.25	3.04
Interpolated hysteresis area	65649	176034	287151	1683193	525486	923229	980057

Note: M_s values were normalized to the total weight of the sample; M_s^* values were normalized to the weight of Fe_2O_3 component alone.

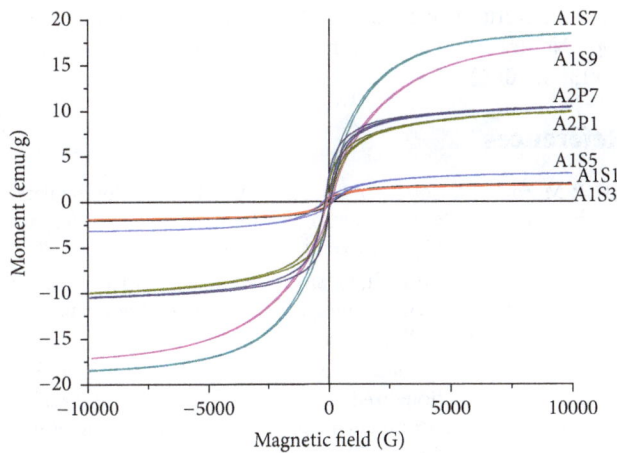

FIGURE 4: *M-H* curves of the glass ceramic samples with different chemical composition.

during the heating process. It is a known fact that P_2O_5 and SiO_2 can easily react with Fe_2O_3 forming nonmagnetic phases at high temperature [18]. Also, Fe could enter into the crystalline structure forming a solid solution [19]. The coercive field was also influenced in a significant way by the crystal dimensions [20]. The SEM results showed that the average crystallite size of A2P1 was smaller than that of A1S7. The smaller crystallite size of the magnetic phase results in a larger coercive force. These observed phenomena were similar to previous reports [21].

Figure 5 shows the relationship between the different magnetic parameters and Fe_2O_3 content of the samples. The integrated hysteresis loop area was calculated for a maximum applied field of 10000 G. The relationship between integrated loop area and the Fe_2O_3 content was similar to that observed in the case of saturation magnetization. It can be seen from Figure 5 that the hysteresis loop area of A1S7 with 30% Fe_2O_3 content is much larger than that of the other samples in group A1, which may be attributed to the higher remanence of A1S7. Theoretically, the energy generated by the glass ceramic samples under low magnetic field can be estimated from the hysteresis loop area [6]. Thus, the above results indicate that A1S7 should be capable of generating a larger amount of heat than the other samples, under a proper alternating magnetic field.

3.4. Induction Heating Ability. To test the heating ability of the samples, the induction heating experiments were performed with the different samples. Hyperthermia heating curves, measured at a fixed current of 2 A for 11 min with 1 g concentrations of the samples, are shown in Figure 6. The heating rates of the samples were basically the same, and the temperature increased rapidly until 10 min. The temperature variation amplitudes of the samples were found to increase gradually with the increase of Fe_2O_3 content of the samples. However, the induction heating ability began to fall when the Fe_2O_3 content was over 30%, which may be caused by the lower magnetite content. Sample A1S7 showed the largest temperature variation amplitudes of 2.7°C in group A1, which is consistent with the results in previous sections. More interestingly, the temperature variation amplitudes of all the

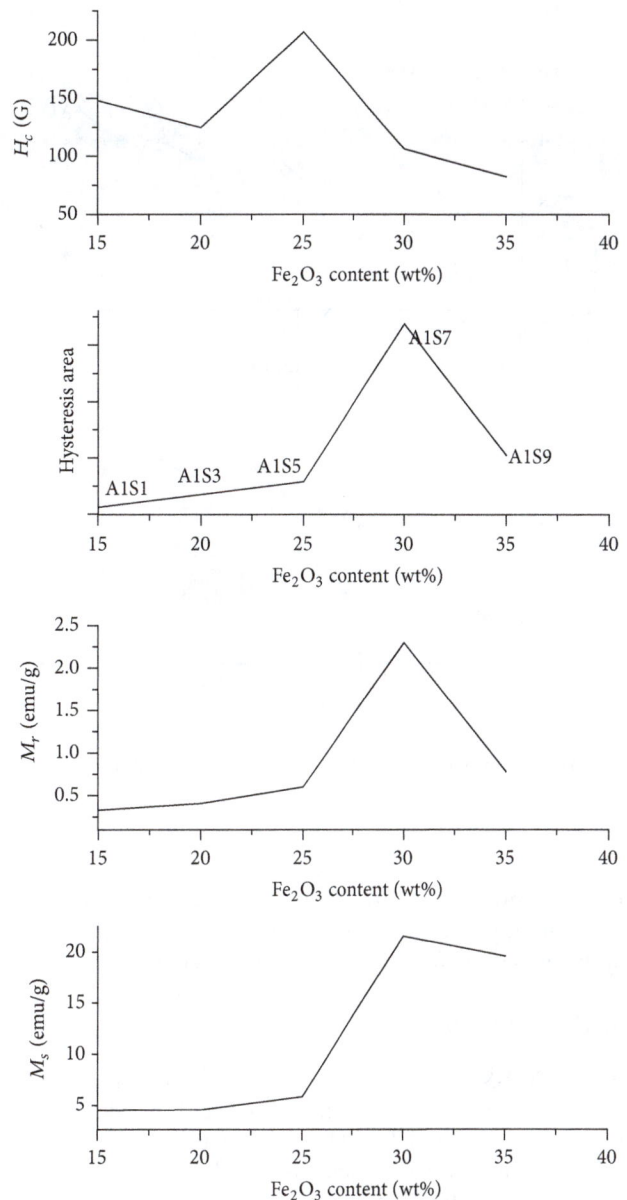

FIGURE 5: Relationship between magnetic properties and the Fe_2O_3 content.

samples in group A1 were about 2°C, while the temperature variation amplitudes of the samples in group A2 with phosphorus were as high as 15°C. The addition of phosphorus not only changed the crystalline phase of the samples but also significantly increased their induction heating ability. These results indicate that P_2O_5-Fe_2O_3-CaO-SiO_2 glass ceramics are more suitable for hyperthermia treatment than Fe_2O_3-CaO-SiO_2 glass ceramics.

In ferromagnetic materials, heat generation is mainly due to hysteresis loss. In addition, other physical parameters such as particle size and distribution of the components, as well as magnetic field intensity, also considerably influence the heating properties under AC magnetic field. The effect of these factors may be the reason for the ability of group A2 samples to generate more heat than group A1 samples.

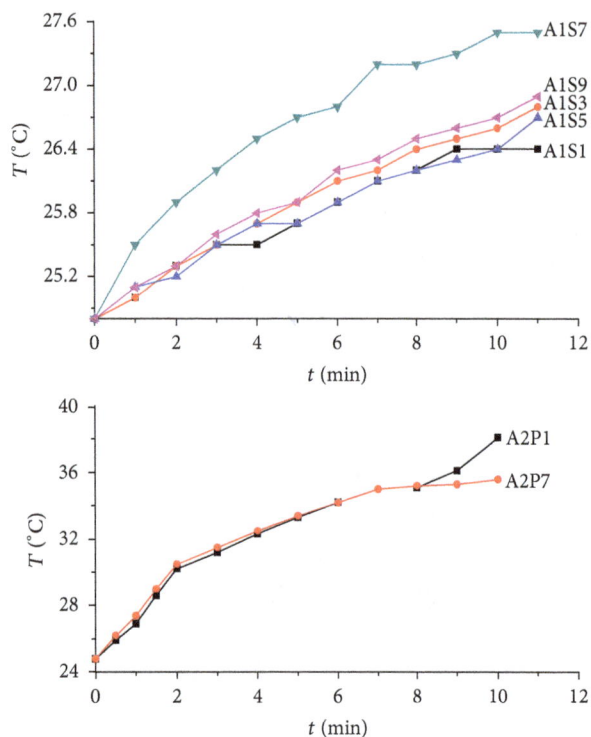

FIGURE 6: Temperature evolution as a function of time in the hyperthermia tests.

Further work is in progress in our laboratory to understand the heating properties of this class of glass ceramics and to optimize the composition for therapeutic application.

4. Conclusions

Fe_2O_3-CaO-SiO_2 glass ceramics of different compositions were prepared by the sol-gel method and tested for their induction heating ability. Magnetic properties of the glass ceramic samples were found to correlate with the Fe_2O_3 content. The saturation magnetization and the integrated loop area increased with the magnetite content in A1 group. Thus, the induction heating ability of A1S7 sample was the strongest among all the prepared Fe_2O_3-CaO-SiO_2 glass ceramics. Moreover, the induction heating ability was significantly improved when a small amount of phosphorus was introduced into the Fe_2O_3-CaO-SiO_2 glass ceramics. This reveals a promising methodology to control the magnetic properties of Fe_2O_3-CaO-SiO_2 glass ceramics through the chemical composition and its potential applications in hyperthermia treatment of cancer.

Competing Interests

The authors declare that they have no competing interests.

Acknowledgments

This research was financially supported by Natural Science Foundation for Youths of Zhejiang Province (LQ16E020001), National Natural Science Foundation of China (no. 51372081), and Natural Science Foundation of Zhejiang Province (LY13E020006).

References

[1] P. Moroz, S. K. Jones, and B. N. Gray, "Magnetically mediated hyperthermia: current status and future directions," *International Journal of Hyperthermia*, vol. 18, no. 4, pp. 267–284, 2002.

[2] P. Wust, B. Hildebrandt, G. Sreenivasa et al., "Hyperthermia in combined treatment of cancer," *The Lancet Oncology*, vol. 3, no. 8, pp. 487–497, 2002.

[3] L. Treccani, T. Yvonne Klein, F. Meder, K. Pardun, and K. Rezwan, "Functionalized ceramics for biomedical, biotechnological and environmental applications," *Acta Biomaterialia*, vol. 9, no. 7, pp. 7115–7150, 2013.

[4] Y. Ebisawa, F. Miyaji, T. Kokubo, K. Ohura, and T. Nakamura, "Bioactivity of ferrimagnetic glass-ceramics in the system FeO-Fe_2O_3-CaO-SiO_2," *Biomaterials*, vol. 18, no. 19, pp. 1277–1284, 1997.

[5] E. Ruiz-Hernández, M. C. Serrano, D. Arcos, and M. Vallet-Regí, "Glass-glass ceramic thermoseeds for hyperthermic treatment of bone tumors," *Journal of Biomedical Materials Research A*, vol. 79, no. 3, pp. 533–543, 2006.

[6] R. K. Singh, A. Srinivasan, and G. P. Kothiyal, "Evaluation of CaO-SiO_2-P_2O_5-Na_2O-Fe_2O_3 bioglass-ceramics for hyperthermia application," *Journal of Materials Science: Materials in Medicine*, vol. 20, no. 1, pp. S147–S151, 2009.

[7] Y. Y. Wang, B. Li, and Y. Y. Wang, "Characterization of Fe_2O_3-CaO-SiO_2 glass ceramics prepared by sol-gel," *Applied Mechanics and Materials*, vol. 624, pp. 114–118, 2014.

[8] Y.-K. Lee, S.-B. Lee, Y.-U. Kim et al., "Effect of ferrite thermoseeds on destruction of carcinoma cells under alternating magnetic field," *Journal of Materials Science*, vol. 38, no. 20, pp. 4221–4233, 2003.

[9] M. Matsumoto, N. Yoshimura, Y. Honda, M. Hiraoka, and K. Ohura, "Ferromagnetic hyperthermia in rabbit eyes using a new glass-ceramic thermoseed," *Graefe's Archive for Clinical and Experimental Ophthalmology*, vol. 232, no. 3, pp. 176–181, 1994.

[10] A. Matsumine, K. Kusuzaki, T. Matsubara et al., "Novel hyperthermia for metastatic bone tumors with magnetic materials by generating an alternating electromagnetic field," *Clinical and Experimental Metastasis*, vol. 24, no. 3, pp. 191–200, 2007.

[11] P. R. Stauffer, T. C. Cetas, A. M. Fletcher et al., "Observations on the use of ferromagnetic implants for inducing hyperthermia," *IEEE Transactions on Biomedical Engineering*, vol. 31, pp. 76–90, 1984.

[12] S. Mornet, S. Vasseur, F. Grasset, and E. Duguet, "Magnetic nanoparticle design for medical diagnosis and therapy," *Journal of Materials Chemistry*, vol. 14, no. 14, pp. 2161–2175, 2004.

[13] Y. Ebisawa, F. Mijaji, T. Kokubo et al., "Bioactivity of ferromagnetic glass-ceramics in the system FeO-Fe_2O_3-CaO-SiO_2," *Biomaterials*, vol. 18, no. 19, pp. 1277–1284, 1997.

[14] O. Bretcanu, E. Verné, M. Cöisson, P. Tiberto, and P. Allia, "Magnetic properties of the ferrimagnetic glass-ceramics for hyperthermia," *Journal of Magnetism and Magnetic Materials*, vol. 305, no. 2, pp. 529–533, 2006.

[15] Y.-K. Lee, K.-N. Kim, S.-Y. Choi, and C.-S. Kim, "Effect of iron state on crystallization and dissolution in Fe_2O_3-CaO-SiO_2 glasses," *Journal of Materials Science: Materials in Medicine*, vol. 11, no. 8, pp. 511–515, 2000.

[16] O. Bretcanu, S. Spriano, E. Verné, M. Cöisson, P. Tiberto, and P. Allia, "The influence of crystallised Fe_3O_4 on the magnetic properties of coprecipitation-derived ferrimagnetic glass–ceramics," *Acta Biomaterialia*, vol. 1, no. 4, pp. 421–429, 2005.

[17] N. Shankhwar, G. P. Kothiyal, and A. Srinivasan, "Understanding the magnetic behavior of heat treated $CaO–P_2O_5–Na_2O–Fe_2O_3–SiO_2$ bioactive glass using electron paramagnetic resonance studies," *Physica B*, vol. 448, pp. 132–135, 2014.

[18] Z. Zhigang, *Ferrite Magnetic Material*, Science Press, Beijing, China, 1981.

[19] Y. Ohashi and L. W. Finger, "The role of octahedral cations in pyroxenoid crystal chemistry. I. Bustamite, wollastonite, and the pectolite-schizolite-serandite series," *American Mineralogist*, vol. 63, pp. 274–288, 1978.

[20] S. Chikazumi, S. Taketomi, M. Ukita et al., "Physics of magnetic fluids," *Journal of Magnetism and Magnetic Materials*, vol. 65, no. 2-3, pp. 245–251, 1987.

[21] R. K. Singh and A. Srinivasan, "Bioactivity of ferrimagnetic $MgO-CaO-SiO_2-P_2O_5-Fe_2O_3$ glass-ceramics," *Ceramics International*, vol. 36, no. 1, pp. 283–290, 2010.

4

Optimization of Alumina Slurry for Oxide-Oxide Ceramic Composites Manufactured by Injection Molding

Catherine Billotte, Edith Roland Fotsing, and Edu Ruiz

NSERC-Safran Chair on 3D Composites for Aerospace, Department of Mechanical Engineering, École Polytechnique de Montréal, P.O. Box 6079, Centre-Ville Station, Montréal, QC, Canada H3C 3A7

Correspondence should be addressed to Catherine Billotte; catherine.billotte@polymtl.ca

Academic Editor: Marco Rossi

This paper focuses on the rheological study of an alumina suspension intended for the manufacturing of oxide-oxide composites by flexible injection. Given the production constraints, it is required to have stable suspension with low viscosity and a Newtonian behavior. This is achieved with a concentration of nitric acid between 0.08 wt% and 0.2 wt% and amount of 3 wt% of PVA binder. The maximum loading of the suspension of 47 vol% suggests that there is no structure development within the suspension with optimized concentration of acid and PVA.

1. Introduction

In the last decades, ceramic matrix composites (CMC) have been increasingly used in various engineering fields such as aerospace (rocket engine nozzle and gas turbines) and automobile (brake pads). Due to the ceramic based matrix, these materials can withstand high temperatures in oxidizing and corrosive environment. Nowadays, efforts focus on cost reduction and the improvement of CMC durability, notably for application in gas turbine engines. Moreover, CMC are more efficient in terms of energy and the light weight of these materials limits the emission of pollutants. Therefore, CMC are viable replacement for superalloys.

The presence of reinforcement not only guarantees the proper load transfer but reduces also the brittleness which characterizes ceramic materials. In fact, fiber reinforcement in CMC contributes to crack deflection, increasing therefore the hardness of the material. The compatibility between matrix and fiber maximizes the mechanical properties and limits the damage tolerance [1–3]. However, because of the crack propagation mechanism, it is necessary that the interphase between the matrix and the fiber is sufficiently weak to allow debonding and the relaxation of the stress at the crack tip. In oxide/oxide composites, there are two ways to promote the debonding during crack propagation of the

matrix: by the use of a fugitive interphase which induces small gap between the matrix and the fiber or by the use of a porous interphase. This latter leads to lowering the elastic properties and the fracture energy [3, 4]. Also, the use of continuous fibers as the reinforcement helps to increase the inelastic deformation at the fiber-matrix interface, improving strength in the presence of holes and notches [5].

Another major concern in the manufacture of CMC is oxidation during sintering or when it is subjected to high temperature stresses greater than 1000°C. The diffusion of oxygen can potentially be limited at the interface by adding additives to the matrix. However, the use of oxide-based composites can overcome these limitations. Oxide/oxide composites are more resistant to oxidation [1] and exhibit comparable mechanical properties (creep performance) to other CMC such as SiC/carbon.

The manufacturing of oxide/oxide ceramic by flexible injection [6, 7] requires the use of oxide ceramic slurry. The particles in suspension must be able to fill the mold containing the oxide reinforcement. Thus, understanding mechanisms such as the colloid stability, dispersion, and rheology are important. Homogeneous distribution of uniform particles in the matrix can promote CMC with homogeneous distribution of the micropores and minimize the shrinkage during the drying and sintering processes [8, 9]. It is therefore important to adjust the rheology of the slurry to ensure proper infiltration of the powder in strands [2, 5, 10]. This infiltration into the reinforcement requires a suspension with a high density of evenly distributed particles in terms of flocculation. Moreover, this requires the use of the dispersing agent and the binder in proper concentration.

While the viscosity of the suspension is among the most important criteria, the shear stress required to displace the suspension must be also considered. It was shown that, depending on pH of the suspension, the particles content, and the dispersant, flow rheology can change drastically from Newtonian behavior to shear thinning or thickening [8–12].

The main challenge with submicron particles remains the small separation distance, especially if the suspension is highly concentrated. A direct consequence of this small distance (less than 5 microns) is the slow settlement under gravity because Brownian motion keeps particles suspended [8]. However, due to the attractive Van der Waals force, small particles have a larger diffusion constant and thus higher probability of aggregation [8, 11, 13]. Then it is necessary to use a repulsive barrier to prevent aggregation. A wide variety of dispersants is available commercially and the choice depends not only on the type of particles, but also on the manufacturing process of CMC. There are several ways to provide repulsion between particles and overcome Van der Waals forces; however steric stabilization is particularly effective in organic liquids and aqueous solvents [13].

Polyelectrolytes and polymers are largely used as steric stabilizers in the literature [14–16]; however the long chain molecules may induce bridging and increase the viscosity of the suspensions. The bridging effect also increases with the reduction of the particle size. Several studies reported the use of small molecules such as ascorbic acid [11], citric acid [17, 18], or nitric acid [19–21] for the dispersion of alumina particles. The adsorption of the molecules occurs between its ionic functional group and the charged oxide of the alumina particle through a ligand exchange process [8, 11].

In order to increase the strength of the green body and being able to get a high density ceramic, a binder must be added to the suspension. The binder is typically a long chain polymer which is adsorbed at the surface of the particles and that can create bridges between the particles through hydrogen bonding [8, 16, 22]. Literature reports the use of cellulose derivatives such as hydroxyethyl cellulose and polyvinyl alcohol [14, 15, 21, 23–25]. When these types of binder are dispersed into water, a gel structure is formed and its action depends on the concentration of particles and level of dispersion. Depending on its molecular weight and quantity, it can strongly affect the rheology of the suspension. There are several factors to take into account when choosing the binder and its burnout characteristics. The binder should be compatible with the slurry and should not displace the dispersant from the particle surface. Additionally, for oxides, it should be less polar than the dispersant [13]. It is necessary to optimize the amount of the binder to form a ceramic cake that can be easily demolded after processing [14].

This paper deals with the influence of PVA, nitric acid, as well as the particles size, and solid loading on the dispersion of α-Al$_2$O$_3$ slurry. Since the suspension is used as matrix for

oxide/oxide processing by flexible injection, the impact of mixing procedure and shear stress on the viscosity will be also assessed and discussed.

2. Materials and Methods

2.1. Materials and Suspension Preparation. The suspensions used in this work were made of α-Al$_2$O$_3$ powder with a BET specific surface of 10 m^2/g and average particle size of 0.3 μm. The powder has a bulk density of 0.8 g/cm^3 and a tap density of 1.1 g/cm^3. Particles morphology was observed in Scanning Electron Microscope (Field Emission Gun FEG-SEMJSM-7600F, JEOL). Dispersion was achieved by adding nitric acid with a purity of 69 wt%. Polyvinyl alcohol (PVA) diluted in deionized water (35 wt%, 1.08 g/cm^3 at 23°C) was used as the binder.

Suspensions with varying solid concentrations of α-Al$_2$O$_3$ powder were prepared using deionized water and HNO$_3$. The basic suspension used, as reference, for most of the study contained 33 vol% alumina particles. The powder is incorporated and dispersed using magnetic stirring for 1 h and then deagglomerated in ultrasonic bath for 5–15 minutes. PVA was added to the suspension using magnetic stirring. The pH of the α-Al$_2$O$_3$ slurry is adjusted with the quantity of HNO$_3$ and was found to settle between 5 and 6 for an acid concentration of 0.06 mol/L (0.17 wt%).

In order to assess the influence of acid concentration and PVA on dispersion, the amount of HNO$_3$ was varied between 0 and 0.34 wt% and the quantity of PVA added to the slurry between 0 and 16.48 wt%. The maximum packing fraction was also investigated using viscosity data of slurries containing 10 to 46 vol% of Al$_2$O$_3$ particles.

2.2. Rheological Measurements. Rheological measurements were performed using a MCR501 rheometer from Anton-Paar, with concentric cylinders' geometry (27 mm bob diameter, 1.17 mm gap). Preliminary rheology tests have been done on the slurry in order to establish the required conditioning (preshear and rest time) to get repeatable and consistent data. It is required for suspension since flow history has a great impact on rheological measurement [26, 27]. The conditioning defined for the suspensions is a preshear of 10 s^{-1} during 5 min followed by a rest time of 5 min. The preshear and rest time must be applied before and between every test. These conditions were defined with the reference suspension containing 33 vol% of Al$_2$O$_3$ particles and are valid for other particles concentration if the amount of HNO$_3$ is enough to maintain the particles dispersed. Potential wall slip and sedimentation were verified as well and the slurry remained stable under the testing conditions. Structure rebuilt experiments were performed and the suspension retrieves its original properties almost immediately after preshear. Moreover, no thixotropy was observed as successive tests showed curves superimposed. Since the suspension is intended for an injection process, no yield stress is expected for standard slurries, verification has been done, and this point is discussed in the Results. All measurements were conducted at 23°C. In order to prevent evaporation, mineral oil has been poured on top of suspension; this liquid is inert and does not interact with the measurement.

Steady shear experiments have been carried out by running loops from 0.1 to $1000\,s^{-1}$ and back down to $0.1\,s^{-1}$. Small amplitude oscillatory shear (SAOS) tests have been conducted from 0.1 Hz to 100 Hz with strain amplitude of 10%. This amplitude of 10% is inside the linear viscoelastic (LVE) domain.

3. Results and Discussion

The rheological behavior of suspensions depends on a number of variables such as the particle shape, size, and distribution of particles sizes but also the interaction forces. The chemicals used in the slurry preparation have great influence on the stability of the suspension [8]. It is therefore crucial to understand how the chemicals interact with each other and with alumina particles to get an acceptable suspension for the injection process.

In this work, the same size and distribution of alumina particles were used for all slurry preparation. The particle SEM morphology presented in Figure 1 shows a fairly even distribution of irregular 3D flake particles. The morphology and the particle size greatly influence the rheological behavior of the suspension, notably at high volume fraction. The reference suspension is composed of 33 vol% Al_2O_3 particles (so-called B33) and 0.17 wt% HNO_3. The basic quantity of PVA binder solution corresponds to 6% of the mass of the alumina particles, which is 3.79 wt% for the B33 suspension. It should be noted that the PVA used in this study is an aqueous solution diluted to 35% by weight, and therefore the weight percentage reported relates to the solution. The choice of the particle base concentration is justified in terms of flow, mold filling, and compaction for the injection process. The influence of dispersant, binder, and particles concentration are investigated in this study.

3.1. Influence of Acid Content. B33 suspensions with HNO_3 varying from 0 to 0.34 wt% were prepared (Table 1). Figure 2 illustrates the viscosity variation with shear rate. Lower viscosity preparations show the most stable curves in the back-forward loops and their flow response is more Newtonian. Similar results have been observed by other authors with ascorbic acid as dispersant [11]. In addition, there is also an overlap of the curves in Figure 2 for some acid contents (0.17 and 0.08 wt%), suggesting that there is a stability zone for the suspension. An offset between the back-forward loops for the 0.04 wt%, 0.25 wt%, and 0.34 wt% suspensions may be an indication of inadequate dispersion of the alumina particles. For low percentages, the attraction forces between particles are too strong, leading to heterogeneities and possible flocculation. When the repulsive forces are too strong, aggregates can be formed and permanent precipitation will occur.

Figure 3 shows the viscosity values at $0.1\,s^{-1}$ of alumina slurry as function of the acid concentration from 0 wt% up to 0.34 wt%. Suspensions with 0 wt% and 0.01 wt% acid (not shown) were tested in a similar manner and a viscosity of about 1000 and 200 Pa·s was measured at $0.1\,s^{-1}$. However, these suspensions are very different with a foamy aspect,

contrasting with suspensions containing more than 0.04 wt% HNO_3 which have a smooth and homogeneous appearance (see Figure 4). Between percentages of 0.01 and 0.04 wt% HNO_3, there is a change in the structure of the slurry itself making measurements more difficult. These suspensions show immediate separation in a very short time just after preparation. The resulting parabolic profile of Figure 3 suggests that there is a critical range of acid concentration between 0.08 and 0.2 wt% where the viscosity is minimum.

When alumina particles are immersed in water, there is a hydrolysis reaction between alumina and hydroxide, and this is enhanced with pH. In the presence of nitric acid, the hydroxyl group on the alumina surface reacts with the H+ that come from the acid. The addition of nitric acid promotes the production of positive charge on the particles and thus repulsive forces [8]:

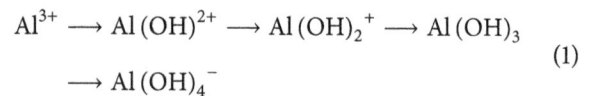

$$Al^{3+} \longrightarrow Al(OH)^{2+} \longrightarrow Al(OH)_2{}^{+} \longrightarrow Al(OH)_3$$
$$\longrightarrow Al(OH)_4{}^{-} \tag{1}$$

The reaction of alumina particles surface (s Al–OH) with H^+ of the nitric acid is

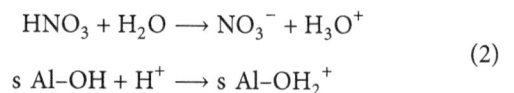

$$HNO_3 + H_2O \longrightarrow NO_3{}^- + H_3O^+$$
$$s\ Al–OH + H^+ \longrightarrow s\ Al–OH_2{}^+ \tag{2}$$

The objective is to obtain electrical stability within the suspension. Below the critical zone, the amount of nitric acid adsorbed on particles surface is insufficient for adequate dispersion and above this zone; the excess of acid may create aggregates and precipitates, both resulting in higher viscosities.

To obtain a uniform green body, it is necessary to disperse adequately the alumina particles in the media and the additives added to the suspension. In this manner, the interactions between the particles will be improved and controlled. Rheology is used to understand the behavior of the suspension and provides information on the interaction between the particles. Strong interactions between particles result in high viscosities and shear-thinning behavior whereas repulsion results in stable suspensions with low viscosities and eases the injection process. On the process level, having a critical acid zone rather than a critical value provides process flexibility. A variation in the critical area will not affect the quality of the final ceramic part. At the critical point, the lowest viscosity is obtained, meaning that the pH of the suspension is below the isoelectric point (IEP) of the oxide powder and above the acidic dissociation constant of the acidic functional group adsorbed [8, 11].

3.2. Influence of PVA Content. B33 slurries were prepared using PVA concentrations varying from 0 to 16.48 wt%. Figure 5 presents the variation of steady shear viscosity with shear rate for different concentrations of PVA. The filled and empty symbols correspond to the back and forth viscosity change with shear rate. For the same PVA content, curves are superimposed and present a Newtonian behavior at low shear rate followed by a slight shear thinning starting around $10\,s^{-1}$.

FIGURE 1: FEG-SEM micrographs of alumina particulates. Scale 1 micron (a) and 100 nm (b).

TABLE 1: B33 suspension preparation with change in PVA and acid content. B33 base concentration contains 0.17 wt% acid and 3.80 wt% PVA.

| | PVA content | | | | | Acid content | | | |
#	Ratio $(\times C_{base})$	wt% acid in H_2O	wt% acid[1]	wt% PVA[1]	#	Ratio $(\times C_{base})$	wt% acid in H_2O	wt% acid[1]	wt% PVA[1]
B33-PVA0	0	0.51	0.17	0	B33-AC0	0	0.00	0.00	3.80
B33-PVA0.5	0.5	0.51	0.17	1.93	B33-AC0.25	0.25	0.13	0.04	3.80
B33 (ref)	1	0.51	0.17	3.80	B33-AC0.5	0.5	0.26	0.08	3.80
B33-PVA2	2	0.51	0.16	7.31	B33 (ref)	1	0.51	0.17	3.80
B33-PVA5	5	0.51	0.15	16.48	B33-AC1.5	1.5	0.76	0.25	3.80
					B33-AC2	2	1.02	0.33	3.80

C_{base}: reference formulation, with base = PVA or acid. Reference formula corresponds to 1 in ratio column; [1]percentage based on total weight of the slurry.

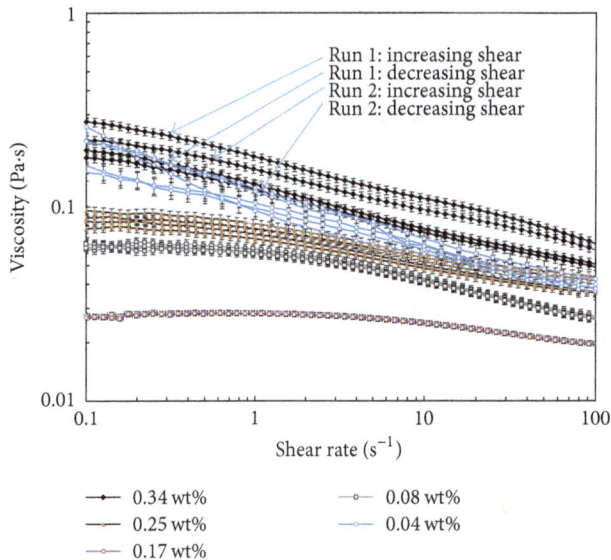

FIGURE 2: Viscosity versus shear rate for two consecutive back-forward loops for B33 slurries with varied acid content from 0.04 to 0.34 wt%.

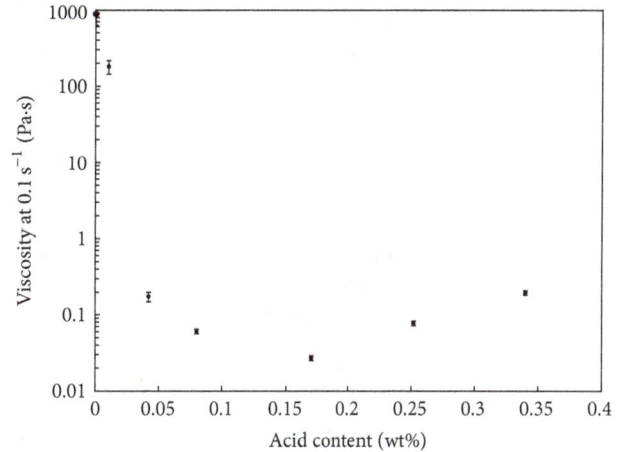

FIGURE 3: Viscosity at $0.1\,s^{-1}$ shear rate versus acid content for B33 slurries with varied acid content from 0 to 0.34 wt%.

As expected, an increase of viscosity with increasing PVA content is observed. For the highest concentration of PVA (16.48 wt%), a bump appears after the Newtonian plateau. It could be associated with a possible change of dispersion state due to an overabundance of PVA, sedimentation, or the beginning of a phase separation. This illustrates the fact that there is a threshold PVA content, which does not affect the dispersion while allowing particles to adhere adequately to form the ceramic cake.

In order to understand the combined effect of PVA and the concentration of the slurry, suspensions containing 25 to 46 vol% of particles with and without PVA were prepared. For

TABLE 2: Different alumina slurries with particulates varying from 10 to 46 vol%. Preparations with and without PVA. Acid content of 0.51 wt% in H_2O for all slurries and PVA content of 6 wt% of alumina particles for slurries with PVA.

#	vol% alumina	wt% acid in H_2O	With PVA (6% wt. alumina)			No PVA		
			wt% acid[1]	wt% alumina[1]	wt% PVA[1]	wt% acid[1]	wt% alumina[1]	wt% PVA[1]
—	Media	0.51	0.51	0	0	0.51	0	0
B10	10	0.51	0.35	29.69	1.78	0.36	30.23	0
B25	25	0.51	0.21	54.67	3.28	0.22	56.52	0
B33	33	0.51	0.17	63.27	3.80	0.18	65.76	0
B40	40	0.51	0.14	69.22	4.15	0.14	72.22	0
B46	46	0.51	0.11	73.48	4.41	0.11	76.86	0

[1]Percentage based on total weight of the slurry.

(a)   (b)

FIGURE 4: Foaming structure for B33 slurries with low acid content of 0 and 0.01 wt% (a) and liquid-like B33 slurries with acid content higher than 0.04 wt% (b).

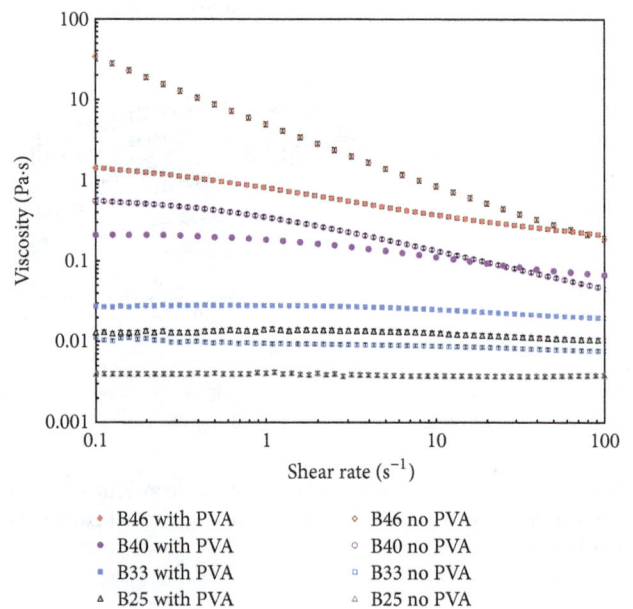

FIGURE 5: Viscosity versus shear rate for back-forward loop for B33 slurries with varied PVA content from 0 to 16.48 wt%.

suspensions containing PVA, the amount of polymer solution corresponds to 6% of the mass of the alumina particles. This ensures that the ratio of adsorption between particles and PVA is constant, assuming that alumina particles are well dispersed into the media. These suspensions are listed in Table 2 and Figure 6 presents the variation of viscosity with shear rate for B10 to B46 slurries. The filled symbols correspond to the suspensions with PVA whereas the empty symbols correspond to slurry without PVA. Generally, for suspensions with PVA, an increase of the particle concentration leads to an increase of the viscosity from 0.01 Pa·s for B25 to 1.3 Pa·s for B46. For low charge suspensions with PVA, the behavior is Newtonian. Towards 40 vol%, the shear-thinning behavior begins to be more pronounced and at 46 vol% the Newtonian plateau disappears and a shear-thinning behavior is observed over the whole range of the shear rate. The B46 suspension with no PVA shows a yield stress. It was evaluated at 1.9 Pa using the Hershel-Bulkley model (not shown here). However, the B46 no PVA suspension is a limit condition and is not suitable for processing. Slurries with lower percentages of alumina particles show a typical power-law flow behavior.

For suspensions B33 and B25, those that do not contain PVA have a lower viscosity than those with PVA. The viscosity

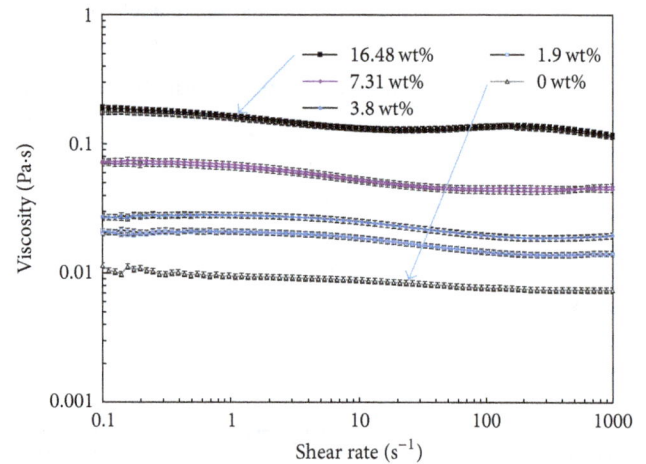

FIGURE 6: Viscosity versus shear rate for slurries containing 25 to 46 vol% of alumina particulates. Slurries with and without PVA. PVA content of 6% of weight of particulates for slurries containing PVA.

of these suspensions with no PVA is low (0.00941 Pa.s for B33 no PVA and 0.00420 Pa·s for B25 no PVA) and close to that of the corresponding suspensive media with PVA (0.00148 Pa·s for B33 and 0.00139 Pa·s for B25). Thus, the viscosity of the media has an influence on the lowest charged suspensions. The increase of viscosity related to the addition of PVA in the media is of the order of 200% for B33 and 240% for B25. For highly charged suspensions B40 and B46, the addition of PVA results in a change in the rheology behavior of the suspension from a strong shear-thinning behavior to a weak shear thinning approaching Newtonian.

The quantity of PVA is adjusted according to the dry mass in alumina (6% of mass of particles); thus the ratio remains constant. Since PVA comes from an aqueous solution 35 wt%, its addition to the slurry slightly lowers the suspension viscosity and decreases the particle loading slightly (in the order of 3-4%). On the other hand, all suspensions were prepared using the same acidified water (same concentration of HNO_3 dispersant). This means that there is probably a small excess of HNO_3 for B25 suspension while B33 suspension exhibits an optimized concentration of nitric acid (Figure 3). For B40 and B46 suspensions, this concentration is less and this leaves room for additional adsorption of PVA on the particles. For alumina suspensions containing both dispersant and PVA binder, it was reported that the dispersant is first adsorbed on particle surface; then the binder coats the rest of the particle [13, 16, 22]. Depending on the dispersant/binder ratio, the rheology of the slurry changes from a more or less shear thinning to Newtonian behavior [16].

Polyvinyl alcohol (PVA) is prepared by hydrolysis of polyvinyl acetate (PVAc), usually leaving a small fraction of unhydrolysed acetate groups in the macromolecule chain structure that contribute to PVA stabilization, preventing depolymerisation [28, 29]. The PVA is adsorbed on the surface of the alumina particles by hydrogen bonding of its –OH functional groups. However, in the case of partially hydrolyzed PVA, which is the PVA used in this study, the adsorption on the alumina particle surface might be preferential for the acetate groups $–OCOCH_3$ rather than the hydroxyl groups [29, 30]. These groups have great influence on the adsorption properties of PVA, providing a chemical conformation of the polymer with a structure with more loops and tails adsorbed better to the surface of the alumina particle [30]. Acetate groups are also sensitive to pH with the level of interactions increasing with pH [29, 31]. Wiśniewska et al. [29] have demonstrated that there is a stability pH range between 3 and 6 for alumina suspension and that beyond the PVA is more stretched and thus the adsorption of the polymer chain is on few particles, causing bridging flocculation of suspension. The alumina suspension in this study has a pH of 5-6, in the limits of stability.

Combining the results from Figure 5 on the impact of PVA concentration for B33 slurry with those of Figure 6 presenting the impact of similar ratio of PVA/alumina for B25 to B46 slurries, it came out that a threshold of 3% of mass of alumina particles can be set for the PVA content. This value corresponds to only 1.9 wt% for B33 slurry suggesting that low viscosity is achievable up to 40 vol% alumina by using the optimal amount of PVA. However, slurry with 33 vol%

of particles with PVA which exhibits a Newtonian behavior combined to a low viscosity represents the best potential with respect to the processability of the oxide/oxide ceramic.

At low shear rate, the interparticles forces and the Brownian motion dominate whereas the hydrodynamic forces predominate and overcome the interparticles interactions at higher shear rates leading to a decrease of the viscosity. For more concentrated suspensions, the separation distance between particles is narrowed and the probability to aggregate increases due to apparition of the Van der Waals attractive forces [8]. This explains the shear-thinning behavior observed for the highly charged suspensions (Figure 6). This rheological behavior was observed for similar systems using oxide nanoparticles in aqueous media, the shear thinning being enhanced by particle loading [12].

The role of PVA is to increase the strength of the green body and it therefore acts as the binder in the alumina suspension. PVA coats the surface of the particles and improves the adhesion between the particles during the formation of the ceramic cake. PVA is water-soluble and is removed from the green body during drying and sintering process. However, its quantity needs to be optimized in order to limit the formation of macroporosities during the sintering process.

3.2.1. SAOS Experiments. In order to understand the combined impact of acid and PVA on the structure of the suspension, Small Amplitude Oscillatory Shear (SAOS) experiments were performed on B33 slurry. Frequency sweep tests were performed on B33 samples with different acid and PVA contents listed in Table 1. Figures 7 and 8 present the change of the loss modulus G'' with frequency for B33 suspensions with different acid and PVA contents, respectively. It can be observed that there is no elastic response, which is consistent with the absence of structure in the suspension. The suspension remains viscous over the entire frequency range tested, and so the energy remains dissipated. The main reason is that slurry remains liquid even with 16.48 wt% PVA solution and there is no development of a gel structure. The loss modulus increases progressively with the frequency and an abrupt change of slope is observed towards 10 Hz (6 rad/s). The friction between the alumina particles at high frequencies can explain this increase of the loss modulus. The friction increases energy dissipation in the slurry and is only related to the particles. A direct influence of nitric acid and PVA was not observed.

3.3. Mixing Equipment on Dispersion. The standard suspension preparation process consists of incorporating different materials in the aqueous media using mechanical agitation with magnetic stirrer for 1 hour followed by an exposure to ultrasound for 5-15 minutes. This type of process deagglomeration of the alumina particles has been widely used by several authors [16, 17, 20, 21, 24]. Preparation with high shear mixer (HSM) is also increasingly used in particular for dispersion of nanoclays [32, 33]. The HSM breaks aggregates by passing the suspension through a small capillary at very high pressure. The B33 suspensions were prepared in the usual manner and using HSM. Figure 9 shows the changes in viscosity as a function of the shear rate for the various

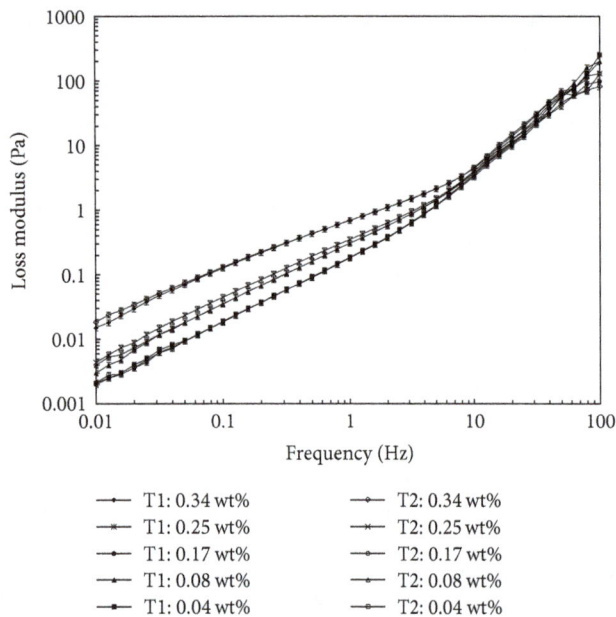

FIGURE 7: Loss modulus versus frequency for B33 slurries containing varied acid content from 0.04 to 0.34 wt%. Two tests illustrated T1 and T2.

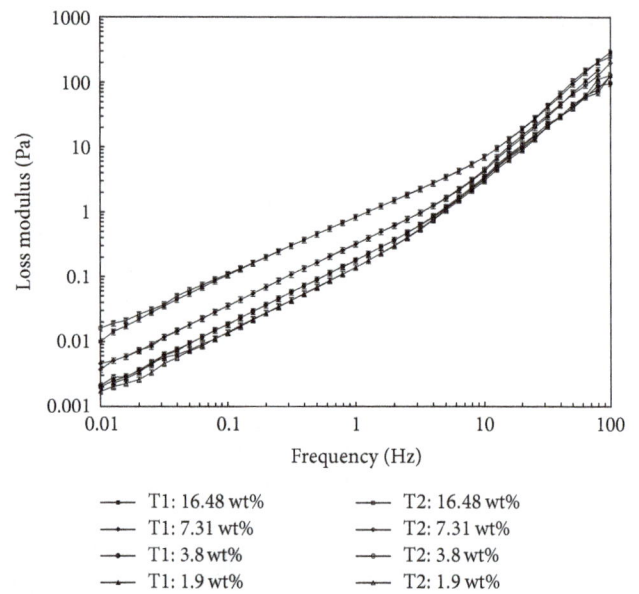

FIGURE 8: Loss modulus versus frequency for B33 slurries containing varied PVA content from 0 to 16.48 wt%. Two tests illustrated T1 and T2.

mixtures. The term HSM 1 p corresponds to 1 pass in the HSM, 5 p to 5 passes, and so on. If 1 and 2 passes in the HSM decrease the viscosity of the B33 slurry, which is an indicator of a better dispersion [8, 11, 14], 5 passes' HSM shows similar results when compared to regular US mix. The viscosity continues to increase after 10 passes. This increase in viscosity can be related to water evaporation associated with overheat in the HSM. However, the viscosity stays in the same decade, varying from 0.025 to 0.04 Pa·s, which is not sufficiently significant at the process level to change of the mixing procedure.

3.4. Maximum Packing Factor. In the suspension, there is a moment where the loading level in particles reaches a maximum. This level defines the maximum packing beyond which the viscosity of the suspension tends to infinite. The shape, particle surface, and the aspect ratio (the ratio between the diameter and the length of the particles) are important aspects to consider when determining the maximum packing. When this ratio increases, the maximum packing tends to decrease. Irregularities on the particle surface tend to decrease the value of the maximum packing, which is the case in this study, with the 3D irregular flake shape of alumina particles (Figure 1). Metzner [34] noted a value of 44 vol% for a rough spherical crystal compared to 68 vol% for smooth spheres.

The maximum packing factor was evaluated for the alumina slurry with and without PVA (Table 2) by using an empirical expression originated by Maron and Pierce:

$$\eta_r = \left(1 - \frac{\Phi}{A}\right)^{-2}, \tag{3}$$

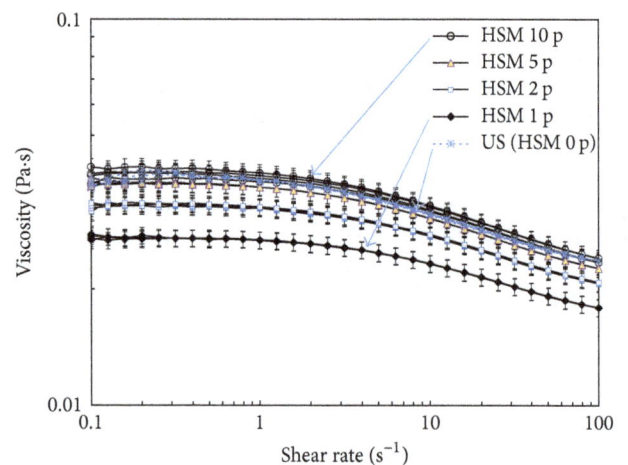

FIGURE 9: Viscosity versus shear rate for B33 base slurry (0.17 wt% acid and 3.80 wt% PVA) preparation using HSM. Consecutive pass in HSM using the same mixture from 1 pass to 10 passes. Slurry with 0 passes corresponds to US.

where η_r is the ratio of the suspension viscosity over the viscosity of the suspending media, evaluated at the same shear stress [35]. This equation allows the determination of an empirical constant A which can be associated with the maximum volumetric packing factor Φ_M of the suspension. Equation (3) can be rewritten as

$$\sqrt{\frac{1}{\eta_r}} = 1 - \frac{\Phi}{\Phi_M}. \tag{4}$$

Relative viscosities of suspensions B10 to B40 are shown in Figure 10. Neither B46 suspensions nor B40 without PVA show Newtonian area, and therefore these slurries were not

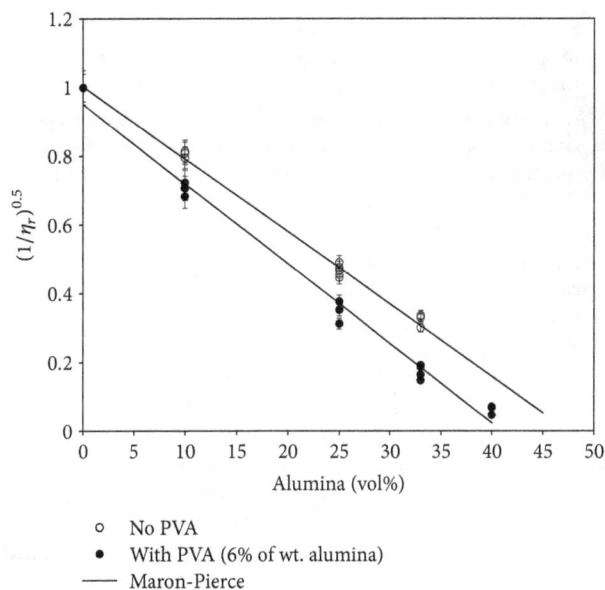

FIGURE 10: Relative viscosity versus alumina concentration for slurries with and without PVA containing 0 to 40 vol% of alumina particulates. Modeling using Maron and Pierce equation.

presented in Figure 10. The maximum packing Φ_M in (4) is calculated from the slope of the linear fitting of the data. The coefficient of determination R^2 between the experimental data and the model is 0.9946 for no PVA suspensions and 0.9844 for suspensions containing PVA. There is a slight difference in slopes for suspension with and without PVA. A maximum packing of 43.1 vol% was calculated for slurries with PVA and 47.4 vol% for slurries without PVA. This confirms that a suspension up to 40 vol% particles can be easily prepared and meet the requirements in terms of packing. At this concentration, the slurry is still liquid and injectable.

The PVA is adsorbed on the surface of the particles, which are already dispersed by nitric acid. The same amount of HNO_3 was used for all the suspensions. It is likely that the PVA has an impact on the dispersion of particles as shown by the difference of 4.3% between Φ_M of suspensions with and without PVA, confirming results discussed above (PVA section). However, this difference is small considering that the PVA content corresponds to 6% of the mass of particles. This confirms that there is probably a large excess of polymer and it is not necessary to put as much for the cohesion of particles during drying and sintering; only a fraction of the polymer is adsorbed on particles surface. These results on the influence of the PVA are added to the previous ones (Figures 5 and 6) showing its influence on the viscosity, the rheology behavior, and the particle loading of the slurry. Given that PVA is the binder that allows the ceramic particles to adhere, a minimum of 3% of the mass of particles is necessary to ensure a ceramic cake that holds. This quantity was confirmed visually by the fact that cakes with less PVA collapse. This percentage of 3% makes it possible to have a stable Newtonian rheological behavior and a low viscosity of 0.03 Pa·s for B33 slurry that ease the injection process and fiber impregnation.

4. Conclusions

Different Al_2O_3 suspensions have been studied with respect to their composition and concentration. The analysis was motivated by the need to provide valuable data to improve the processing of oxide/oxide composites by flexible injection. The formulation is simple and requires the use of an acidic dispersant and a binder. The slurry can be processed by using mechanical stirring and ultrasounds for dispersion. The following main results were obtained:

(i) Based on the rheology analysis, it was found that there is a critical concentration of HNO_3 between 0.08 and 0.2 wt% that ensures a stable dispersion and a lower viscosity.

(ii) A concentration of PVA binder of 3% of particle mass is enough to ensure a Newtonian behavior and low viscosity and ease the injection of suspension charged up to 40 vol% particulates.

(iii) The SAOS measurements show that there is no structure development with increased frequency. Increasing loss modulus is consistent with energy dissipation occurring over the entire frequency range.

(iv) The maximum packing fraction is 47.4 vol% alumina for suspension without PVA and 43.1 vol% for suspensions with PVA content corresponding to 6% mass of particles. The small difference in slope suggests that the polymer is in excess in the suspension. Reducing this quantity by half should be sufficient to reach suitable viscosity for injection molding.

Conflicts of Interest

The authors declare that there are no conflicts of interest regarding the publication of this paper.

Acknowledgments

This research was supported by the Natural Sciences and Engineering Research Council of Canada (NSERC). The authors are also grateful to the industrial partner Groupe Safran for supporting this project.

References

[1] M. Parlier and M. H. Ritti, "State of the art and perspectives for oxide/oxide composites," *Aerospace Science and Technology*, vol. 7, no. 3, pp. 211–221, 2003.

[2] P. Colomban and G. Gouadec, "The ideal ceramic-fibre/oxide-matrix composite: how to reconcile antagonist physical and chemical requirements ?" *Annales de Chimie: Science des Materiaux*, vol. 30, no. 6, pp. 673–688, 2005.

[3] F. W. Zok, "Developments in oxide fiber composites," *Journal of the American Ceramic Society*, vol. 89, no. 11, pp. 3309–3324, 2006.

[4] J. H. Weaver, J. Rannou, M. A. Mattoni, and F. W. Zok, "Interface properties in a porous-matrix oxide composite," *Journal of the American Ceramic Society*, vol. 89, no. 9, pp. 2869–2873, 2006.

[5] C. G. Levi, J. Y. Yang, B. J. Dalgleish, F. W. Zok, and A. G. Evans, "Processing and performance of an all-oxide ceramic composite," *Journal of the American Ceramic Society*, vol. 81, no. 8, pp. 2077–2086, 1998.

[6] M. Podgorski, C. Billotte Cabre, B. Dambrine, L. Molliex, E. Ruiz, and S. Turenne, "Method for manufacturing part made of composite material," 2016.

[7] M. Podgorski, C. Billotte Cabre, B. Dambrine, L. Molliex, E. Ruiz, and S. Turenne, "Method for manufacturing a refractory part made of composte material," 2016.

[8] G. V. Franks and Y. Gan, "Charging behavior at the alumina-water interface and implications for ceramic processing," *Journal of the American Ceramic Society*, vol. 90, no. 11, pp. 3373–3388, 2007.

[9] P. Tomasik, C. H. Schilling, R. Jankowiak, and J.-C. Kim, "The role of organic dispersants in aqueous alumina suspensions," *Journal of the European Ceramic Society*, vol. 23, no. 6, pp. 913–919, 2003.

[10] M. Parlier, M. H. Ritti, and A. Jankowiak, "Potential and perspectives for oxide/oxide composites," *Journal Aerospace Lab*, vol. 3, pp. 1–11, 2011.

[11] S. Çinar and M. Akinc, "Ascorbic acid as a dispersant for concentrated alumina nanopowder suspensions," *Journal of the European Ceramic Society*, vol. 34, no. 8, pp. 1997–2004, 2014.

[12] S. Çinar, D. D. Anderson, and M. Akinc, "Influence of bound water layer on the viscosity of oxide nanopowder suspensions," *Journal of the European Ceramic Society*, vol. 35, no. 2, pp. 613–622, 2015.

[13] M. N. Rahaman, "Science of colloidal processing," in *Ceramic Processing and Sintering*, pp. 181–245, Marcel Dekker, Inc., New York, NY, USA, 2003.

[14] N. Kiratzis and P. F. Luckham, "Rheological behavior of stabilized aqueous alumina dispersions in presence of hydroxyethyl cellulose," *Journal of the European Ceramic Society*, vol. 18, no. 7, pp. 783–790, 1998.

[15] N. E. Kiratzis and P. F. Luckham, "The rheology of aqueous alumina suspensions in the presence of hydroxyethylcellulose as binder," *Journal of the European Ceramic Society*, vol. 19, no. 15, pp. 2605–2612, 1999.

[16] M. R. Ben Romdhane, S. Baklouti, J. Bouaziz, T. Chartier, and J.-F. Baumard, "Dispersion of Al_2O_3 concentrated suspensions with new molecules able to act as binder," *Journal of the European Ceramic Society*, vol. 24, no. 9, pp. 2723–2731, 2004.

[17] Y.-K. Leong, "Role of molecular architecture of citric and related polyacids on the yield stress of α-alumina slurries: inter- and intramolecular forces," *Journal of the American Ceramic Society*, vol. 93, no. 9, pp. 2598–2605, 2010.

[18] K. Sato, H. Yilmaz, A. Ijuin, Y. Hotta, and K. Watari, "Acetic acid mediated interactions between alumina surfaces," *Applied Surface Science*, vol. 258, no. 8, pp. 4011–4015, 2012.

[19] L. F. G. Setz, A. C. Silva, S. C. Santos, S. R. H. Mello-Castanho, and M. R. Morelli, "A viscoelastic approach from α-Al_2O_3 suspensions with high solids content," *Journal of the European Ceramic Society*, vol. 33, no. 15-16, pp. 3211–3219, 2013.

[20] L. Stappers, L. Zhang, O. Van der Biest, and J. Fransaer, "The effect of electrolyte conductivity on electrophoretic deposition," *Journal of Colloid and Interface Science*, vol. 328, no. 2, pp. 436–446, 2008.

[21] F. Chabert, D. E. Dunstan, and G. V. Franks, "Cross-linked polyvinyl alcohol as a binder for gelcasting and green machining," *Journal of the American Ceramic Society*, vol. 91, no. 10, pp. 3138–3146, 2008.

[22] S. Baklouti, T. Chartier, and J. F. Baumard, "Binder distribution in spray-dried alumina agglomerates," *Journal of the European Ceramic Society*, vol. 18, no. 14, pp. 2117–2121, 1998.

[23] A. L. Ahmad, M. R. Othman, and N. F. Idrus, "Synthesis and characterization of nano-composite alumina-titania ceramic membrane for gas separation," *Journal of the American Ceramic Society*, vol. 89, no. 10, pp. 3187–3193, 2006.

[24] M. A. Huha and J. A. Lewis, "Polymer effects on the chemorheological and drying behavior of alumina-poly(vinyl alcohol) gelcasting suspensions," *Journal of the American Ceramic Society*, vol. 83, no. 8, pp. 1957–1963, 2000.

[25] K. Livanov, H. Jelitto, B. Bar-On, K. Schulte, G. A. Schneider, and D. H. Wagner, "Tough alumina/polymer layered composites with high ceramic content," *Journal of the American Ceramic Society*, vol. 98, no. 4, pp. 1285–1291, 2015.

[26] C. Billotte, P. J. Carreau, and M.-C. Heuzey, "Rheological characterization of a solder paste for surface mount applications," *Rheologica Acta*, vol. 45, no. 4, pp. 374–386, 2006.

[27] G. P. Citerne, P. J. Carreau, and M. Moan, "Rheological properties of peanut butter," *Rheologica Acta*, vol. 40, no. 1, pp. 86–96, 2001.

[28] S. Baklouti, J. Bouaziz, T. Chartier, and J.-F. Baumard, "Binder burnout and evolution of the mechanical strength of dry-pressed ceramics containing poly (vinyl alcohol)," *Journal of the European Ceramic Society*, vol. 21, no. 8, pp. 1087–1092, 2001.

[29] M. Wiśniewska, S. Chibowski, T. Urban, and D. Sternik, "Investigation of the alumina properties with adsorbed polyvinyl alcohol," *Journal of Thermal Analysis and Calorimetry*, vol. 103, no. 1, pp. 329–337, 2011.

[30] S. Chibowski, M. Paszkiewicz, and M. Krupa, "Investigation of the influence of the polyvinyl alcohol adsorption on the electrical properties of Al_2O_3-solution interface, thickness of the adsorption layers of PVA," *Powder Technology*, vol. 107, no. 3, pp. 251–255, 2000.

[31] D. Santhiya, S. Subramanian, K. A. Natarajan, and S. G. Malghan, "Surface chemical studies on the competitive adsorption of poly(acrylic acid) and poly(vinyl alcohol) onto alumina," *Journal of Colloid and Interface Science*, vol. 216, no. 1, pp. 143–153, 1999.

[32] F. Bensadoun, N. Kchit, C. Billotte, S. Bickerton, F. Trochu, and E. Ruiz, "A study of nanoclay reinforcement of biocomposites made by liquid composite molding," *International Journal of Polymer Science*, vol. 2011, Article ID 964193, 10 pages, 2011.

[33] F. Bensadoun, N. Kchit, C. Billotte, F. Trochu, and E. Ruiz, "A comparative study of dispersion techniques for nanocomposite made with nanoclays and an unsaturated polyester resin," *Journal of Nanomaterials*, vol. 2011, Article ID 406087, 2011.

[34] A. B. Metzner, "Rheology of suspensions in polymeric liquids," *Journal of Rheology*, vol. 29, no. 6, pp. 739–775, 1985.

[35] A. J. Poslinski, M. E. Ryan, R. K. Gupta, S. G. Seshadri, and F. J. Frechette, "Rheological behavior of filled polymeric systems I. Yield stress and shear-thinning effects," *Journal of Rheology*, vol. 32, no. 7, pp. 703–735, 1988.

5

Tuning, Impedance Matching, and Temperature Regulation during High-Temperature Microwave Sintering of Ceramics

Sylvain Marinel ⓘ**,[1] Nicolas Renaut,[1] Etienne Savary,[1,2] Rodolphe Macaigne,[1] Guillaume Riquet,[1] Christophe Coureau,[3] Thibault Gadeyne,[4] and David Guillet[4]**

[1]*Laboratoire de Cristallographie et Sciences des Matériaux, Normandie Univ, ENSICAEN, UNICAEN, CNRS, CRISMAT, 14000 Caen, France*
[2]*Université de Valenciennes et du Hainaut-Cambrésis, Boulevard du Général de Gaulle, 59600 Maubeuge, France*
[3]*SOLCERA, ZI n°1 rue de l'industrie, 27000 Evreux, France*
[4]*SAIREM, 12 Porte du Grand Lyon, 01702 Neyron, France*

Correspondence should be addressed to Sylvain Marinel; sylvain.marinel@ensicaen.fr

Academic Editor: Joon-Hyung Lee

Over the years, microwave radiation has emerged as an efficient source of energy for material processing. This technology provides a rapid and a volumetric heating of material. However, the main issues that prevent microwave technology from being widespread in material processing are temperature control regulation and heating distribution within the sample. Most of the experimental works are usually manually monitored, and their reproducibility is rarely evaluated and discussed. In this work, an originally designed 915 MHz microwave single-mode applicator for high-temperature processing is presented. The overall microwave system is described in terms of an equivalent electrical circuit. This circuit has allowed to point out the different parameters which need to be adjusted to get a fully controlled heating process. The basic principle of regulation is then depicted in terms of a block function diagram. From it, the process has been developed and tested to sinter zirconia- and spinel-based ceramics. It is clearly shown that the process can be successfully used to program multistep temperature cycles up to ~1550°C, improving significantly the reproducibility and the ease of use of this emerging high-temperature process technology.

1. Introduction

In recent years, microwave energy has been successfully used for processing many materials, including the fast synthesis of various inorganic compounds and the sintering of ceramics. Kitchen et al. [1] have recently published an overview paper describing different works related to the solid state synthesis of widely spread oxides (BaTiO$_3$, spinel phase, BaZrO$_3$, etc.), carbides (TiC, WC, Mo$_2$C, etc.), or nitrides (AlN, Ta$_2$N, etc.) using microwave energy. Microwave energy has also been successfully used to sinter various materials. For instance, Brosnan et al. [2] have sintered alumina in using a 2.45 GHz hybrid microwave multimode cavity. Many other works have reported similar results on various materials, mostly oxide ceramics [3–6]. All these processing methods obviously imply a high-temperature stage (>1200°C) and, in comparison to conventional technologies involving infra-red radiation heating sources (electrical or gas furnaces), the microwave technology is known to be faster and greener, with less energy consumption. Bykov et al. [7] have written an interesting overview underlining the specific advantages of microwave sintering over conventional processes. Most especially, it is well established that taking into consideration the penetration depth of the microwave electrical field (E), which is a few centimeter in most oxide materials, the heat is volumetric generated. Starck et al. [8] indeed reported that the penetration depth of microwaves at 2.45 GHz into alumina is around 1.5 cm at room temperature. This is a great advantage over infra-red conventional heating, for which the penetration depth of the radiation is a few μm. As a result, the energy is transferred through the material by direct microwave-matter coupling, not as a thermal heat flux. Consequently, the rate of heating is often high, and the uniformity of heat distribution is greatly improved [9]. All these features undeniably contribute to the high efficiency of the microwave thermal process. However, up to now, the main issues that prevent the microwave technology from being widespread used at a larger scale, including the industrial scale, is temperature control regulation, which also means its measurement and its distribution within the sample. This is even truer when high temperature processes (>1200°C) are involved, and the final microstructure must be carefully controlled. Regarding the temperature measurement, Macaigne et al. [10] have recently published a paper describing an original method to measure the thermal emissivity of a material being microwave heated, allowing to measure with high confidence the actual material temperature by a infra-red single-color pyrometer. It is worthy to mention that this method takes into account the emissivity temperature dependence, which is of primary importance for precisely measuring the temperature by a pyrometer. This method is nowadays used in our laboratory. However, up to now, very few methods have been reported to automatically regulate the temperature of samples which are being heat treated in a microwave single-mode applicator. Most of the time the temperature is measured by an infra-red pyrometer, and the regulation is performed manually: the operator usually adjusts both the incident microwave power and the resonance mode of the cavity so that the sample temperature follows the desired thermal cycle. This later method works but is obviously not appropriate when high robustness and reliability are required. In this work, the goal was to be able to program a multistep thermal cycle in a large temperature range (from room temperature to ~1600°C). In the literature, only few articles deal with automatic temperature regulation during high-temperature microwave processing of materials. Vogt et al. [11] from Los Alamos Laboratory (Defense National USA laboratory) have reported an interesting method implying a rapid feedback control system operating with a traveling wave tube amplifier for the regulation of the sintering temperature of bars and tubes in a single-mode microwave cavity. This method was tested to sinter NICALON, which is a multifilament silicon carbide fiber, at roughly 1100°C. The described method was based on automatically adjusting the microwave generator power and frequency (centered around 2.94 GHz) to follow-up the desired temperature. It is important to mention that as temperature increases, the cavity load impedance changes (since the physical properties of the load depend on the temperature), and therefore, the conditions to get a good microwave material absorption vary. However, this method is not really applicable since most of the microwave generators used in the laboratory and industry are working at a fixed frequency (usually 2.45 GHz or 915 MHz). Although the importance of the impedance matching and cavity tuning

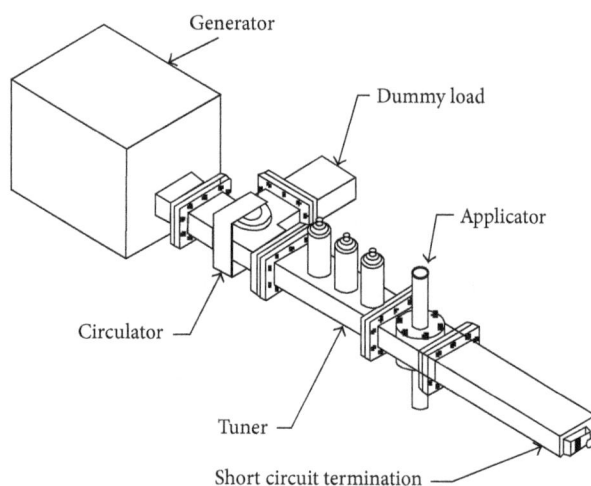

FIGURE 1: A typical laboratory set-up used for material processing at high temperature in a single-mode microwave cavity.

has been already found and discussed in medicine for microwave exposure to biological tissue [12] (at, obviously, a low temperature), those aspects are rarely evoked in high-temperature microwave processes. Thus, in this work, an original process will be described which allows to automatically regulate the temperature of a material being heat treated by a microwave up to 1600°C. This process uses a common fixed frequency, that is, 915 MHz. First of all, the microwave system will be described in terms of an equivalent electrical circuit. Based on this circuit, the various impedance components and relevant parameters will be defined. Afterwards, the overall basic principle of temperature regulation will be described in terms of a block function diagram. Finally, the process will be tested to sinter at high temperatures (up to 1600°C) on some widely spread oxides, including zirconia and spinel.

2. Experimental and Results

2.1. Experimental Microwave Set-Up. Figure 1 shows a typical laboratory set-up used for material processing at high temperature in a single-mode microwave cavity. The microwave generator delivers a microwave power towards a rectangular wave guide directly connected to a circulator. The latter is used for the protection of the generator from the damaging effects of the reflected microwave power. The circulator absorbs the reflected power through a dummy load. This circulator is followed by a conventional rectangular wave guide equipped with a tuner.

Coupling microwave power to a load requires the respective complex impedance between the load and the microwave power source to be matched. Manually adjusted tuners, as the one shown in Figure 1, are by far the most common for industrial and laboratory heating applications. A simple design indeed, widely used, employs three threaded stubs screwed directly through the broad wall of the wave guide. Still, operating a manual tuner is mostly a practiced work that can only be learned from experience. In this work, an automatic tuner will be used in order to keep

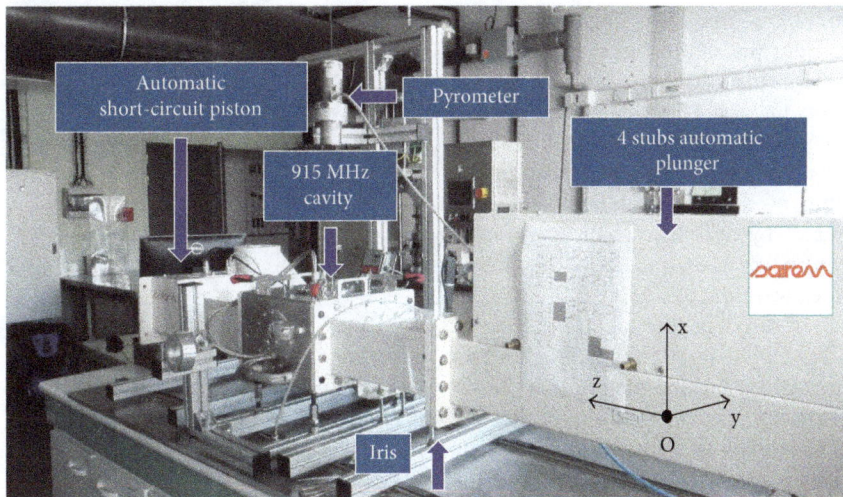

FIGURE 2: Photograph showing the overall 915 MHz microwave process. The short circuit motorization and the 4 stubs automatic plunger are connected to a personal computer and can be monitored by it.

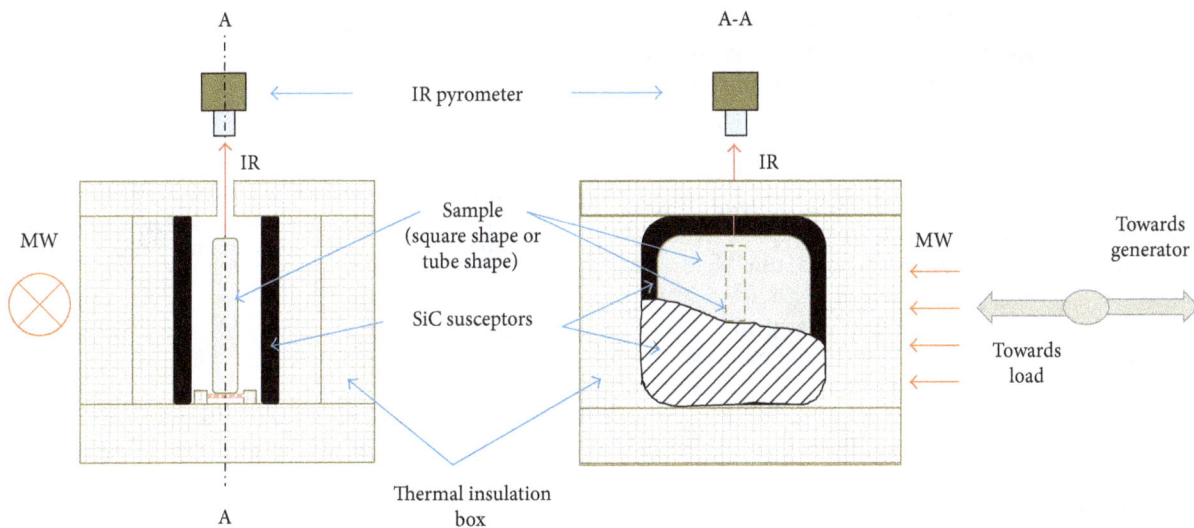

FIGURE 3: Scheme showing the assembly used to heat up the ceramic green sample into the 915 MHz single-mode cavity.

up with the dynamics of the heating process. Otherwise, with an automatic system, a better impedance match can be achieved within milliseconds while the manual impedance process usually takes a few seconds. For more information regarding the function of each microwave components, Gerling et al. [13] have published an interesting technical review. Following the tuner, there is the resonant cavity, in which the sample will be heat treated. Finally, to tune the resonance mode, a movable short circuit termination is used at the end of the line as shown in Figure 1. Taking into account our goal to get a fully automatic heating process, the basic principle depicted on Figure 1 has been modified in first, using an automatic 4 stubs impedance plunger (AI4S from SAIREM) and second, using a motorized short circuit piston/termination. Figure 2 shows a photograph of the overall microwave process that includes the 915 MHz–5 kW generator from SAIREM (not shown on the photograph), an automatic impedance tuner based on a 2×2 tuning plungers (AI4S from SAIREM)),

and a resonant applicator, delimited by a fixed iris and a short motorized circuit termination.

2.2. Microwave Heating Assembly. The microwave assembly designed for our purpose is shown in Figure 3. The sample having a square or a cylindrical shape is vertically mounted onto an alumina support piece. In case of a square shape sample, the larger side of the sample is aligned along the microwave propagation direction (see the "Z" direction). In case of a cylindrical shape, the cylinder axis is parallel to the X direction (vertical direction). Two SiC pieces are used as susceptors to initiate the sample heating especially for the low loss dielectric material. These two pieces are schematically positioned as shown in Figure 3. The susceptors allow to rapidly heat up the samples. The susceptors are two hot surfaces in front of the large side of the sample that prevent it from being too rapidly cooled by thermal radiation. This geometry was demonstrated to be efficient to homogenize the temperature within the sample. Finally, the overall assembly, made from the sample and susceptors, is introduced into a thermal insulation box made of a microwave transparent refractory material (alumina-silicate porous foam from RATH). The transverse electric (TE 105) mode is tuned so that the sample, located in the center of the cavity, is subjected to a maximum of the electrical field (E field). Temperature is measured by a single-color pyrometer (IRCON modline 5 G) which targets on the edge of the sample and for which the temperature range is $350°C < \theta < 2000°C$. Single-color pyrometer was used instead of a 2-color pyrometer since the apparent emissivity of the material and its environment can be measured using the method previously developed in the laboratory by Macaigne et al. [10]. By this method, the apparent emissivity was found to be 0.7 at ~1500°C for spinel, and a similar value was measured for zirconia. It must be understood that the thermal radiation received by the IR pyrometer comes from the material and its environment (thermal radiation from the susceptors). The environment being the same for both materials (Figure 3), as well as the maximum temperature (~1500°C), the apparent emissivity remains nearly constant regardless the material (zirconia or spinel). Consequently, the value of the emissivity was fixed at 0.7 for all experiments.

2.3. Analogous Electrical Circuit: Microwave Heating Process.

Based on our experimental set-up, an analogous electrical circuit of the overall microwave process can be drawn (Figure 4(a)). This analogous circuit is well known in microwave engineering, and more particularly in telecommunication science [14], but is practically not used by chemists working with microwaves for high-temperature processing and power applications.

The microwave generator is represented by a voltage source (E_0) with its internal resistance (R_0). The 4 stubs tuning plunger with the iris is modeled as an "n turn" transformer. The resonant cavity is represented by a resonant LC circuit with a resonant angular velocity ($\omega = (1/\sqrt{LC})$). It is worthwhile mentioning that the resonance conditions of the microwave cavity are properly adjusted by moving the short circuit piston position to match with the fixed frequency (in doing so, L and C change). In an (L, C) electrical conventional circuit, the resonance conditions are tuned by adjusting the frequency of the electrical signal. Finally, Joule losses within the cavity wall and into the sample which is being heated is represented by a series resistance (r). It is essential to mention that considering the sample temperature increasing from room temperature to high temperature (above 1500°C), a large variation of its physical properties is expected during the heating process. For instance, Heiroth et al. [15] have measured the electrical conductivity of yttrium-stabilized zirconia varying from 4.10^{-5} S·cm^{-1} at 675 K to 3.10^{-3} S·cm^{-1} at 952 K. This shows that, from the equivalent electrical circuit aspect, a variable resistor (r), depending on the temperature, must be taken into account. It also means that the load impedance is going to change during the heating cycle progress. Thereafter, the electrical circuit can be simplified as shown on Figure 4(b). On it, the load

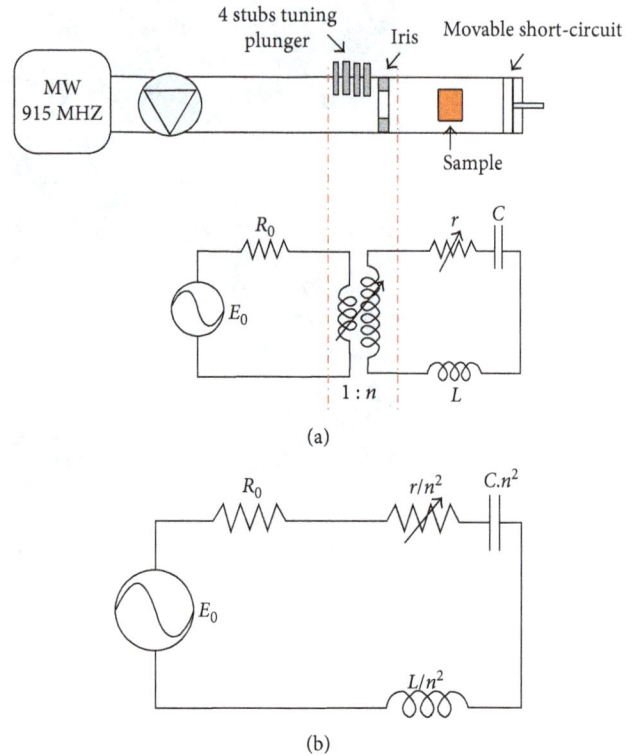

FIGURE 4: (a) Basic equivalent circuit of the microwave heating process. (b) Equivalent electrical circuit with load impedance referred to the primary side of the transformer.

impedance is referred to the primary side of the transformer, so that the transformer is removed from the electrical circuit. From this simple analogous electrical circuit, the optimal conditions to improve the power transfer from the source to the load, and to define a satisfactory strategy for designing the temperature regulation block function diagram, can be found.

2.4. Optimal Power Transfer Conditions and Block Function Diagram.

From the equivalent electrical circuit (Figure 4(b)), the coupling coefficient (β) can be defined as follows:

$$\beta = \frac{R_0 n^2}{r}. \tag{1}$$

The optimal power transfer will be achieved when the power losses into the load equals the power losses into the input impedance (R_0). This condition means

$$R_0 = \frac{r}{n^2}. \tag{2}$$

Therefore, the coupling coefficient equals $\beta = 1$. It is useful to define the module of the reflection coefficient (ρ) which can be expressed as follows:

$$\rho = \frac{\beta - 1}{\beta + 1}. \tag{3}$$

The reflection coefficient is the amplitude of the reflected voltage wave normalized to the amplitude of the incident voltage wave. When the load impedance ideally matches to

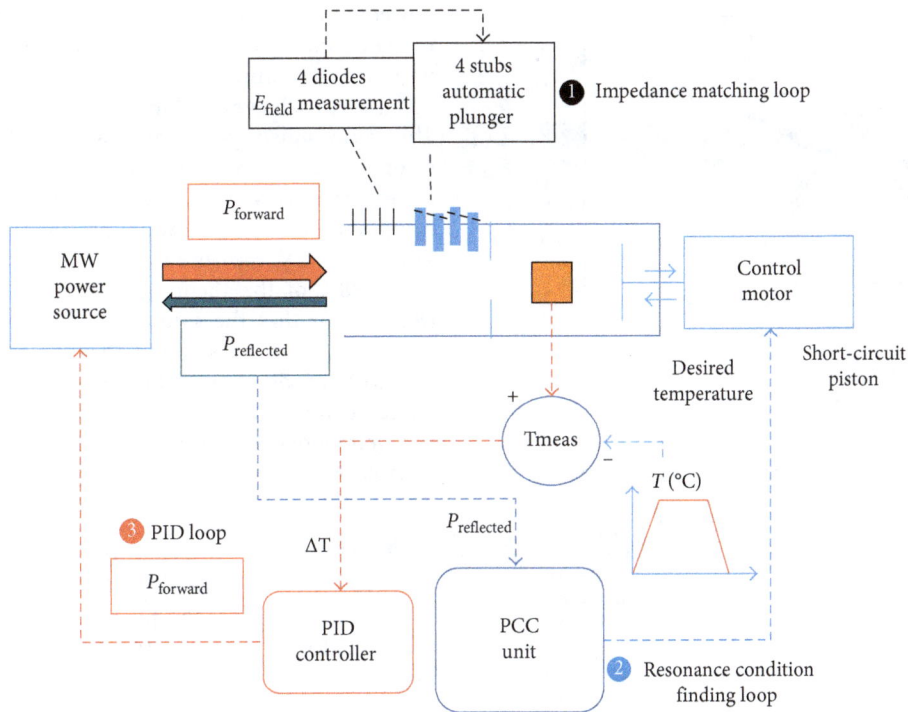

FIGURE 5: Block function diagram of the overall 915 MHz microwave process designed to fully automatize the temperature regulation with resonance tuning and impedance matching.

the input impedance, the reflected voltage equals zero and the reflection coefficient becomes zero. The last useful expression is the ratio of the reflected microwave power ($P_{reflected}$) to the forwarded microwave power ($P_{forward}$), linked to each other by the following equation:

$$\frac{P_{reflected}}{P_{forward}} = \rho^2. \tag{4}$$

In order to fully automatize the 915 MHz microwave heating process, the block function diagram (Figure 5) has been implemented. An independent loop, labeled ①, was first implemented to make sure that the load impedance matches with the line impedance. Considering the analogous electrical circuit, this means that the "n" turn value changes automatically so that the β value remains close to one. This loop needs to be running during the entire thermal cycle as the load impedance varies with temperature (the r value is obviously changing, see (1)). Once the impedance matching is accommodated, the resonance conditions need to be tuned as well. To do so, a resonance finding loop has been implemented as well (see the loop labeled ②). The latter is working in adjusting the cavity length during the thermal cycle. The transverse electric (TE 105) mode is tuned so that the sample, located in the center of the cavity, is subjected to a maximum of the electrical field (E field). Another loop, labeled ③, was implemented to calculate and command the microwave forward power ($P_{forward}$) needed to follow-up the desired temperature cycle, in using an adaptive PID module (proportional, integer, and derivative). The whole system is controlled by a personal computer (PCC unit).

2.5. Applications of the As-Developed Microwave Process

2.5.1. Microwave Sintering of MgO-Stabilized ZrO$_2$-Based Ceramic Tube.
In order to test our process, a green zirconia-based sample made of an MgO (~2-3 weight %) stabilized zirconia material has been sintered using a programmed multistep heating cycle shown in Figure 6(a). The green samples (green density ~2.9 g/cm^3 or ~51% of the theoretical density value) were provided by the SOLCERA company and shaped by thermoplastic injection. After subjecting the sample to a thermal cycle in a conventional furnace for the removal of the organic binders and plasticizers, the cylinder (diameter 5.6 mm, height 31.8 mm, and thickness 0.9 mm) has been sintered in air in our 915 MHz cavity, using the thermal cycle shown in Figure 6(a). For confidential reasons, no additional details can be given regarding the preparation of the samples. However, the validation of the process being our goal, those details are unnecessary. The sintering thermal cycle has involved four heating ramps, respectively, at 30°C/min (from 400°C to 1000°C), 10°C/min (from 1000°C to 1400°C), 5°C/min (from 1400°C to 1500°C), and finally at 2.5°C/min from 1500°C to 1515°C. This multistep thermal cycle has been chosen in order to evaluate the dynamics of the system and to observe how the different ramps are managed. In order to evaluate the P, I, and D parameters (loop ③), a fixed microwave forward power of ~900 W is, first of all, applied. The temperature increase—as seen on the curve (see the dashed line)—is recorded and analyzed so that the initial PID parameters are calculated (thereafter, a self-tuning adaptive PID has been used through the use of an adaptive PID controller from the Eurotherm 2408 model). Afterwards, the system works by itself and the actual

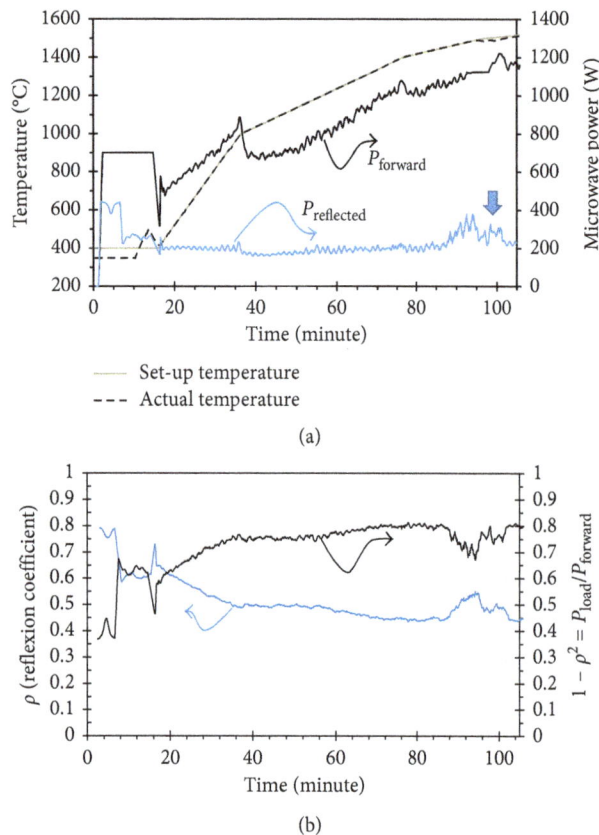

FIGURE 6: (a) Curves of the set-up and actual temperatures versus time and curves of the forwarded and reflected microwave power versus time during the microwave sintering of MgO-stabilized zirconia. (b) Curves of the reflexion coefficient (ρ) and ($1-\rho^2$) versus time during the microwave sintering of zirconia-based material.

temperature automatically follows up the programmed temperature cycle. It can be seen that the actual temperature fits well with the programmed cycle and that the changes of the heating ramps are well managed. On the curve shown in Figure 6(a), it can be seen that the forward power continuously varies from 400 W at the beginning of the process to 1200 W when the highest temperature is reached. During the heating process, the reflected power does not vary much and its value remains close to 200 W.

At the highest temperature, its value slightly increases up to 400 W before decreasing to 250 W at the end of the process (the arrow), whereas the actual temperature still follows up the desired cycle. It shows that the dynamic response is good enough to follow-up, even at a high temperature, a specific temperature cycle. From the forwarded and reflected powers, the reflexion coefficient (ρ) has been calculated and plotted versus time (Figure 6(b)). Its initial value is ~0.8, but it rapidly decreases to ~0.45, showing that the tuning process (including both the impedance matching (loop①) and the resonance condition finding loop (loop②)) are properly working. From the ρ value, the ratio of the loaded power into the cavity over the forwarded power was plotted versus time (Figure 6(b)). This information gives the amount of the incident microwave

power stored into the applicator, that is, the useful power. It is a good measure of the efficiency of the process. Once the program is running, it can be seen that this ratio rapidly increases to achieve a stable value around 80% (Figure 6(b)). This ratio shows good impedance matching, satisfactory resonance conditions, and high process efficiency.

Figure 7 shows the green zirconia sample and the one being sintered by microwave heating. After sintering, the sample density, measured by the Archimedes method reaches ~98% of the theoretical value. Moreover, it can be clearly seen that the sample mechanical integrity is very satisfactory (no cracks), and that the microwave sintering has led to produce a very homogeneous piece. As expected, a fine microstructure (Figure 7(b)) composed of both cubic and tetragonal phases is observed by Scanning Electron Microscopy (SEM Zeiss EVO MA 15).

2.5.2. Microwave Sintering of Square-Shaped MgAl₂O₄ Ceramic. Green square shape ($65 \times 65 \times 12 \text{ mm}^3$) $MgAl_2O_4$ (Baikowski S30CR, >99% pure) samples were prepared by the SOLCERA company. The green density was approximately ~52% of the theoretical one. To evaluate the reproducibility of the as-developed MW process, several pieces were microwave sintered using quite the same experimental conditions as those previously described. For this material, the programmed thermal cycle is given in Figure 8(a). It resembles the one used for zirconia, apart from the last heating ramp @ 2.5°C/min, which is being replaced by a dwell @ 1500°C for 5 minutes. Three samples were sintered using exactly the process previously described and using the thermal cycle (Figure 8(a)). In each case, homogeneous sintered pieces were obtained (Figure 8(b)) with similar densities around 96-97% (measured by the Archimedes method). The samples microstructures were observed by the Scanning Electron Microscopy, and a typical microstructure is shown in Figure 8(c). This observation clearly shows a homogeneous and dense microstructure which confirms the satisfactory reproducibility of the as-developed 915 MHz microwave process.

3. Conclusion

This study describes the development of an original 915 MHz single-mode microwave cavity which is fully instrumented and automatized in order to regulate multistep temperature heating cycles up to 1600°C for material processing and sintering. At first, the whole system was described using a simple analogous electrical circuit which allows to clearly point out the different parameters that need to be controlled (tuning of the resonance conditions and impedance matching). From this electrical circuit, a block function diagram has been designed and implemented using automatic impedance-matching tuning plungers and a motorized short circuit termination. The as-developed process was successfully used to sinter different ceramic pieces, including zirconia tubes and square-shaped spinel materials by subjecting pieces to a programmed multistep thermal cycle. The process has

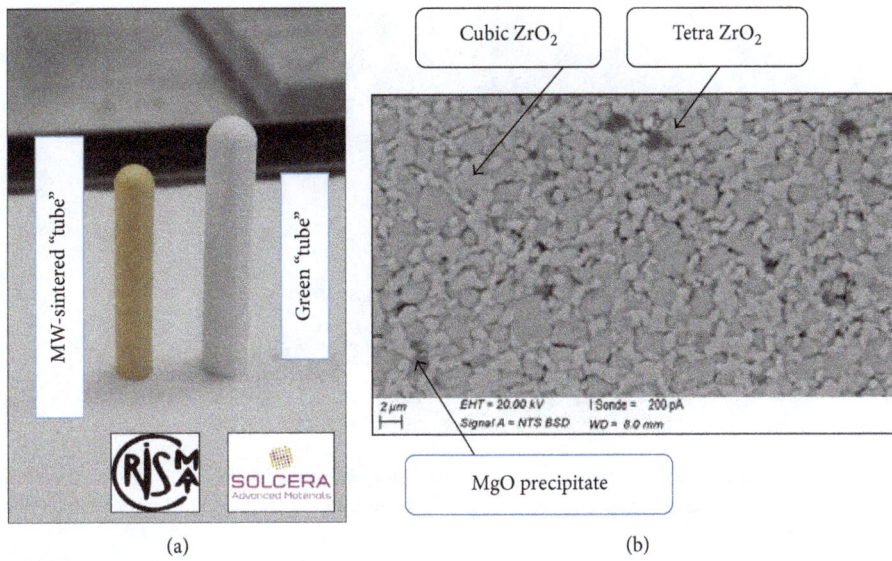

FIGURE 7: (a) Photograph showing the zirconia material tube before and after microwave sintering. (b) SEM microstructure of MgO-stabilized microwave-sintered zirconia.

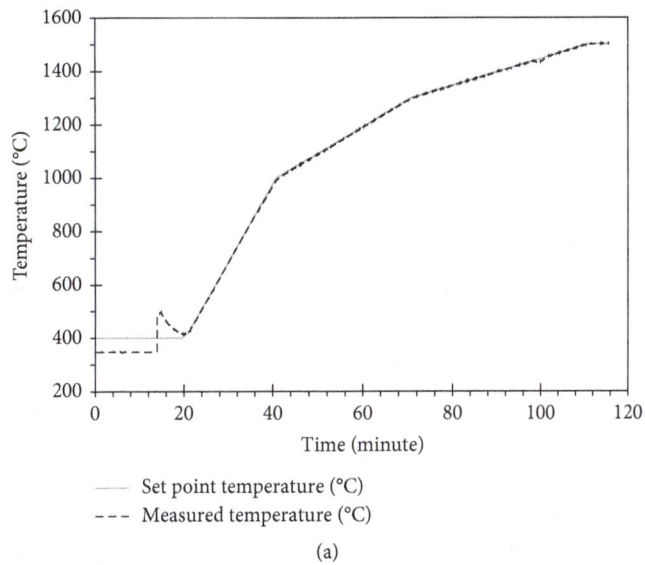

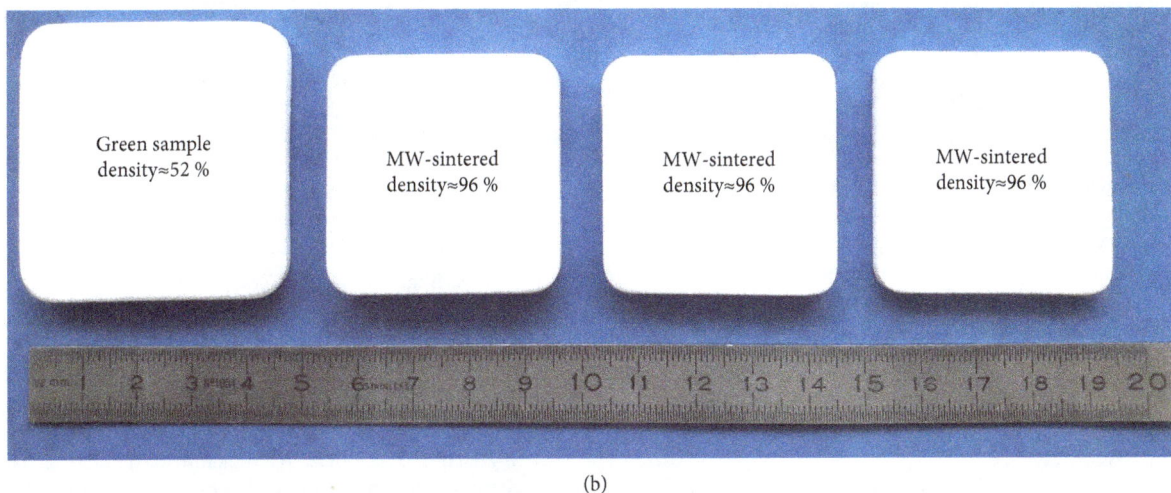

FIGURE 8: Continued.

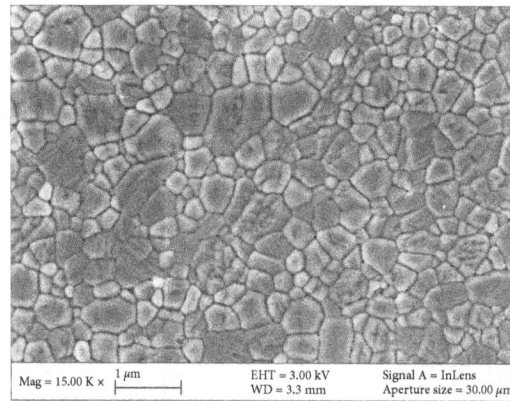

(c)

FIGURE 8: (a) Curves of the set-up and actual temperatures versus time for sintering $MgAl_2O_4$ square shape samples, (b) photograph showing the spinel-based material samples before and after microwave sintering, and (c) typical SEM microstructure recorded on $MgAl_2O_4$ microwave sintered .

proven its reliability, efficiency, and autonomy, opening up the path to industrial applications of microwave processing of materials at high temperature.

Conflicts of Interest

The authors declare that there are no conflicts of interest regarding the publication of this paper.

Acknowledgments

The authors acknowledge Christelle Bilot and Anthony Bilot for their valuable assistance in conducting the experimental work. This work was supported by the Direction Générale de l'Armement (DGA) through the projects BAMBI (Ref. ANR-11-ASTR-0025) and MAMBO (Ref. ANR-16-ASMA-0003). The authors would also like to thank the Solcera Advanced Material Company (Evreux) and the "Conseil Régional de Normandie" for supporting this work.

References

[1] H. J. Kitchen, S. R. Vallance, J. L. Kennedy et al., "Modern microwave methods in solid-state inorganic materials chemistry: from fundamentals to manufacturing," *Chemical Reviews*, vol. 114, no. 2, pp. 1170–1206, 2014.

[2] K. H. Brosnan, G. L. Messing, and D. K. Agrawal, "Microwave sintering of alumina at 2.45 GHz," *Journal of the American Ceramic Society*, vol. 86, no. 8, pp. 1307–1312, 2003.

[3] R. R. Menezes and R. H. G. A. Kiminami, "Microwave sintering of alumina–zirconia nanocomposites," *Journal of Materials Processing Technology*, vol. 203, no. 1–3, pp. 513–517, 2008.

[4] R. R. Mishra and A. K. Sharma, "Microwave–material interaction phenomena: heating mechanisms, challenges and opportunities in material processing," *Composites Part A:*

Applied Science and Manufacturing, vol. 81, pp. 78–97, 2016.

[5] M. Oghbaei and O. Mirzaee, "Microwave versus conventional sintering: a review of fundamentals, advantages and applications," *Journal of Alloys and Compounds*, vol. 494, no. 1-2, pp. 175–189, 2010.

[6] J. Croquesel, D. Bouvard, J.-M. Chaix, C. P. Carry, and S. Saunier, "Development of an instrumented and automated single mode cavity for ceramic microwave sintering: application to an alpha pure alumina powder," *Materials & Design*, vol. 88, pp. 98–105, 2015.

[7] Y. V. Bykov, K. I. Rybakov, and V. E. Semenov, "High-temperature microwave processing of materials," *Journal of Physics D: Applied Physics*, vol. 34, no. 13, pp. R55–R75, 2001.

[8] A. V. Starck and A. Mühlbauer, *Handbook of Thermoprocessing Technologies*, Vulkan-Verlag GmbH, Essen, Germany, 2005.

[9] J. Sun, W. Wang, and Q. Yue, "Review on microwave-matter interaction fundamentals and efficient microwave-associated heating strategies," *Materials*, vol. 9, no. 4, p. 231, 2016.

[10] R. Macaigne, S. Marinel, D. Goeuriot, C. Meunier, S. Saunier, and G. Riquet, "Microwave sintering of pure and TiO_2 doped $MgAl_2O_4$ ceramic using calibrated, contactless in-situ dilatometry," *Ceramics International*, vol. 42, no. 15, pp. 16997–17003, 2016.

[11] G. J. Vogt, A. Regan, A. Rohlev, and M. Curtin, "Microwave process control through a traveling wave tube source," *MRS Proceedings*, vol. 430, 1996.

[12] G. Atanasova and N. Atanasov, "Tuning, coupling and matching of a resonant cavity in microwave exposure system for biological objects," *Electromagnetic Biology and Medicine*, vol. 32, no. 2, pp. 218–225, 2013.

[13] J. F. Gerling, *Waveguide Components and Configurations for Optimal Performance in Microwave Heating Systems, Technical Document*, Gerling Applied Engineering, Inc., Modesto, CA, USA, 2000.

[14] A. Das and S. K. Das, *Microwave Engineering 2E*, Tata McGraw-Hill Education, New York, NY, USA, 2009.

[15] S. Heiroth, T. Lippert, A. Wokaun et al., "Yttria-stabilized zirconia thin films by pulsed laser deposition: microstructural and compositional control," *Journal of the European Ceramic Society*, vol. 30, no. 2, pp. 489–495, 2010.

6

Synergistic Sintering of Lignite Fly Ash and Steelmaking Residues towards Sustainable Compacted Ceramics

V. G. Karayannis,[1] A. K. Moutsatsou,[2] A. N. Baklavaridis,[1] E. L. Katsika,[2] and A. E. Domopoulou[1]

[1]*Department of Environmental Engineering, Technological Education Institute of Western Macedonia, Kila, 50100 Kozani, Greece*
[2]*School of Chemical Engineering, National Technical University of Athens, Zografou Campus, 15773 Athens, Greece*

Correspondence should be addressed to V. G. Karayannis; vkarayan@teiwm.gr

Academic Editor: Peter Majewski

The development of value-added ceramic materials deriving only from industrial by-products is particularly interesting from technological, economic, and environmental point of views. In this work, the synergistic sintering of ternary and binary mixtures of fly ash, steelmaking electric arc furnace dust, and ladle furnace slag for the synthesis of compacted ceramics is reported. The sintered specimens' microstructure and mineralogical composition were characterized by SEM-EDS and XRD, respectively. Moreover, the shrinkage, apparent density, water absorption, and Vickers microhardness (HV) were investigated at different sintering temperatures and raw material compositions. The characterization of the sintered compacts revealed the successful consolidation of the ceramic microstructures. According to the experimental findings, the ceramics obtained from fly ash/steel dust mixtures exhibited enhanced properties compared to the other mixtures tested. Moreover, the processing temperature affected the final properties of the produced ceramics. Specifically, a 407% HV increase for EAFD and a 2221% increase for the FA-EAFD mixture were recorded, by increasing the sintering temperature from 1050 to 1150°C. Likewise, a 972% shrinkage increase for EAFD and a 577% shrinkage increase for the FA-EAFD mixture were recorded, by increasing the sintering temperature from 1050 to 1150°C. The research results aim at shedding more light on the development of sustainable sintered ceramics from secondary industrial resources towards circular economy.

1. Introduction

Management and valorization of huge amounts of waste by-products (solid residues), recovered from industrial activities, still remain a major problem in many countries. In steel industry, the production of 1 ton of steel results in the generation of 2–4 tons of various types of wastes [1]. Therefore, the valorization of solid by-products recovered from steelmaking and metal industries would induce significant economic and environmental benefits [2]. However, the suitable disposal and handling of wastes of this category remain both dangerous and expensive [3, 4]. Three types of solid residues mainly derive from the steel industries, namely, blast furnace slag, steel slag, and electric arc furnace dust. Blast furnace slag and steel slag are being used as competitive raw materials in the mineral and cement industry [1]. Specifically, steel slag

is widely used in road and pavement construction materials, while blast furnace slag usage in the cement industry is currently increasing due to the reduction of production cost that is achieved in the final cementitious products. The recycling of electric arc furnace dust (EAFD) is also very important due to the fact that it is one of the major steel industry by-products produced in large quantities worldwide and considered as potentially hazardous [3, 5]. In that sense, various efforts aiming at the effective recycling of EAFD have been proposed. It has been reported that the addition of up to 15% EAFD in the ceramic clay body leads to the production of ceramic specimens with good physical and mechanical properties while the leachability of EAFD lies within acceptable limits [6]. Moreover, clay mixtures with up to 5% EAFD could be effectively used for the production of bricks and roof tiles [7].

TABLE 1: The solid residue compositions of the specimens.

S/N	Sample name	EAFD (mass%)	FA (mass%)	LFS (mass%)
1	FA	0	100	0
2	EAFD	100	0	0
3	FA-EAFD	50	50	0
4	FA-EAFD-LFS	33.33	33.33	33.33

On the other hand, coal fly ash (FA) is a fine powder obtained by the electrostatic precipitation of dust-like particles from the flue gases of lignite-fed boilers. In Greece, more than 8 million tons of ash are produced annually from lignite combustion power stations. Specifically, 80% of the total amount of ash derives from the Northern Greece (Region of Western Macedonia), where the main lignite deposits are located. Nowadays, limited amounts of FA are being reused, while more than 80% of the overall FA output is directly deposited into ponds and landfills. However, the contamination from FA deposits can easily occur, because the FA particles usually contain potentially hazardous trace elements (heavy metals) due to their high surface area [8, 9]. Subsequently, there are serious concerns that this situation will possibly cause severe and irreversible long-term environmental effects. In order to avert this environmental impact, the FA valorization in construction and other civil engineering applications should be considered. Several studies were already focused upon the addition of various loadings of FA in clayey mixtures, for the production of conventional extruded or sintered ceramics or even composite materials, destined for construction applications [10–15]. At the same time, several studies reported the usage of FA for the production of advanced ceramics and composite materials [16, 17].

As aforementioned, fly ash (FA) and electric arc furnace dust (EAFD) are two of the most abundant industrial by-products. Since FA and EAFD consist of several oxides, compositionally appropriate mixtures could be used as raw (starting) materials in the ceramic and construction industries. On the other hand, it has been reported that various mixtures including several solid industrial wastes (containing SiO_2, Al_2O_3, and CaO) could be used for the production of ceramics, glass-ceramics, and cement-based materials destined for construction applications [18–21]. Moreover, residues generated from galvanization and metal finishing processes and containing toxic heavy metals can be thermally treated in the presence of fly ash and/or clay for the immobilization of hazardous agents into novel ceramics to be used in building, construction, and other applications [3, 20–24]. However, limited works are adverted to the development of sintered ceramics by synergistically using FA and steel industry powdery residues 100% as the raw materials. Karamberi and Moutsatsou reported the production of glass and glass-ceramics from steel and Fe-Ni production slags by vitrification [18].

In the current research, the development of sintered ceramic compacts consisting of fly ash (FA), electric arc furnace dust (EAFD), and ladle furnace slag (LFS) is reported. The physicomechanical properties of the ceramics produced are studied at different sintering temperatures and raw material mixture compositions.

2. Experimental

2.1. Raw Materials. EAFD and LFS were obtained from a Greek steel industry, while FA was collected from a lignite-fed power station situated in Northern Greece. All materials were used as-received without any further treatment. The chemical composition of these materials was determined using a Spectro X-Lab 2000 X-ray Fluorescence (XRF) spectrometer. The analysis shows that EAFD contains Fe and Zn oxides, in quantities higher than 50 wt%. Several other oxides of Pb, Ca, Na, Si, Mn, and Mg were also detected in EAFD. LFS mainly contains Fe, Ca, and Si oxides in quantities higher than 80 wt%. Moreover, CaO concentrations higher than 50 wt% were found in LFS, while SiO_2 and Fe and Al oxides are also present. The FA used was a highly calcareous one. The CaO and SiO_2 content (wt%) exceeded 60%, while Al_2O_3 and Fe_2O_3 were also detected.

2.2. Preparation of Ceramic Compacts. Mixtures of the as-received industrial residues (in equal compositions) were prepared. The residue compositions of the specimens prepared for this study are shown in Table 1.

The initial mixtures were then compressed into disc-shaped green compacts (13 mm in diameter) by uniaxial cold-pressing in a stainless steel die (SPECAC, 15011). The compaction pressure has been optimized, prior to the fabrication of the final samples, so as to obtain sample having sufficient green density and strength for subsequent handling and thermal treatment. The green specimens were heated at 10°C/min up to the designated processing (sintering) temperature of 1050°C or 1150°C. The samples were then kept at the peak temperature for 2 h. The thermal treatment was carried out in a laboratory chamber furnace, under controlled conditions. Finally, the sintered samples were gradually cooled down to ambient temperature, inside the furnace (Thermoconcept, KL06/13). Lower sintering temperature (950°C) was also tested, but the produced ceramic samples were not successfully consolidated (SEM micrograph is provided in Results and Discussion). Photographs of the initial (nonsintered) and sintered disc-shaped ceramic samples are shown in Figure 1.

2.3. Characterization of Ceramic Compacts. Phase identification of the ceramic samples was carried out via X-ray diffraction (XRD) measurements (Siemens, Diffractometer D-5000). Microstructural examination of the specimens was realized by Scanning Electron Microscopy (SEM-Jeol,

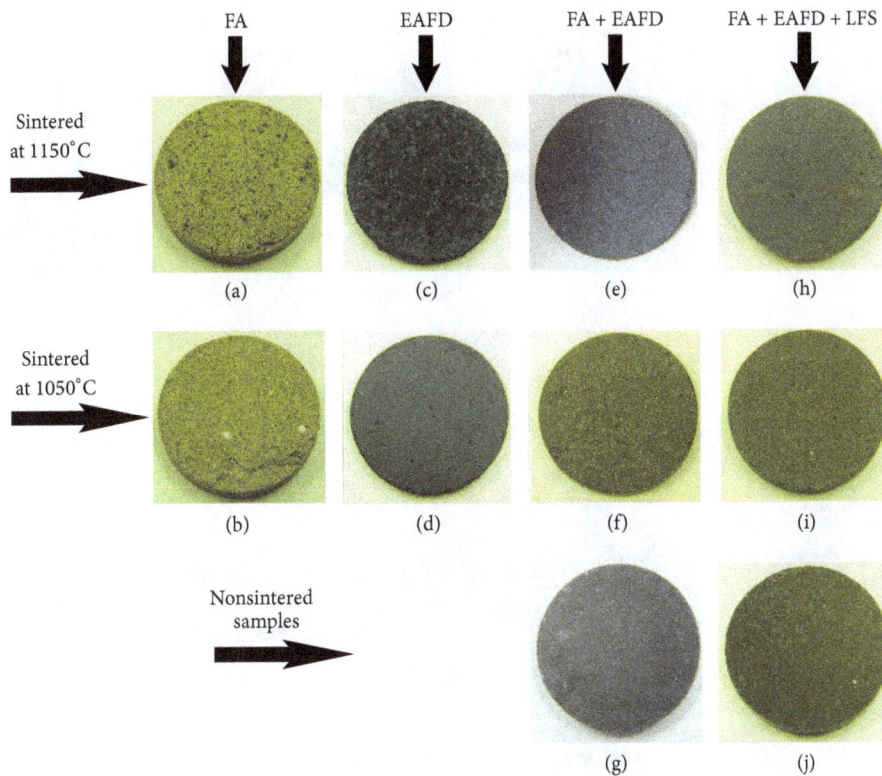

FIGURE 1: Selected photographs of the disc-shaped ceramic specimens (with diameter 13 mm) consisting of 100 wt.% FA (a, b), 100 wt.% EAFD (c, d), 50-50 wt.% FA + EAFD mixture (e, f, g), and 33,3-33,3-33,3 wt.% FA + EAFD + LFS mixture (h, i, j). Sintered samples at 1150°C (a, c, e, h), 1050°C (b, d, f, i), and nonsintered samples (g, j) are shown from top to bottom, respectively.

JSM-6400) coupled with Energy Dispersive X-ray Spectroscopy (EDX) analysis. The apparent density was determined by means of a specific apparatus (Shimadzu, SMK401-AUW220V) according to the Archimedes principle. Microhardness was measured using a Vickers indenter (Shimadzu, HMV-2T) by applying a load of 200 g (dwell time of 15 s) on the specimen surface. The mean hardness values were calculated over five valid indentations per specimen.

3. Results and Discussion

XRD measurements were conducted in order to identify the mineralogical phases that are present in the raw materials as well as in the mixtures and the sintered samples as well. In Figure 2, the XRD patterns of the sintered samples (at 1050°C) containing only EAFD, FA, LFS, and the mixture EAFD-FA-LFS are provided. It can be seen from Figure 2 that the sintered samples are composed of several crystalline phases. The main crystalline phases identified are magnetite (Fe_3O_4), hematite (Fe_2O_3), zinc oxide (ZnO), gehlenite ($Ca_2Al[AlSiO_7]$), $MgAl_2Si_3O_{10}$, quartz (SiO_2), cyanite (Al_2SiO_5), lime (CaO), and portlandite ($Ca(OH)_2$). Fe_3O_4 and ZnO should mainly originate from EAFD. Gehlenite derives from carbonate-silicate-spinel reactions. Both the gehlenite and even the $MgAl_2Si_3O_{10}$ formation should be associated with the presence of FA in the sample [25, 26]. It should be noted that most of the crystalline phases identified were also present in the starting raw materials (as-received

industrial by-products). However, the thermal treatment of the compacted mixture leads to changes in the intensity of the peaks corresponding to the aforementioned mineralogical phases. The sintered mixture (FA-EAFD-LFS) also contains an amorphous phase, which becomes evident from a hump in the pattern recorded between $2\theta = 30°$ and $2\theta = 35°$ (Figure 2). The occurrence of amorphous phase may be mainly ascribed to bloating of LFS particles and secondarily to the aluminosilicate glass which is present in FA [27, 28].

The role of sintering temperature in the mineralogical phase composition of the samples was also studied with XRD. Specifically, the effect of sintering temperature on the phase composition of the ceramic samples was studied for the FA-EAFD mixture. In Figure 3, XRD patterns of the sintered (at 1050°C) FA, EAFD, and its mixture sintered at two different peak temperatures: 1050 and 1150°C. No significant mineralogical phase changes are observed in the XRD patterns of Figure 3, with increasing the sintering temperature from 1050 to 1150°C.

SEM-EDS examination sheds light into the surface morphology and the elemental composition of the ceramic samples. In Figures 4(a) and 4(b), SEM images of the binary EAFD-FA mixture sintered at 1150°C are provided, while images of the ternary EAFD-FA-LFS mixture sintered at 1150°C are shown in Figures 4(c) and 4(d). The EAFD-FA sample exhibits a better consolidated morphology compared to EAFD-FA-LFS.

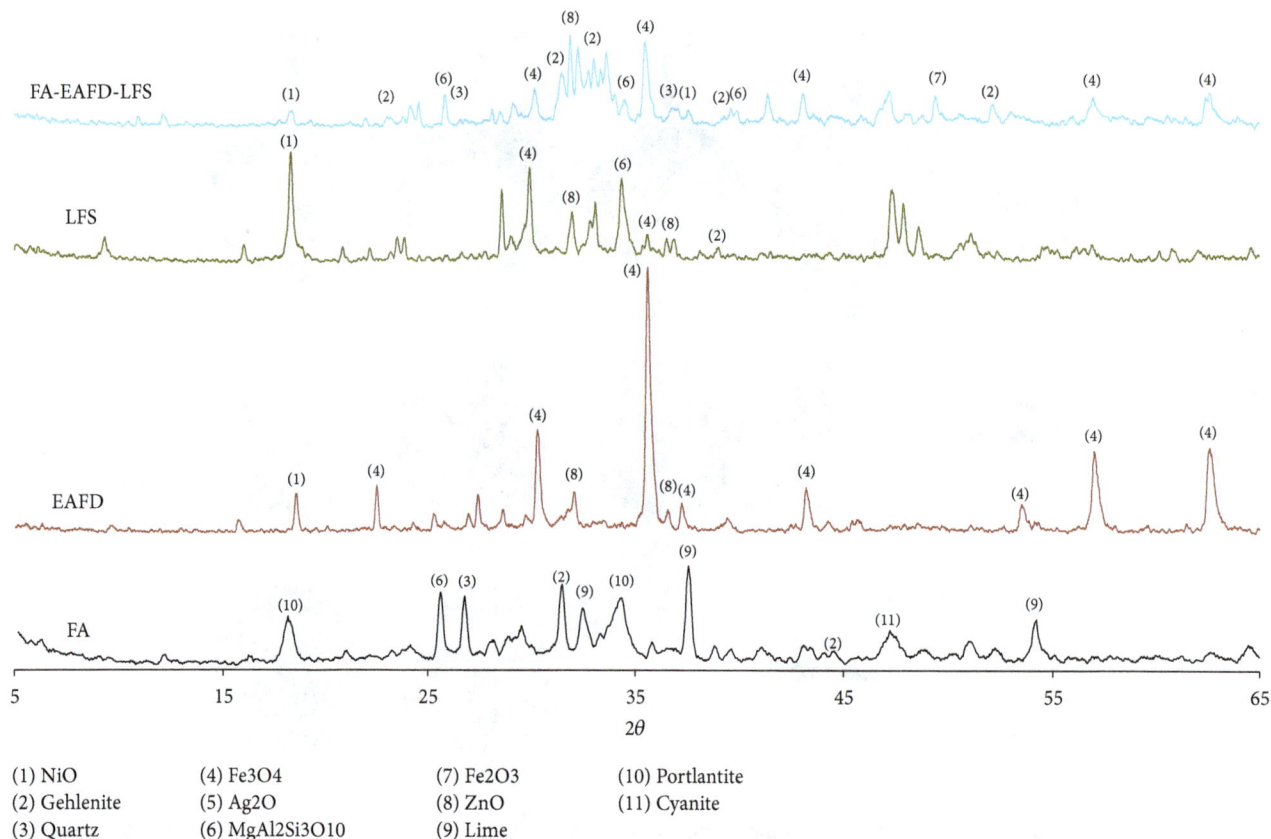

FIGURE 2: XRD patterns of the sintered samples containing only FA, EAFD, LFS, and their ternary mixture. The numbered peaks correspond to the identified phases, shown in the legend (below the figure).

(1) NiO   (4) Fe3O4   (7) Fe2O3   (10) Portlantite
(2) Gehlenite   (5) Ag2O   (8) ZnO   (11) Cyanite
(3) Quartz   (6) MgAl2Si3O10   (9) Lime

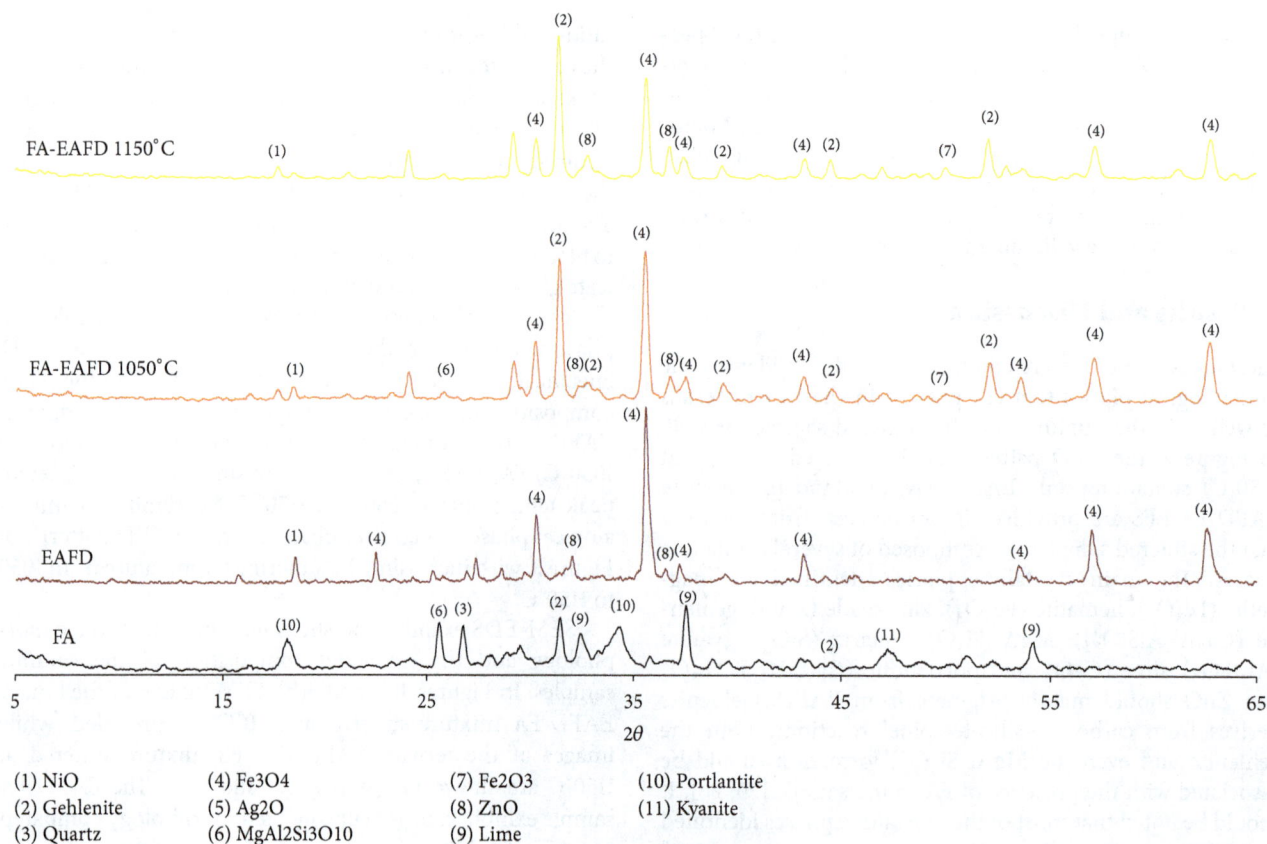

(1) NiO   (4) Fe3O4   (7) Fe2O3   (10) Portlantite
(2) Gehlenite   (5) Ag2O   (8) ZnO   (11) Kyanite
(3) Quartz   (6) MgAl2Si3O10   (9) Lime

FIGURE 3: XRD patterns of the FA, EAFD (sintered at 1050°C), and its mixture, sintered in two different temperatures: 1050 and 1150°C.

(a)

(b)

(c)

(d)

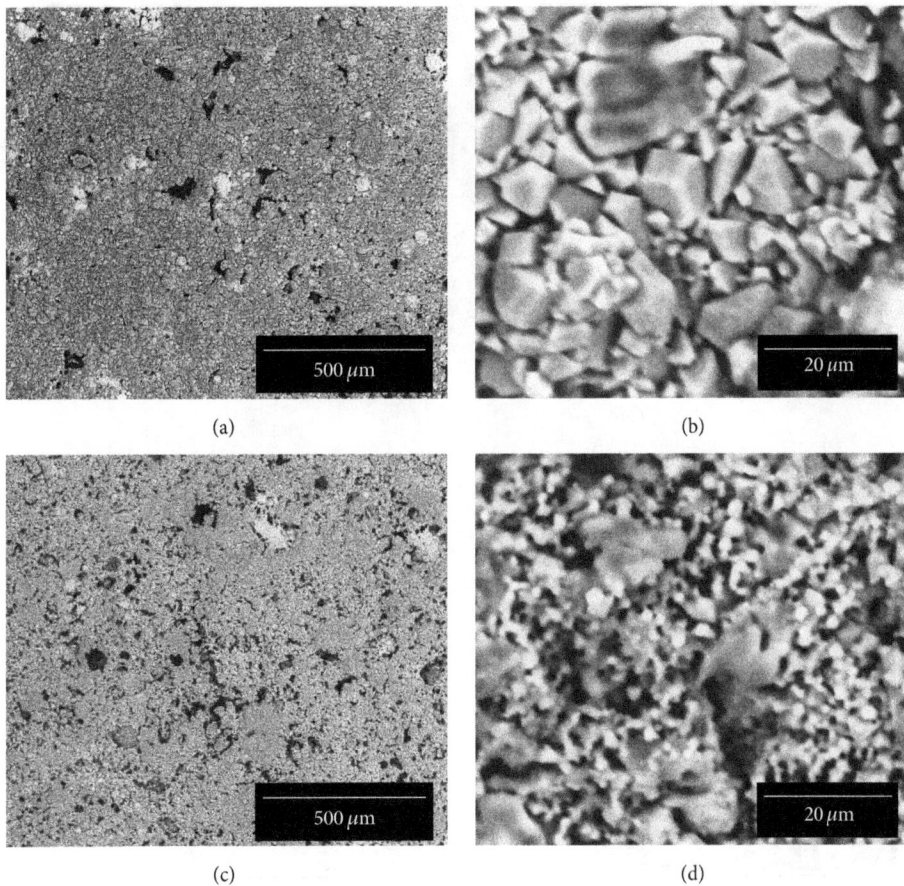

FIGURE 4: SEM images of the EAFD-FA mixture (a and b) as well as of the EAFD-FA-LFS mixture (c and d), both sintered at 1150°C.

SEM micrographs of EAFD-FA-LFS samples sintered at 950°C and 1150°C are shown in Figures 5(a) and 5(b), respectively. The morphology of both samples reveals the effective densification of the ceramic mixtures. Particularly in the ceramic sample sintered at 1150°C (Figure 5(b)), a more interconnected network with sintering necks can be observed. Some glassy structures are also found especially in this sample (sintered at 1150°C) (see large areas in Figure 5(b) without well-specified grain boundaries). This is in good accordance with the occurrence of the amorphous phase in the XRD patterns. Separate FA cenospheres (approximately 1-2 μm) were also found inside the ceramic bodies, and both samples seem to be porous (Figures 5(a) and 5(b)). In Figure 5(c), an EDS spectrum, taken from the sample area included in Figure 5(a), is shown. All elements that are present in the EDS spectrum were also represented in the phase composition of the sample identified from the XRD measurements. Similar EDS spectra were obtained from the rest of the ceramic samples.

Various properties were studied in order to assess the performance of the ceramic compacts produced. Volume change, apparent density, water absorption, and Vickers microhardness, are shown in Figures 6(a), 6(b), 6(c), and 6(d), respectively. FA exhibits a highly positive volume change while EAFD has a highly negative volume change. The

positive volume change (expansion) of FA can be attributed to the collapse of the amorphous and crystalline aluminosilicate phases (e.g., $MgAl_2Si_3O_{10}$) contained in FA [29]. The collapse of the silicate phases occurs at temperatures around 950°C [30]. The positive volume change of FA is more pronounced in the temperature of 1050°C, while in higher temperatures (1150°C) this variation is not that evident. On the other hand, the negative volume change (shrinkage) of EAFD can be attributed to the increased densification capability of EAFD due to the sintering of the metal oxides. The negative volume change in EAFD is more pronounced (−30%) at higher temperatures (1150°C), which is in good agreement with the literature [31]. In the produced mixtures (FA-EAFD and FA-EAFD-LFS), an intermediate behavior (shrinkage or expansion) is observed depending on the FA and EAFD composition in the mixture. The volume change on the mixtures is rather not significant due to the antagonistic effect of FA and EAFD on the volume change.

On the other hand, the apparent density (Figure 6(b)) of the materials produced lies in the range of 2.8–3.0 g/cm^3. Except for FA, by increasing the processing temperature from 1050 to 1150°C, the apparent density slightly increases. This is also consistent with the SEM observations (Figures 6(a) and 6(b)), from which a higher densification degree was observed for the mixture processed at the highest temperature (1150°C).

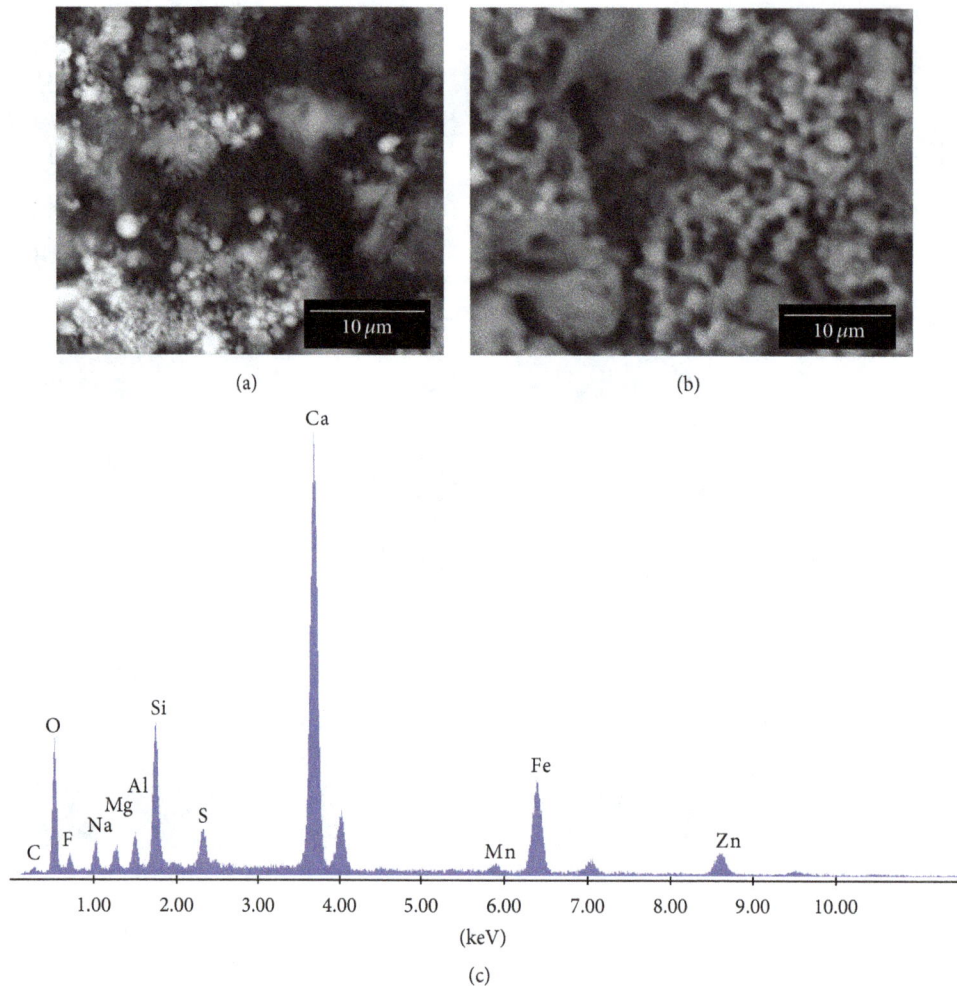

FIGURE 5: SEM images of two EAFD-FA-LFS samples sintered at 950°C (a) and at 1150°C (b). The EDS spectrum (c) was taken from the sample area included in image (a).

FA is a high specific surface area material, with enhanced absorption capabilities due to its high porosity and surface area [29]. The water absorption (Figure 6(c)) of FA is much higher than the one of EAFD. Furthermore, the water absorption of FA increases significantly (almost doubles) with a temperature increase from 1050 to 1150°C. This may be attributed to the removal of preexisting volatile (burnable) matter at higher temperatures [32]. The water absorption of the FA-EAFD and FA-EAFD-LFS mixtures exhibits intermediate values, in the range of 20–30%. However, by increasing the processing temperature (to 1150°C), the water absorption slightly reduces. The latter should be ascribed to the higher densification degree achieved at higher temperatures and the blockage of sorption sites (pores) which exist in FA.

The Vickers (HV) testing results reveal the important role of EAFD loading in the microhardness of the ceramic compacts. Actually, EAFD should be much harder than the other solid residues used in this study due to its metal oxides content. In that sense, the highest HV values were obtained for the binary mixture (FA-EAFD), as it contains higher EAFD loadings. Much lower HV values were recorded for the ternary mixture (FA-EAFD-LFS). The aforementioned behavior becomes more pronounced at higher temperatures (1150°C) due to the higher densification degree achieved. Indeed, the hardness of the binary mixture (FA-EAFD) increases from 10 HV to 234 HV by increasing the temperature from 1050 to 1150°C.

4. Conclusions

Sustainable ceramic materials deriving from 100% industrial waste by-product mixtures were successfully produced in this study. The elemental and structural analyses (XRF, XRD) indicated that the waste by-products consist in various valuable mineralogical phases (quartz, metal oxides, etc.) that are useful for the production of advanced materials. However, the recorded differences in chemical and mineralogical composition of the secondary resources being used as the raw materials for the production of ceramics should be taken into consideration. Morphological observations via SEM and apparent density measurements denote the successful densification of waste by-products towards the formation

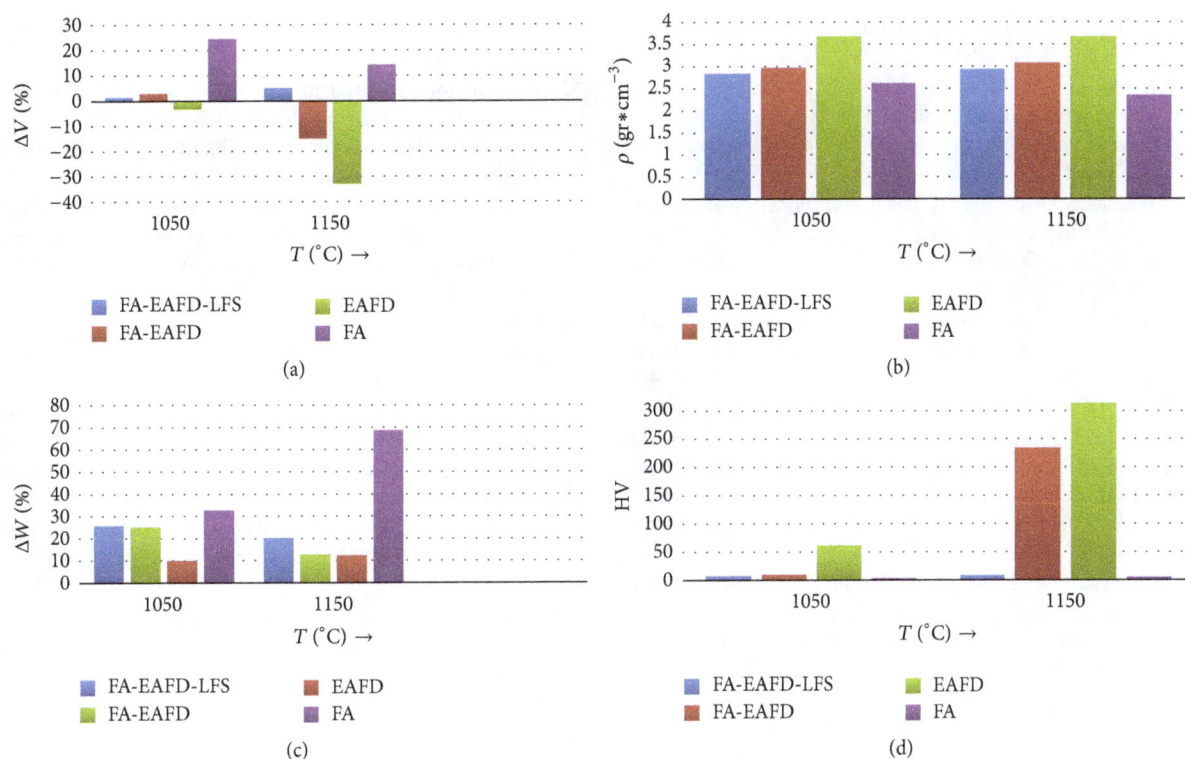

FIGURE 6: Volume change (a), apparent density (b), water absorption (c), and Vickers microhardness (d), for all the samples produced in this work.

of sintered ceramic compacts. Moreover, volume change, water absorption, and microhardness measurements conducted contribute to assess the performance of the ceramics produced. FA and EAFD ratio in the compacted ceramic mixtures seems to play a crucial role in the attained properties. An antagonistic effect of FA and EAFD on the volume change leads to intermediate volume change values for the ceramic mixtures. Furthermore, the addition of FA plays a significant role in the water absorption of the ceramic mixtures. Both the binary (FA-EAFD) and the ternary mixture (FA-EAFD-LFS) exhibit water absorption values lower than 20%, acceptable for traditional ceramics. Nevertheless, the addition of EAFD mainly affected the Vickers microhardness of the compacted ceramic mixtures. Rather significant microhardness values (234 HV) were recorded for the binary (FA-EAFD) ceramic mixture at higher processing temperatures. The processing temperature also influences the other properties examined and the quality of the compacted ceramic mixtures. Consequently, the results presented here indicate that value-added ceramic products can be produced mainly from compacted FA-EAFD by-product mixtures, because these ceramic materials exhibit enhanced physicomechanical properties. Finally, additional properties testing is underway so as to further explore potential fields of technological applications (e.g., insulation), in which the produced ceramics may be used.

Conflicts of Interest

The authors declare that there are no conflicts of interest regarding the publication of this paper.

Acknowledgments

This research has been cofinanced by the European Union (European Social Fund (ESF)) and Greek National Funds through the Operational Program "Education and Lifelong Learning" of the National Strategic Reference Framework (NSRF) Research Funding Program ARCHIMEDES III (Scientific Coordinator: Associate Professor Vayos G. Karayannis): Investing in Knowledge Society through the European Social Fund.

References

[1] B. Das, S. Prakash, P. S. R. Reddy, and V. N. Misra, "An overview of utilization of slag and sludge from steel industries," *Resources, Conservation and Recycling*, vol. 50, no. 1, pp. 40–57, 2007.

[2] V. G. Karayannis and A. K. Moutsatsou, "Synthesis and characterization of nickel-alumina composites from recycled nickel powder," *Advances in Materials Science and Engineering*, vol. 2012, Article ID 395612, 2012.

[3] I. B. Singh, K. Chaturvedi, R. K. Morchhale, and A. H. Yegneswaran, "Thermal treatment of toxic metals of industrial hazardous wastes with fly ash and clay," *Journal of Hazardous Materials*, vol. 141, no. 1, pp. 215–222, 2007.

[4] M.-T. TANG, J. PENG, B. PENG, D. YU, and C.-B. TANG, "Thermal solidification of stainless steelmaking dust," *Transactions of Nonferrous Metals Society of China (English Edition)*, vol. 18, no. 1, pp. 202–206, 2008.

[5] T. Sofilić, A. Rastovčan-Mioč, Š. Cerjan-Stefanović, V. Novosel-Radović, and M. Jenko, "Characterization of steel mill

electric-arc furnace dust," *Journal of Hazardous Materials*, vol. 109, no. 1-3, pp. 59–70, 2004.

[6] A. T. Machado, F. R. Valenzuela-Diaz, C. A. C. de Souza, and L. R. P. de Andrade Lima, "Structural ceramics made with clay and steel dust pollutants," *Applied Clay Science*, vol. 51, no. 4, pp. 503–506, 2011.

[7] P.-T. Teo, A. S. Anasyida, P. Basu, and M. S. Nurulakmal, "Recycling of Malaysia's electric arc furnace (EAF) slag waste into heavy-duty green ceramic tile," *Waste Management*, vol. 34, no. 12, pp. 2697–2708, 2014.

[8] V. Mymrin, R. A. C. Ribeiro, K. Alekseev, E. Zelinskaya, N. Tolmacheva, and R. Catai, "Environment friendly ceramics from hazardous industrial wastes," *Ceramics International*, vol. 40, no. 7, pp. 9427–9437, 2014.

[9] P. S. Polie, M. R. Ilic, and A. R. Popovic, "Environmental impact assessment of lignite fly ash and its utilization products as recycled hazardous wastes on surface and ground water quality," *Handbook of Environmental Chemistry, Vol. 5: Water Pollution*, vol. 5, pp. 61–110, 2006.

[10] M. Arsenovic, Z. Radojević, Ž. Jakšić, and L. Pezo, "Mathematical approach to application of industrial wastes in clay brick production - Part II: Optimization," *Ceramics International*, vol. 41, no. 3, pp. 4899–4905, 2015.

[11] V. Karayannis, X. Spiliotis, A. Domopoulou, K. Ntampegliotis, and G. Papapolymerou, "Optimized synthesis of construction ceramic materials using high-Ca fly ash as admixture," *Romanian Journal of Materials*, vol. 45, no. 4, pp. 358–363, 2015.

[12] V. G. Karayannis, A. K. Moutsatsou, and E. L. Katsika, "Synthesis of microwave-sintered ceramics from lignite fly and bottom ashes," *Journal of Ceramic Processing Research*, vol. 14, no. 1, pp. 45–50, 2013.

[13] B. Kim and M. Prezzi, "Compaction characteristics and corrosivity of Indiana class-F fly and bottom ash mixtures," *Construction and Building Materials*, vol. 22, no. 4, pp. 694–702, 2008.

[14] A. Zimmer and C. P. Bergmann, "Fly ash of mineral coal as ceramic tiles raw material," *Waste Management*, vol. 27, no. 1, pp. 59–68, 2007.

[15] S. Rai, D. H. Lataye, M. J. Chaddha et al., "An alternative to clay in building materials: Red mud sintering using fly ash via taguchi's methodology," *Advances in Materials Science and Engineering*, vol. 2013, Article ID 757923, 2013.

[16] J. Huang, M. Fang, Z. Huang et al., "Preparation, microstructure, and mechanical properties of spinel-corundum-sialon composite materials from waste fly ash and aluminum dross," *Advances in Materials Science and Engineering*, vol. 2014, Article ID 789867, 2014.

[17] V. G. Karayannis, A. K. Moutsatsou, and E. L. Katsika, "Recycling of lignite highly-calcareous fly ash into nickel-based composites," *Fresenius Environmental Bulletin*, vol. 21, no. 8B, pp. 2375–2380, 2012.

[18] A. Karamberi and A. Moutsatsou, "Vitrification of lignite fly ash and metal slags for the production of glass and glass ceramics," *China Particuology*, vol. 4, no. 5, pp. 250–253, 2006.

[19] A. Olgun, Y. Erdogan, Y. Ayhan, and B. Zeybek, "Development of ceramic tiles from coal fly ash and tincal ore waste," *Ceramics International*, vol. 31, no. 1, pp. 153–158, 2005.

[20] P. Porreca, E. Furlani, L. Fedrizzi et al., "Sintered ceramics from special waste incinerator ashes and steelmaking slag," *Industrial Ceramics*, vol. 27, no. 3, pp. 197–203, 2007.

[21] L. F. Vilches, C. Fernández-Pereira, J. Olivares del Valle, and J. Vale, "Recycling potential of coal fly ash and titanium waste as new fireproof products," *Chemical Engineering Journal*, vol. 95, no. 1, pp. 155–161, 2003.

[22] E. Bernardo, L. Esposito, E. Rambaldi, A. Tucci, Y. Pontikes, and G. N. Angelopoulos, "Sintered esseneite-wollastonite-plagioclase glass-ceramics from vitrified waste," *Journal of the European Ceramic Society*, vol. 29, no. 14, pp. 2921–2927, 2009.

[23] G. Bantsis, C. Sikalidis, M. Betsiou, T. Yioultsis, and A. Bourliva, "Ceramic building materials for electromagnetic interference shielding using metallurgical slags," *Advances in Applied Ceramics*, vol. 110, no. 4, pp. 233–237, 2011.

[24] L. S. Pioro and I. L. Pioro, "Reprocessing of metallurgical slag into materials for the building industry," *Waste Management*, vol. 24, no. 4, pp. 371–379, 2004.

[25] P. Asokan, M. Saxena, and S. R. Asolekar, "Coal combustion residues—environmental implications and recycling potentials," *Resources, Conservation and Recycling*, vol. 43, no. 3, pp. 239–262, 2005.

[26] M. Ilic, C. Cheeseman, C. Sollars, and J. Knight, "Mineralogy and microstructure of sintered lignite coal fly ash," *Fuel*, vol. 82, no. 3, pp. 331–336, 2003.

[27] V. Karayannis, A. Moutsatsou, N. Koukouzas, and C. Vasilatos, "Valorization of CFB-combustion fly ashes as the raw materials in the development of value-added ceramics," *Fresenius Environmental Bulletin*, vol. 22, no. 12C, pp. 3873–3879, 2013.

[28] N. Koukouzas, C. Vasilatos, G. Itskos, I. Mitsis, and A. Moutsatsou, "Removal of heavy metals from wastewater using CFB-coal fly ash zeolitic materials," *Journal of Hazardous Materials*, vol. 173, no. 1-3, pp. 581–588, 2010.

[29] R. Aiello, F. Testa, and G. Giordano, "Impact of zeolites and other porous materials on the new technologies at the beginning of the new millennium," in *Proceedings of the 2nd International FEZA Conference*, Elsevier Science, Taormina, Italy, September, 2002.

[30] J. Kovac, A. Trnik, I. Medved, I. Stubna, and L. Vozar, "Influence of fly ash added to a ceramic body on its thermophysical properties," *Thermal Science*, vol. 20, no. 2, pp. 603–612, 2016.

[31] C. Sikalidis and M. Mitrakas, "Utilization of electric arc furnace dust as raw material for the production of ceramic and concrete building products," *Journal of Environmental Science and Health - Part A Toxic/Hazardous Substances and Environmental Engineering*, vol. 41, no. 9, pp. 1943–1954, 2006.

[32] K. Pimraksa, M. Wilhelm, and W. Wruss, "A new approach to the production of bricks made of 100% fly ash," *Tile & brick International*, vol. 16, no. 6, pp. 428–435, 2000.

7

The Study of Microwave and Electric Hybrid Sintering Process of AZO Target

Ling-yun Han, Yong-chun Shu, Yang Liu, Lu Gao, Qing-tong Wang, and Xiao-na Zhang

The Key Laboratory of Weak Light Nonlinear Photonics, Ministry of Education, Nankai University, Tianjin 300457, China

Correspondence should be addressed to Yong-chun Shu; shuyc@nankai.edu.cn

Academic Editor: Philip Eisenlohr

We simulated the microwave sintering of ZnO by 3D modelling. A large-size Al-doped ZnO (AZO) green ceramic compact was prepared by slurry casting. Through studying the microwave and electric hybrid sintering of the green compact, a relative density of up to 98.1% could be obtained by starting microwave heating at 1200°C and increasing the power 20 min later to 4 kW for an AZO ceramic target measuring 120 × 240 × 12 mm. The resistivity of AZO targets sintered with microwave assistance was investigated. The energy consumption of sintering could be greatly reduced by this heating method. Until now, few studies have been reported on the microwave and electric hybrid sintering of large-size AZO ceramic targets. This research can aid in developing sintering technology for large-size high-quality oxide ceramic targets.

1. Introduction

Because of the high optical transmittance and low resistivity of transparent conducting oxide (TCO), TCO films are applied in many electronic devices such as solar cells, flat displays, and touch panels [1–4]. Indium tin oxide (ITO) is currently the most popular industrial TCO film material. However, because of the shortage and high cost of In as a resource, an alternative must be found. Some researchers have found Al-doped ZnO (AZO) film to be a potential alternative to ITO, with optical and electrical properties approaching those of the conventional material [5–7].

Magnetron sputtering is the main fabrication technology used for industrial TCO films. The method requires oxide ceramic feed materials with the same components and quality as the sputtering target. Sintering, as the most important step in preparing the target, has been studied by many researchers, but most studies have investigated electricity as the sintering power [8–12]. The obvious drawbacks of this method include power wastes and the possibility of cracking the ceramic target, as the temperature distribution is not homogeneous during sintering.

Microwave energy is a new sintering power that may increase sintering speed and homogeneity while decreasing the energy cost [13–15]. The working theory is that microwave radiation causes the electric dipole moment rotation of the polar molecules in the dielectric materials, converting the electromagnetic energy to heat energy instantaneously, rapidly, and efficiently. Gunnewiek and Kiminami studied the effect of microwave heating rate on the sintering of ZnO, indicating that the relative density of microwave sintered ZnO was independent of the heating rate, but the grain growth could be restricted by heating rates with a lower time and energy requirement [16]; Birnboim et al. studied the influence of microwave frequency on ZnO sintering, indicating that the temperature gradients could be reduced by changing the microwave frequency [17], while the research of Janney et al. revealed microwave radiation enhanced sintering of ceramic compacts, including lower sintering temperatures and reduced activation energies [18, 19]. However, the limit of microwave sintering research above is obvious: the compacts in the experiments are all in millimeter size, which is too small for industry. The size of AZO targets required for industry is usually beyond 100 × 200 × 10 mm.

This study addresses the technology of microwave and electric hybrid sintering to prepare large-size (≥100 × 200 × 10 mm) AZO ceramic targets with high density (≥98%) for magnetron sputtering.

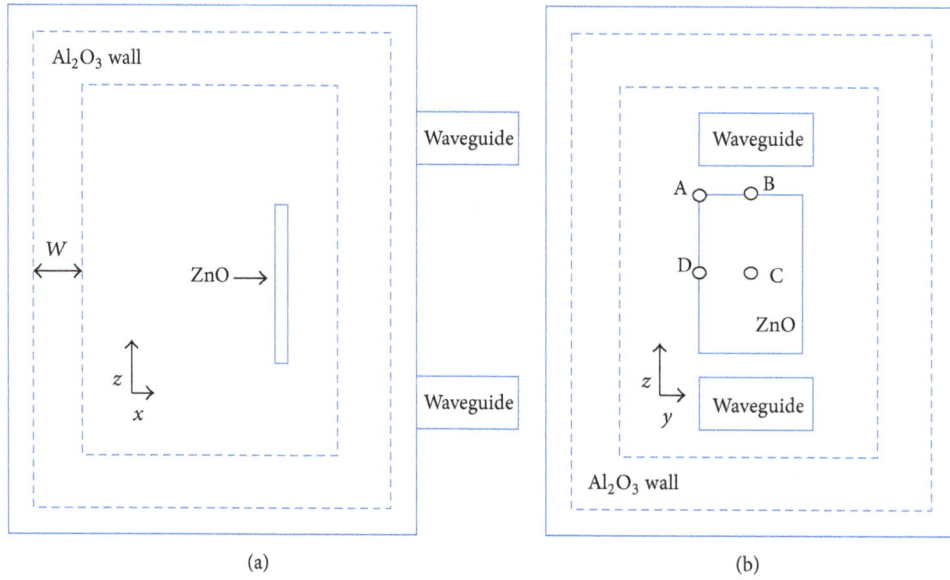

FIGURE 1: Cross-sectional view of the microwave sintering apparatus, (a) the x-z plane view, (b) the y-z plane view.

2. Simulation of the Microwave Sintering Process

In this study, the finite element method (FEM) [20] is applied to simulate microwave sintering using COMSOL software. By setting the physical parameters of the microwave transmitter and the ZnO samples and investigating the sintering process with single- and double-source microwave transmitters, we obtain the distributions of temperature and electric field during sintering. Electromagnetic field equation (1), heat transport equation (2), and electromagnetic power equation (3) are as follows:

$$\nabla^2 E + k^2 \cdot E = 0, \tag{1}$$

$$\rho(T) C(T) \frac{\partial T}{\partial \tau} = \lambda(T) \nabla^2 T + P, \tag{2}$$

$$P = \frac{1}{2} 2\pi f \varepsilon_0 \varepsilon''(T) |E|^2. \tag{3}$$

In the equations ρ, C, and λ are density, capacity, and thermal conductivity, respectively; f is the working frequency, the complex permittivity $\varepsilon = \varepsilon' - j * \varepsilon''$, and the propagation vector of the electromagnetic wave $k = 2\pi f (\mu_0 \varepsilon_0 \varepsilon)^{1/2}$.

2.1. Model and Simulation.
Figure 1 depicts the configuration of the microwave furnace including the sample, as used in our study. The cavity volume of the microwave furnace is $420 \times 420 \times 420$ mm. The 86×43 mm waveguide ports are excitation sources in the TE_{10} mode, and the operating frequency is 2.45 GHz. The total output power of the microwave is 10 kW. Two equally sized ports are placed in the y-z plane of the cavity. To reduce heat loss and protect the wall of the cavity, a porous Al_2O_3 sheet is introduced as insulation; the gap between the outer boundary of the sheet and the cavity wall is 20 mm. The thickness W is 40 mm, and the ZnO sample

in the furnace has a volume of $10 \times 10 \times 100$ mm. During the simulation, the sample is perpendicular to the half-height surface and the locations of the two microwave ports have mirror symmetry relative to the half-height surface. Labels A, B, C, and D designate reference points at four different positions. The parameters of the porous Al_2O_3 and ZnO including complex permittivity ε (a.u.), density (g/cm^3), specific heat capacity C_p (J/(g·K)), and heat transfer coefficient K_p (W/(m^2 · °C)) are shown in Table 1.

We selected air as the sintering atmosphere in the cavity and copper as a perfect cavity wall. The penetration of the electromagnetic field into the sample is characterized by the skin depth δ, given by (4) [21]. The permittivity and δ of the ZnO sample vary with temperature, as shown in Figure 2. Consider

$$\delta = \frac{c}{\sqrt{2}\pi f \sqrt{\sqrt{(\varepsilon')^2 + (\varepsilon'')^2} - \varepsilon'}}, \tag{4}$$

where c is the speed of light.

2.2. Results and Discussion of Simulation

2.2.1. Electric Field Distribution. The total output power was 10 kW. Each port for the double-source microwave had an output power of 5 kW in the simulation. By comparing the output power of the single- and double-source microwave ports, we obtain the electric field distributions from the single- and double-source microwave transmitter ports, as shown in Figure 3. Although the dimension of the electric field does not differ between the double- and single-source microwaves, the intensity of the field produced by the double-source microwave is significantly higher and the cross-sectional symmetric conditions in both the y and z directions are satisfied by the double-source microwave, which greatly reduces the load on memory and computation.

TABLE 1: Relevant parameters of ZnO and Al_2O_3.

	Al_2O_3	ZnO
ε (a.u.)	$4.5 - j \cdot 0.0001$	See details in Figure 2
ρ (g/cm^3)	0.24	$\rho/\rho_{th} = 0.52$ (for $T \leq 600°C$) $\rho/\rho_{th} = 1 - \exp[-0.00586(T - 463.7)]$ (for $600°C < T \leq 1150°C$) $\rho/\rho_{th} = 0.776 + 0.1792 * 10^{-3}T$ (for $T > 1150°C$)
C_p (J/(g·K))	1.09	$C_p = -9880/T^2 + 7.34 * 10^{-3}T + 0.58$
K_p (W/(m^2·°C))	$8 * 10^{-4} + 8 * 10^{-7}(T - 500)$ for $T \leq 500°C$ $8 * 10^{-4} + 16 * 10^{-7}(T - 500)$ for $T > 500°C$	Low temperature stage: $K_p \geq 2.67 * 10^{-2}$ High temperature stage: $K_p = 5.34 * 10^{-3}(T/520)^{2.5}$ $T > 1100°C$: $K_p = 0.0348$

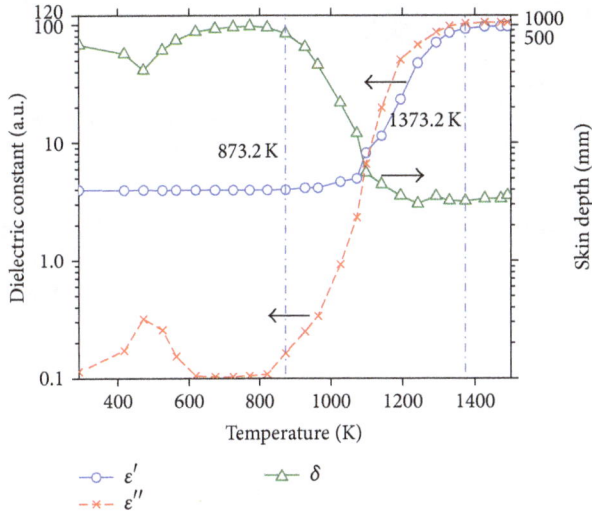

FIGURE 2: Real and imaginary parts ε' and ε'' of the complex dielectric constant and the skin depth δ of ZnO versus temperature T at 2.45 GHz.

2.2.2. Temperature Field Distribution. The single- and double-source microwave power was set to 10 kW and 5 kW, respectively, which ensured the same total output power. The initial temperature and density of the ZnO sample was 25°C and 2.917 g/cm^3, respectively.

When the temperatures of the four reference points all reached or exceeded 1200°C, we terminated the simulation. The temperature changes of the four locations are plotted in Figure 4. Obvious differences in temperature occur among the four reference points at temperatures below 1000°C. However, these differences decrease gradually as the temperature increases. Eventually, the temperatures of all points equalize above 1000°C. This indicates that microwave sintering may be more homogenous at high temperatures. Additionally, single-source microwave sintering requires more time to reach 1100°C than the double-source system. With respect to energy utilization, double-source microwave sintering will be more effective and practical than single-source.

The simulation was the first time to use a 3D model to illustrate the microwave sintering process. It was closer to practical sintering process with the insulation and dynamic parameters introduced in the simulation. Though the simulation has some difference with the experiment because of the limit of computational ability, the results have shown us the possibility of getting a bigger homogeneous sintering area for solving the cracking problem in the preparation of big-size ceramic target by setting multisource microwave. Part of the furnace in the experiments was designed by the simulation.

3. Experimental

The ZnO and Al_2O_3 powders used in the experiment were nanosized particles; ammonium polyacrylate, polycarboxylate superplasticizer, and arabic gum were used as the dispersant, surface active agent, and binder, respectively. All the materials used were analytical reagent (AR). The AZO ceramic targets were prepared by sintering the AZO green compacts. The process consisted of several steps: (1) the ZnO and Al_2O_3 powders were mixed in mass ratio of 98 : 2 by ball-milling for 2 h and then poured into a solution of distilled water, dispersant, surface active agent, and blinder in the volume ratio of 50 : 1.2 : 0.8 : 0.5. (2) After ball-milling in vacuum for 2 h, white slurries with proper viscosity and fluidity were obtained. (3) The slurries were poured into the plaster mold and allowed to consolidate for 48 h at 25°C. (4) The wet green compacts were placed into a drying oven for 1 week at 80°C. After the thermal treatment at 500°C to remove the water and chemical reagents, we obtained dried AZO green compacts of 170 × 340 × 17 mm.

The green compacts were then electrically heated at 1°C/min to the setting temperatures (1000, 1100, 1200, and 1300°C), after which microwave heating was started. The output power of the microwave was varied from 1 to 5 kW and the change rate of power varied from 0.16 kW/min to 0.8 kW/min (5–25 min). The rectangular bulk AZO at the size of 120 × 240 × 12 mm was obtained after heating by electricity and microwave for 3 h at 1460°C.

Figure 5 shows a schematic view of the furnace, which was self-designed and made by Gaoge Thermal Treatment, Inc., Hefei, China. Resistance rods were located around the inner wall of the furnace, which produced heat by electricity. Eight microwave transmitters of the same standard were distributed symmetrically at the top of the furnace, which provided a stable, uniform, and freely controlled electromagnetic field. The frequency was 2.45 GHz and the maximum power of each transmitter was 1 kW. The theoretical density of AZO target in the experiment was 5.56 g/cm^3. The densities

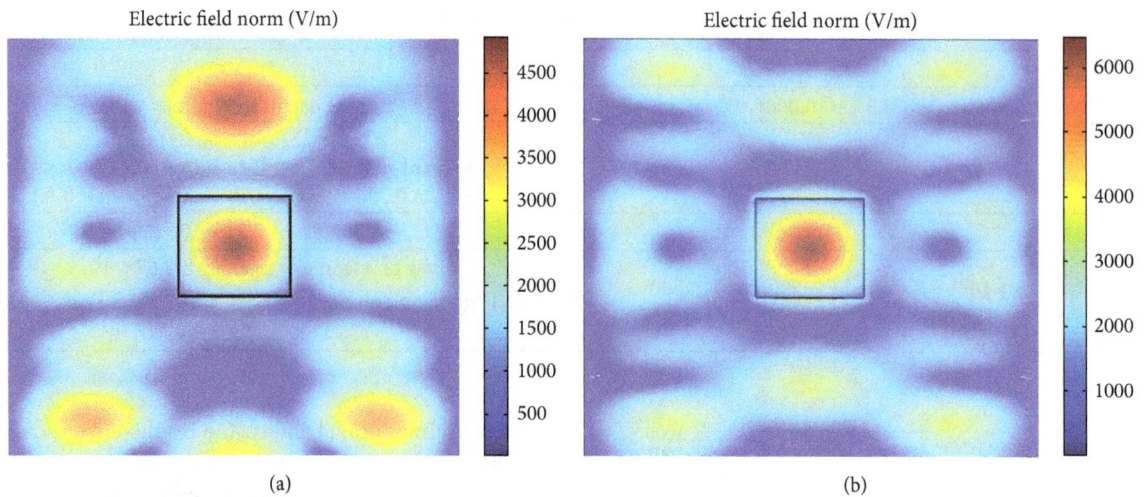

FIGURE 3: Surface plots of microwave electric field distribution for (a) single-source and (b) double-source microwave ports.

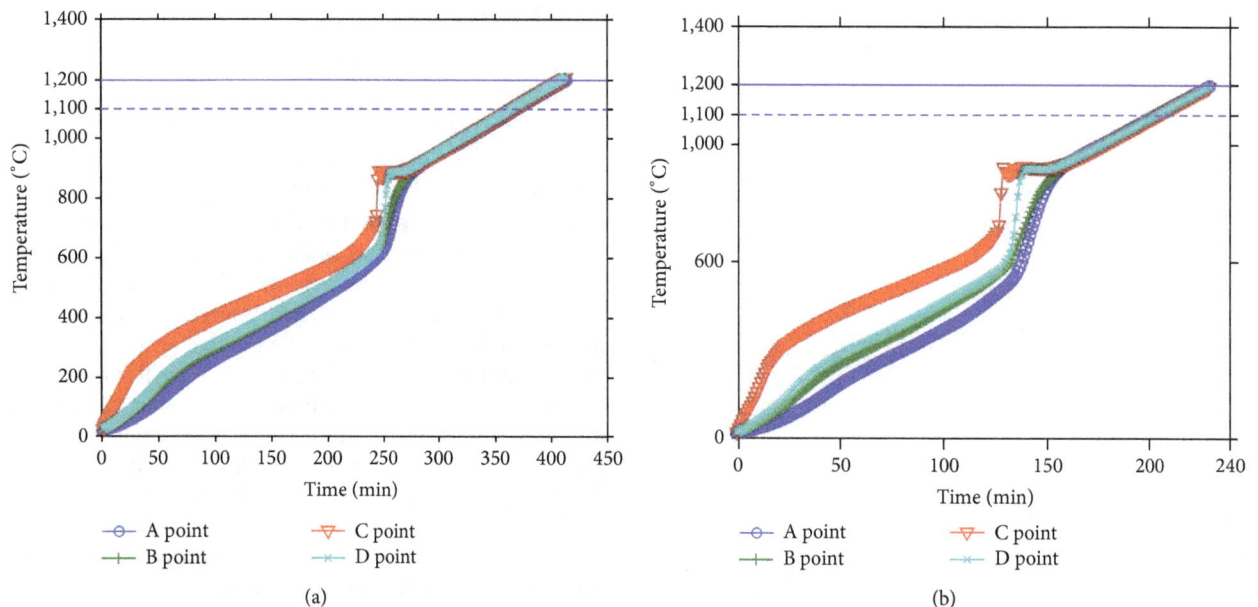

FIGURE 4: Plots of temperatures of four reference points versus heating time in the microwave heating process for (a) single-source and (b) double-source microwave output.

of the targets were measured based on Archimedes' principle. The microstructures and morphologies were observed by a Leo 1530vp scanning electron microscope (SEM). The resistivity of the sintered targets was tested by a four-point probe method (ST2258A, Jingge Electronic Co., Suzhou, China).

4. Results and Discussion

4.1. Effect of Microwave Starting Temperature on the Microstructure and Density of AZO Target. Figure 6 shows X-ray diffraction patterns for the samples prepared through conventional sintering (1) and that prepared through microwave sintering at 1200°C (2), 1300°C (3) as starting temperature. It is evident that both samples exhibit ZnO phase (Z) with $ZnAl_2O_4$ phase (ZA) as the second phase from (1) and (2). But at 1300°C the XRD analysis shows only ZnO phase as shown in (3). The absence (presence in very low intensity) of the characteristic peaks of $ZnAl_2O_4$ may be explained by the melting of $ZnAl_2O_4$. The huge heat induced by microwave radiation and electricity at 1300°C turned $ZnAl_2O_4$ phase to amorphous state. Also, we can see that the intensity of the diffraction peaks for the sample prepared through microwave sintering is lower than that prepared through conventional sintering, which can be explained as the fact that the rapid heating of microwave prevented the growth of grain size.

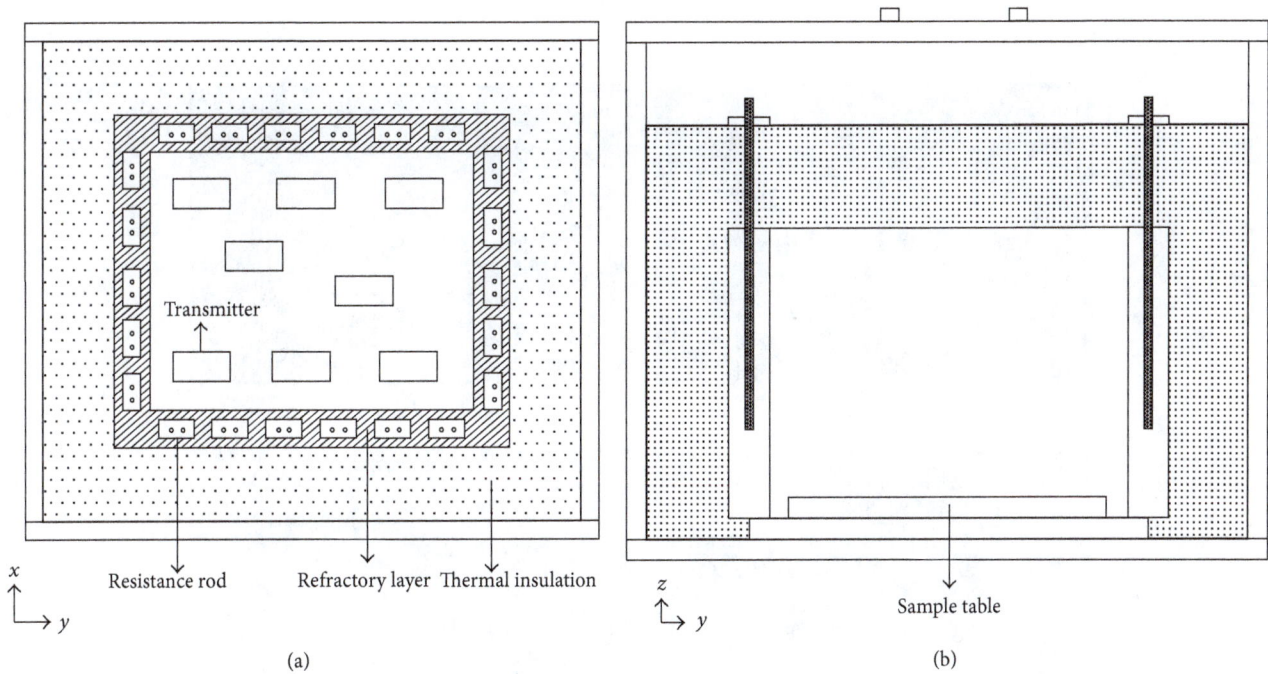

FIGURE 5: A schematic view of the furnace: (a) in x-y view; (b) in y-z view.

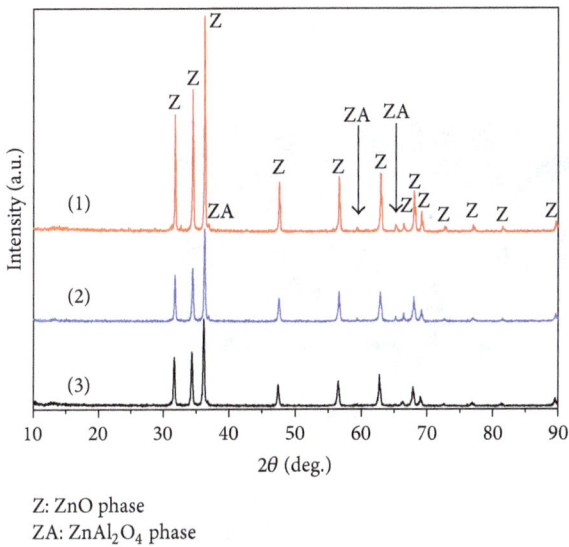

Z: ZnO phase
ZA: ZnAl$_2$O$_4$ phase

FIGURE 6: X-ray diffraction patterns for the AZO sample prepared through (1) conventional sintering, (2) (3) microwave sintering started at 1200°C, 1300°C.

TABLE 2: The grain size of microwave sintered samples.

Sample	Grain size (μm)
a	8.3
b	12.1
c	22.5
d	28.4

that the grains grow with increases in the starting temperature. The pictures also show that sample (c), with a start temperature of 1200°C, has fewer pores and a smaller average grain size than the other samples do.

Figure 8 reveals the fracture morphology of the AZO targets sintered at different microwave starting temperatures. The pore sizes in samples (a), (b), and (c) are greater than those in sample (d). Comparing samples (a), (b), (c), and (d) in Figure 8 we note the absence of the second-phase ZnAl$_2$O$_4$, which is confirmed by XRD study in Figure 6.

Figure 9 shows the variation of the relative density of the target in different microwave starting temperatures. The relative density could increase to 98% when the microwave starting temperature was raised to 1200°C. After this maximum, the relative density decreased with increases in the microwave starting temperature. According to the computational simulation, this result may be explained by the development of a high-intensity electromagnetic field around the sample. The temperature was raised rapidly by a large amount of heat generated from interactions between the electromagnetic field and the sample, which would increase the surface evaporation of ZnO. The reduction of ZnO would decrease the relative density of the target.

Figure 7 shows the microstructures of the AZO targets sintered at different starting temperatures for microwave heating. The average grain size of the sintered samples was estimated from the pictures using the linear intercept method [22]. The results are summarized in Table 2. The grain size increased from sample (a) to sample (d) with increasing microwave heating start temperatures. The starting temperature of microwave heating could alter the grain growth, such

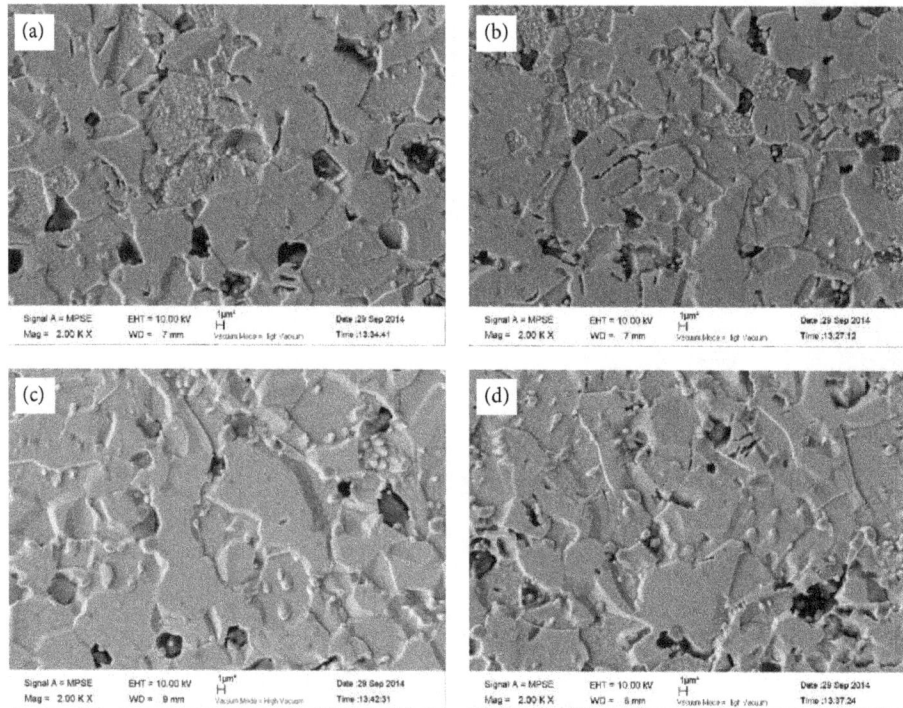

FIGURE 7: Microstructures of AZO targets sintered at different microwave starting temperatures: (a) 1000, (b) 1100, (c) 1200, and (d) 1300°C.

FIGURE 8: Fracture morphologies of AZO targets at different microwave starting temperatures: (a) 1000°C, (b) 1100°C, (c) 1200°C, and (d) 1300°C.

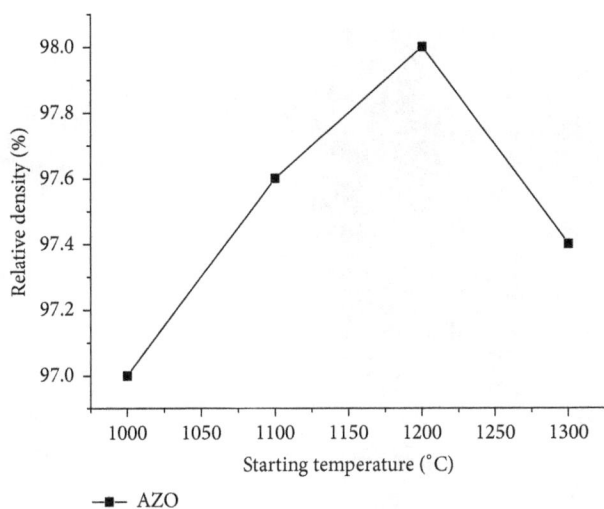

FIGURE 9: Relative densities of AZO targets as a function of microwave starting temperature.

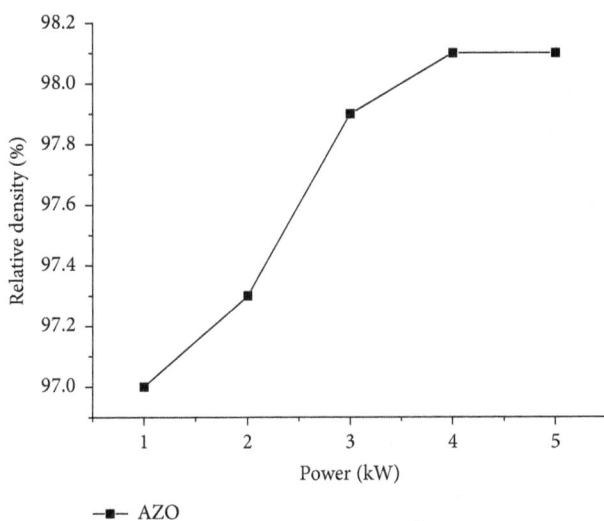

FIGURE 11: Relative densities of AZO targets as a function of change rate of microwave power.

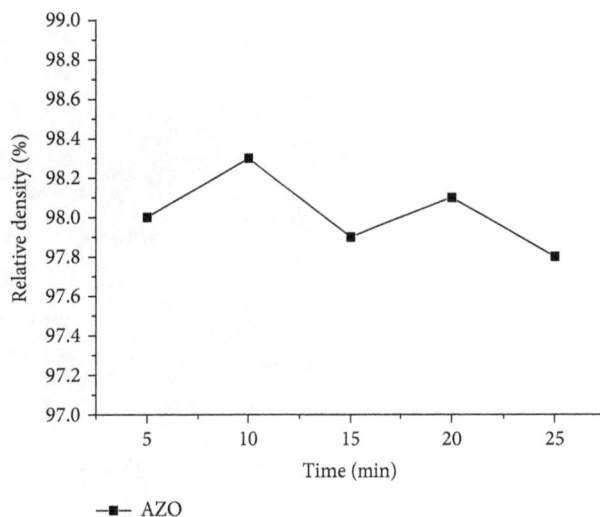

FIGURE 10: Relative densities of AZO targets as a function of microwave power.

All the above data show that 1200°C is the optimal temperature at which microwave heating should begin during the sintering of the AZO target.

4.2. Effect of Microwave Power on the Sintering of AZO Ceramic Target.

As shown in Figure 10, the relative density of AZO target increased with increasing microwave power. When the microwave power was 4 kW, the relative density was 98.1% and remaining constant with further power increases. This may be caused by the increased energy of the electromagnetic field with increased microwave power, which would produce heat, promoting powder aggregation and decreasing the size of pores in the ceramic in rapid time. As the active energy for powder aggregation increases as a function of the sintering temperature, once the energy required for powder aggregation exceeded the microwave radiation energy, the powder aggregation process would be prevented by the high active energy; thus the ceramic density would remain constant.

4.3. Effect of Change Rate of Microwave Power during the Sintering Process.

Figure 11 shows the relative densities of the different samples as a function of change rate of microwave power. The change in relative density was very small with the increase in change rate of microwave power, which correlated with the observation that the microwave heating rate had little effect on the density [16]. According to the study of Rahaman and Chu et al. [23, 24], the ceramic density was not affected by factors other than the sintering temperature.

Figure 12 displays photos of the AZO targets after sintering under different change rates of microwave power. Some obvious cracks occurred at the border of sample (a) and severe oversintering was visible on the surface. No cracks were found on the surfaces of sample (b) or (c). However, the sintering of (b) and (c) was inhomogeneous, as indicated by different surface colours. Sample (d) revealed homogeneous sintering without cracks on the sample surface. A comparison of these samples indicated that sintering was inhomogeneous and the target could crack when the microwave power changed too quickly. This may be because rapid increases in microwave power caused similarly rapid increases in temperature, resulting in a large amount of heat at the sintering area of the AZO target. The dramatic temperature gradient in the area could cause thermal stress in the target. Thermal stress was the main reason for the inhomogeneous sintering. So taking control of change rate of microwave power is necessary for a high-quality target.

4.4. Electrical Properties.

Table 3 shows the resistivities of AZO targets prepared with microwave assistance at the starting temperatures of 1000–1300°C. The results indicated that the resistivities of AZO targets sintered with microwave assistance decreased from 7.2×10^{-2} Ω·cm to 2.8×10^{-3} Ω·cm

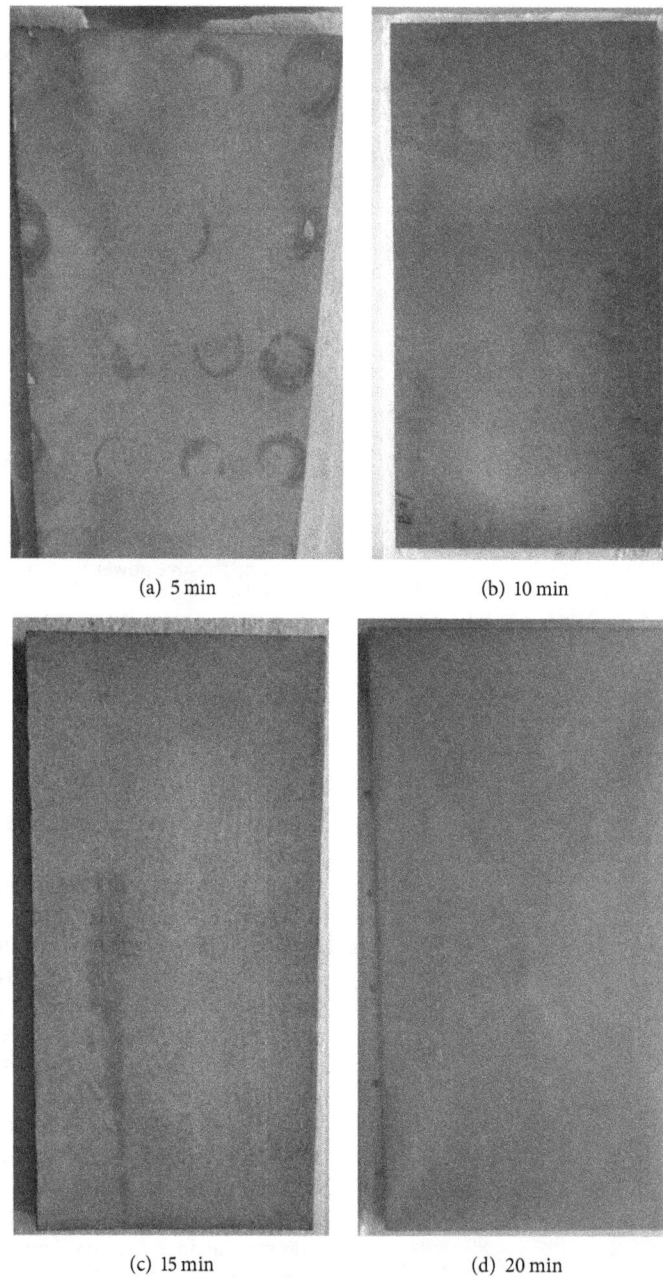

(a) 5 min

(b) 10 min

(c) 15 min

(d) 20 min

FIGURE 12: Photographs of AZO targets after sintering with different change rates of microwave power.

TABLE 3: Resistivities of AZO targets.

Sample	Microwave starting temperature (°C)	Resistivity (Ω·cm)
1	1000	7.2×10^{-2}
2	1100	5.1×10^{-2}
3	1200	2.8×10^{-3}
4	1300	5.3×10^{-3}

with the starting temperature raised from 1000°C to 1200°C. When the temperature was raised to 1300°C, the resistivity increased to 5.3×10^{-3} Ω·cm. The variation of resistivities

can be attributed to the sintered densities as shown in Figure 9. The lower sintered density led to serious scattering of electrons by the large amounts of pores in the targets. When the microwave starting temperature was increased from 1000°C to 1200°C, the increase in sintered density and the change in pores alleviated the scattering of electrons, resulting in the increase of the resistivity. Moreover, as the formation of $ZnAl_2O_4$ phase, the Al dopant may gradually become more active as a conducting carrier between 1000°C and 1200°C. But with the absence of $ZnAl_2O_4$ phase and the decrease in sintered density after the starting temperature was raised to 1300°C, the resistivity increased slightly.

5. Conclusions

(1) FEM simulation indicated that a homogeneous electric field distribution for sintering could be achieved by designing a microwave furnace with proper constraints. The sintering field increased in homogeneity with a multisource microwave relative to that achieved by a single-source microwave at high temperature. The sintering time could be reduced by using a multisource microwave instead of a single-source microwave.

(2) By preparing the AZO target of $120 \times 240 \times 12$ mm with microwave and electric hybrid sintering, the microwave starting temperature was found to affect the densification of the AZO target. Increasing the microwave power could increase the density of the AZO target. The change rate of microwave power had no obvious effect on the density of the target but significant effects on the integrity and uniformity of the large-sized AZO target.

(3) Based on the experimental results, an AZO target of $120 \times 240 \times 12$ mm with a relative density of 98.1% can be obtained by introducing microwave power when the sample has reached 1200°C by electrical heating, with the output power increased to 4 kW in 20 min.

(4) The resistivity of AZO targets sintered with microwave started at 1200°C could be as low as $2.8 \times 10^{-3}\ \Omega \cdot$cm.

Competing Interests

The authors declare that there are no competing interests regarding the publication of this paper.

Acknowledgments

The authors are grateful for financial support from the National High Technology Research and Development Program of China (2012AA03030315).

References

[1] X. T. Hao, F. R. Zhu, K. S. Ong, and L. W. Tan, "Hydrogenated aluminium-doped zinc oxide semiconductor thin films for polymeric light-emitting diodes," *Semiconductor Science and Technology*, vol. 21, no. 1, pp. 48–54, 2006.

[2] C. G. Granqvist, "Transparent conductors as solar energy materials: a panoramic review," *Solar Energy Materials and Solar Cells*, vol. 91, no. 17, pp. 1529–1598, 2007.

[3] T. Minami, "Present status of transparent conducting oxide thin-film development for Indium-Tin-Oxide (ITO) substitutes," *Thin Solid Films*, vol. 516, no. 17, pp. 5822–5828, 2008.

[4] X. Yu, X. M. Yu, J. J. Zhang, and H. J. Pan, "Gradient Al-doped ZnO multi-buffer layers: effect on the photovoltaic properties of organic solar cells," *Materials Letters*, vol. 161, pp. 624–627, 2015.

[5] B. Hwang, Y.-K. Paek, S.-H. Yang, S. Lim, W.-S. Seo, and K.-S. Oh, "Densification of Al-doped ZnO via preliminary heat treatment under external pressure," *Journal of Alloys and Compounds*, vol. 509, no. 27, pp. 7478–7483, 2011.

[6] T. Minami, "Substitution of transparent conducting oxide thin films for indium tin oxide transparent electrode applications," *Thin Solid Films*, vol. 516, no. 7, pp. 1314–1321, 2008.

[7] G. Fang, D. Li, and B.-L. Yao, "Fabrication and vacuum annealing of transparent conductive AZO thin films prepared by DC magnetron sputtering," *Vacuum*, vol. 68, no. 4, pp. 363–372, 2002.

[8] M.-W. Wu, "Two-step sintering of aluminum-doped zinc oxide sputtering target by using a submicrometer zinc oxide powder," *Ceramics International*, vol. 38, no. 8, pp. 6229–6234, 2012.

[9] H. Cheng, X. J. Xu, H. H. Hng, and J. Ma, "Characterization of Al-doped ZnO thermoelectric materials prepared by RF plasma powder processing and hot press sintering," *Ceramics International*, vol. 35, no. 8, pp. 3067–3072, 2009.

[10] X.-M. Wang, X. Bai, H.-Y. Duan et al., "Preparation of Al-doped ZnO sputter target by hot pressing," *Transactions of Nonferrous Metals Society of China*, vol. 21, no. 7, pp. 1550–1556, 2011.

[11] J. Zhang, W. Zhang, E. Zhao, and H. J. Jacques, "Study of high-density AZO ceramic target," *Materials Science in Semiconductor Processing*, vol. 14, no. 3-4, pp. 189–192, 2011.

[12] H. Bouhamed and S. Baklouti, "Synthesis and characterization of Al_2O_3/Zno nanocomposite by pressureless sintering," *Powder Technology*, vol. 264, pp. 278–290, 2014.

[13] D. E. Clark and W. H. Sutton, "Microwave processing of materials," *Annual Review of Materials Science*, vol. 26, no. 1, pp. 299–331, 1996.

[14] J. D. Katz and R. D. Blake, "Microwave sintering of multiple alumina and composite components," *American Ceramic Society Bulletin*, vol. 67, pp. 1304–1308, 1991.

[15] Z. Xie, J. Yang, X. Huang, and Y. Huang, "Microwave processing and properties of ceramics with different dielectric loss," *Journal of the European Ceramic Society*, vol. 19, no. 3, pp. 381–387, 1999.

[16] R. F. K. Gunnewiek and R. H. G. A. Kiminami, "Effect of heating rate on microwave sintering of nanocrystalline zinc oxide," *Ceramics International*, vol. 40, no. 7, pp. 10667–10675, 2014.

[17] A. Birnboim, D. Gershon, J. Calame et al., "Comparative study of microwave sintering of zinc oxide at 2.45, 30, and 83 GHz," *Journal of the American Ceramic Society*, vol. 81, no. 6, pp. 1493–1501, 1998.

[18] M. A. Janney and H. D. Kimrey, "Microstructure evolution in microwave sintered alumina," in *Ceramic Powder Science, Vol. II*, pp. 919–924, American Ceramic Society, 1988.

[19] M. A. Janney, C. L. Calhoun, and H. D. Kimrey, "Microwave sintering of zirconia-8 mol% yttria," *Ceramic Transactions*, vol. 21, pp. 311–318, 1991.

[20] D. C. Dibben and A. C. Metaxas, "Finite element time domain analysis of multimode applicators using edge elements," *Journal of Microwave Power & Electromagnetic Energy*, vol. 29, no. 4, pp. 242–251, 1994.

[21] A. K. Shukla, A. Mondal, and A. Upadhyaya, "Numerical modeling of microwave heating," *Science of Sintering*, vol. 42, no. 1, pp. 99–124, 2010.

[22] J. C. Wurst and J. A. Nelson, "Lineal intercept technique for measuring grain size in two-phase polycrystalline ceramics," *Journal of the American Ceramic Society*, vol. 55, no. 2, p. 109, 1972.

[23] M. N. Rahaman, *Ceramic Processing and Sintering*, Taylor & Francis, New York, NY, USA, 2nd edition, 2003.

[24] M. Y. Chu, N. Rahaman, L. C. Jonghe, and R. J. Brook, "Effect of heating rate on sintering and coarsening," *Journal of the American Ceramic Society*, vol. 74, no. 6, pp. 1217–1225, 1991.

8

Effect of Recasted Material Addition on the Quality of Metal-Ceramic Bond: A Macro-, Micro-, and Nanostudy

Karolina Beer–Lech ⓘ,[1] Krzysztof Pałka,[2] Anna Skic ⓘ,[1] Barbara Surowska,[2] and Krzysztof Gołacki[1]

[1]Department of Mechanical Engineering and Automation, Faculty of Production Engineering,
 University of Life Sciences in Lublin, Głęboka Street 28, 20-612 Lublin, Poland
[2]Department of Materials Engineering, Faculty of Mechanical Engineering, Lublin University of Technology,
 Nadbystrzycka Street 36, 20-618 Lublin, Poland

Correspondence should be addressed to Anna Skic; anna.skic@up.lublin.pl

Academic Editor: Patrice Berthod

Using the recasted alloys in dental prosthetics could affect the quality of the metal-ceramic bond. However, scientists, alloys producers, and prosthetists are still of different opinions. The purpose of this study was to estimate the influence of recasting of the NiCrMo alloy on the metal-ceramic bond quality. The research was carried out on macro-, micro- and nanoscales using the three-point bending test procedure and hardness tests as well as atomic force microscopy and SEM analyses. The SEM analyses showed good integrity of the metal-ceramic bond. The τ_b index of all test samples was greater than 45 MPa. The highest values were recorded for the samples made of 50% and 100% of a brand new material. SEM analysis made after the bending test confirmed good metal-ceramic bond and exhibited adhesive-cohesive fracture. The largest hardness of metal plates was found for the samples containing 50% of the recycled material. Atomic force microscopy studies showed that the alloy containing 50% of the recycled material was characterized by the highest values of surface roughness parameters.

1. Introduction

The metal-ceramic dentures are still widely used due to good combination of their durability—high strength of metal substructures and esthetics of porcelain. They are also more economical than implants. Despite great popularity of precious metal alloys, the nonprecious Co-Cr and Ni-Cr ones are often used interchangeably because of their good mechanical properties and low costs.

Despite questioning the use of nickel alloys as biomaterials, due to the potential harmful effects on human tissues, nickel-base alloys are still widely used to manufacture the substructures of ceramic crowns and bridges. These alloys have larger elasticity modulus than the gold ones and thus thinner cross section of the alloy can be used to reduce destruction of the healthy tooth during crown manufacturing [1]. Moreover, the coefficient of thermal expansion of nickel alloys is compatible with that of thermal expansion of conventional ceramics that are used to produce dental veneers which provides a good metal-ceramic bond [2].

The final quality and reliability of partial dentures is influenced mainly by properties of denture and their design [3], the casting method [4, 5], and the use of already melted materials for casting its metal substructures [6]. The practice of using recasting materials is a very common way to reduce the cost of prosthetic components manufactured in dental laboratories. However, producer's and researcher's points of view are different in this case. Some manufacturers of dental alloys permit the use of once melted alloys but not less than the 50% addition of the brand new material. They also require that the material must come from the same batch. Another group of manufacturers do not allow using remelted material (e.g., Heraeus Kulzer Co.) or do not provide any information on the use of dental alloys obtained from recycling.

Recasting of Ni-Cr dental alloys is the current topic in the literature. Researchers examine the effect of using recasted materials on properties such as chemical composition and precipitate formation and microstructure and its influence on mechanical properties, including testing of metal-ceramic bond strength.

Characterization of microstructure and metal-ceramic bond quality is generally made using several methods, for example, scanning electron microscopy (SEM) and atomic force microscopy (AFM). These methods have also been used to evaluate the surface morphology [7]. AFM provides three-dimensional detailed topographical images of surface roughness on a nanometer scale and has been used in dental research [8–14]. Roughness could be defined as a complex role of irregularities or little indentations that characterize a surface and influence on, inter alia, metal-ceramic bond quality [15]. Determination of surface roughness plays an important role in materials used for dental prostheses.

In spite of that the topic of recycling of nickel-based dental alloys is frequently discussed by research institutions, there is still no clear answer whether and how the use of the recycled material affects the quality of the metal-ceramic bond. There is also a lack of considering this issue from the perspective of different scales, which would allow a more complete identification of the causes of possible differences in the results of individual tests.

The purpose of this study was to investigate the influence of recycled material addition on the quality of the metal-ceramic bond (for the NiCrMo alloy), basis on tests in the micro- and nanoscales (surface properties) as well as in the macroscale (bond strength).

2. Materials and Methods

2.1. The Alloy and Casting Method. Specimens for all tests were made of the Heraenium NA alloy (Heraeus Kulzer GmbH, Germany), whose chemical compositions were as follows (by weight): Ni-59.3% Cr-24% Mo-10% Fe-1.5%, Mn-1.5%, and Ta-1.5% Si-1.5% Nb-1.0%, according to the manufacturer's data [16]. The abovementioned alloy is used for casting crowns and bridges. The castings were made using a disposable alloy with a starting composition of 100%, 50%, and 0% of the brand new material made up to 100% by the recasted material according to Table 1. All materials came from the same batch. Casting and dental ceramic coating were carried out in a professional prosthetic laboratory according to the procedures applied to manufacture dental crowns and bridges. The vacuum pressure casting method (Nautilus, Bego Co.) and the ceramic crucibles (Nautilus) were used.

The specimens were purified from the investment material using first the coarse (250 μm) and then the precision (100 μm) abrasive cleaning (sandblasting). Sandblasting time for all samples was equal (15 s for each particle size). After sandblasting and cleaning the casting channels, from specimens "100" were used to manufacture the second and third groups of test specimens (foundry cones were not reused).

TABLE 1: Marking of the specimens.

Specimen marking	Composition
100	Castings made of 100% of the brand new alloy
50	Castings made of 50% of the brand new alloy and 50% of the already melted alloy comes from material 100
0	Castings made of 0% of the brand new alloy (100% of the already melted alloy comes from material 100)

2.2. Samples and Porcelain Coating Method. Shapes and dimensions of the samples were in agreement with the ISO 9693 standard [17]. The samples had the shape of rectangular plates with dimensions of $25 \times 3 \times 0.5$ mm. They were cast from patterns cutout from wax sheets of 0.5 mm thick. To make foundry molds, the Bellavest SH investment bond (Bego Co.) was applied. The 10 minutes oxidation program was applied for all tested samples (1223.15 K).

The castings were coated with d.Sign porcelain (opaque, dentin, glaze, and Ivoclar Vivadent) into the rectangular area (8×3 mm) in the center of one side of each metal specimen according to the manufacturer's specification. The samples were checked for dimensional correctness and the presence of potential external casting defects. Defected specimens were removed from the tests. The image of prepared specimens is presented in Figure 1(a).

2.3. SEM and AFM Structure Characterization. Characterization of specimen's surface was made by SEM and atomic force microscopy (NTEGRA Prima, NT-MDT, Moscow, Russia). Images of ceramic and metal surfaces were obtained using the scanning electron microscopy Nova NanoSEM 450 (FEI, Eindhoven, Netherlands) before and after the bending test. For the preparation for SEM observations sputtering of the samples with the gold layer of thickness about 7 nm was applied. Additionally, chemical composition analyses using the EDS method were carried out. Analyses were made using the Octane Pro EDS detector (EDAX) with the excitation voltage of 15 kV. Each specimen was analyzed three times, and the signal was detected from the area of about 600×800 μm. The metal surface roughness parameters (R_q is the root-mean-square deviation of the roughness profile, R_a is the arithmetical mean deviation of the roughness profile, R_p is the maximum peak height of the roughness profile, R_v is the maximum valley depth of the roughness profile, and R_z is the maximum height of roughness profile) were determined with NT-MDT software application [13]. In order to determine surface roughness of the alloy, the areas of 10×10 μm were scanned with 512×512 data points in the semicontact mode using a silicon cantilever (NSG30; NT-MDT) 125 μm long, 4.0 μm thick with a tip radius of 10 nm, and the average resonance frequency of 300 kHz. The scan rate was 1 Hz. Regions of interest were defined under the optical microscope. For each of the samples, three different areas, not closely located, were chosen. Obtained height images were used for determination of surface roughness parameters. All parameters were calculated for 15 lines per one image, and

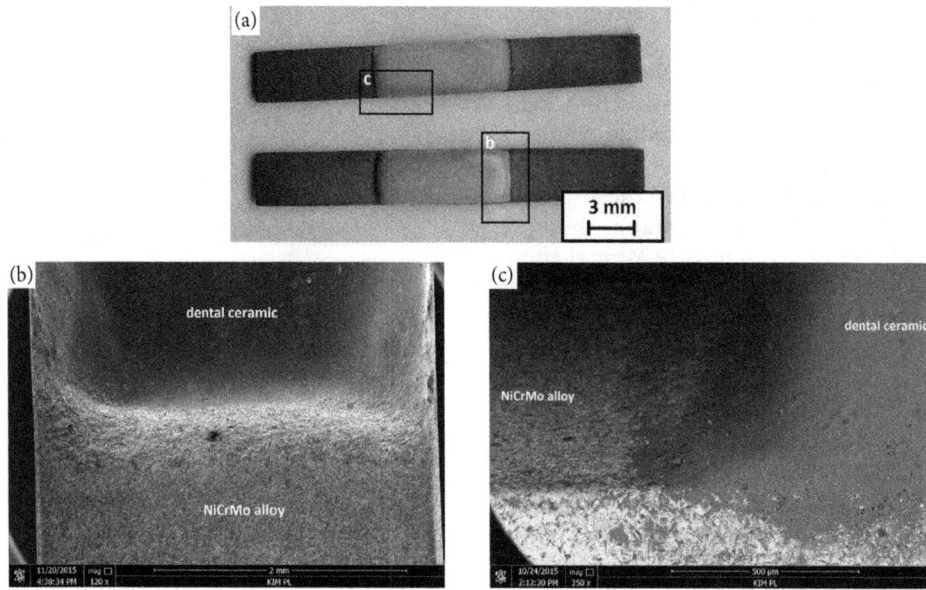

FIGURE 1: The exemplary groups (SEM images): (a) specimens used for the bending test; (b) metal-ceramic bond surface (view from the top); (c) metal-ceramic bond surface (lateral view) .

finally, a set of 540 data per experimental group was taken for further analysis.

The roughness parameters were checked for normal distribution and variances homogeneity (Levene's test). As the data were not normally distributed, the Kruskall–Wallis analysis of variance was applied to assess significant effects of recasted NiCrMo alloy addition on the determined parameters. Multiple comparisons of mean ranks were used to determine significant differences among the experimental groups (at the significance level of 0.05).

2.4. The Bending Test and Hardness Measurements. After the preliminary bending tests, using the power test analysis from the Statistica package, the number of samples needed for testing was determined. The test power was equal to 0.8, the value of the significance level was $\alpha \leq 0.05$, and the measurement error was equal to 5%.

Twelve samples were selected from each batch for the bending test, according to the ISO 9693-1:2012 standard [17]. The three-point bending test was performed on a Instron 8801 machine equipped with a 0.1 N accuracy load cell. The distance between the supports was 20 mm, and the diameter of the rollers supporting the sample was ⌀1 mm. The test was conducted at the crosshead rate of 1.5 mm/min. The measurements were made up to a significant reduction in the value of the force appeared. This indicated the destruction of the metal-porcelain connection. The maximum recorded force was taken as the fracture force (F_{fail}). The bond strength was calculated as the bonding compatibility index τ_b [17] using the following equation:

$$\tau_b = k_1 \times F_{fail}. \tag{1}$$

The fail force Ffail was multiplied by the coefficient k_1, which depends on the thickness of the metal substrate, dM

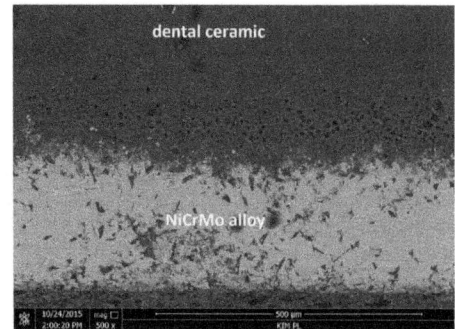

FIGURE 2: Lateral view of the metal-ceramic bond surface (specimen 100): porosity in dental porcelain (SEM image).

d_M ((0.5 ± 0.05) mm), and the Young's modulus, E_M and the Young's modulus value, EM, of the used metallic materials. Young's modulus was provided by the alloy manufacturer (for Heraenium Na, the value was equal to 222 GPa). Hardness measurements (on the metal) were performed using the Vickers hardness tester FM-800 (Future-Tech Corp.). The applied load was 4.9 N which corresponds to the scale HV0.5. The duration of action of force was 10 s.

3. Results and Discussion

Microstructural (SEM) analysis showed a good integrity of the metal-ceramic bonding in each case of tested specimens, regardless of the content of recycled material. Metal surface is characterized by a homogeneous structure, without any cracks and porosity (Figures 1(b) and 1(c)). On the ceramic surface, cracks were not observed either (Figure 1(b)). In a lateral view, the presence of microporosity in the ceramic layer was observed in most specimens (Figure 2). However, this fact cannot be

(a)

(b)

(c)

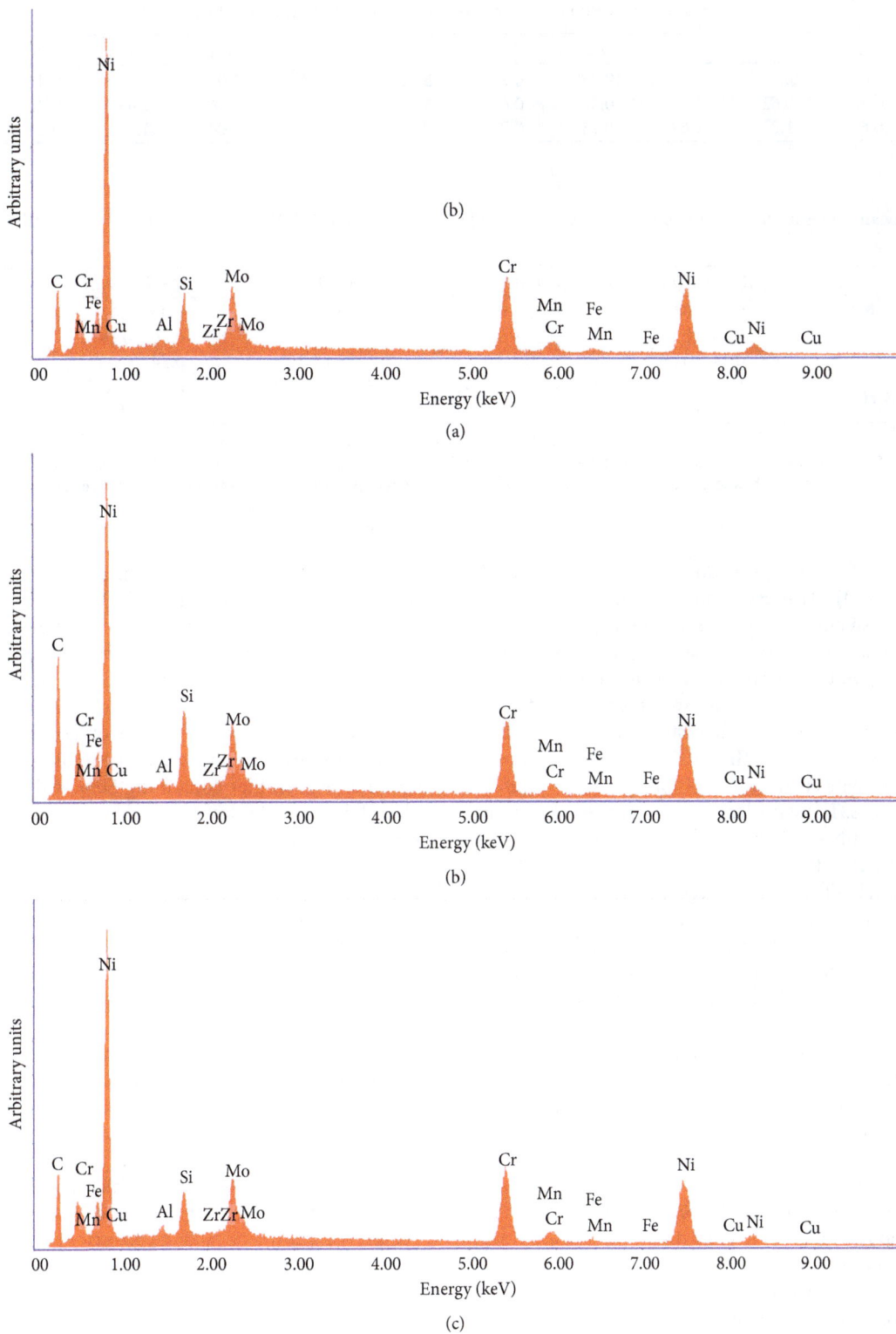

FIGURE 3: The results of chemical compositions analyses (EDS). (a) Sample 0. (b) Sample 50. (c) Sample 100.

associated with the content of the recycled material in metal castings. It would rather be due to precision of porcelain application on the sample surface. It is noteworthy that the precise of the application of the ceramic layer was the same in each case.

The results of chemical composition analyses (EDS) are presented in Figure 3 and Table 2 and contain the mean values. There were no significant differences observed in the amount of metallic elements for each type of the specimen. The differences were observed only for the carbon

TABLE 2: The results of chemical composition analyses (%) (EDS).

Sample	Al	Ta	Si	Zr	Nb	Mo	Cr	Mn	Fe	Cu	Ni
0	1.06	0.71	2.68	0.08	0.74	8.91	23.36	1.08	1.47	0.48	Balanced
50	0.86	0.62	2.43	0.40	0.68	8.54	24.73	1.46	1.49	0.46	Balanced
100	0.61	1.27	2.67	0.13	0.78	9.15	24.27	1.04	1.54	0.53	Balanced

TABLE 3: The mean values of surface roughness parameters (with SD values) and Kruskal–Wallis H_{K-W} statistics results calculated for each parameter.

Specimen	$R_q \pm SD$ (μm)	$R_a \pm SD$ (μm)	$R_p \pm SD$ (μm)	$R_v \pm SD$ (μm)	$R_z \pm SD$ (μm)
100	0.27 ± 0.08^a	0.23 ± 0.07^a	0.50 ± 0.14^a	0.49 ± 0.11^a	1.04 ± 0.17^a
50	0.38 ± 0.13^b	0.27 ± 0.10^b	0.68 ± 0.27^b	0.80 ± 0.19^b	1.37 ± 0.39^b
0	0.24 ± 0.06^a	0.20 ± 0.03^a	0.33 ± 0.12^c	0.34 ± 0.05^c	0.75 ± 0.15^c
Kruskal–Wallis H test	$H_{K-W} = 27.42$; $P < 0.05$	$H_{K-W} = 12.41$; $P < 0.05$	$H_{K-W} = 25.99$; $P < 0.05$	$H_{K-W} = 77.27$; $P < 0.05$	$H_{K-W} = 49.71$; $P < 0.05$

R_q, root-mean-square deviation of the roughness profile; R_a, arithmetical mean deviation of the profile; R_p, maximum roughness profile peak height; R_v, maximum roughness profile valley depth; R_z, ten-point height of irregularities; the mean values with dissimilar letters in the same column are significantly different at $P < 0.05$. The letters a, b, and c indicate the existence of significant differences between the mean values of surface roughness parameters.

content with the highest peak obtained for specimen 50 (Figures 3(a)–3(c)). However, due to the EDS method limitations, the quantitative results were not presented.

The results of surface roughness analysis are shown in Table 3. The highest values of all calculated roughness parameters were observed for the specimens marked 50, whereas the 0 specimens were characterized by the smallest surface roughness (Table 3). This is also shown in Figure 4, which presents typical 2D and 3D topography images of investigated alloys. The statistical analysis showed significant effects of recasted NiCrMo alloy addition on R_q, R_a, R_p, R_v, and R_z parameters ($P < 0.05$). Despite the fact that the roughness was significantly the largest for specimen 50, its average value was still small (R_q below 0.5μm). The low roughness is very important to avoid bacterial adhesion (size of about 1μm) on the alloy surface when parts of the crown, not coated with the porcelain veneer, may remain in contact with the oral environment [18–21].

The τ_b index of all tested samples, determined at the three-point bending test, was higher than 45 MPa (Table 4). This indicates good connection of the pair metal-dental ceramic. According to Craig [22], a proper metal-ceramic bond exists when the τ_b index reaches over 25 MPa. The highest values were recorded for the samples cast at 50% and 100% of the brand new material (approximately 48-49 MPa). As the data did not meet the normal distribution conditions, the Kruskall–Wallis analysis of variance was applied to estimate significant effects of recasted NiCrMo alloy addition on the bonding compatibility index. The test did not show significant differences between values of τ_b obtained for casting composed of 100%, 50%, and 0% of the brand new material (Table 4).

The macro- and microscopic observations of the samples after the bending test are shown in Figure 5. The crack appearing between the dental ceramic and metal substrate is shown in Figure 5(a). It is visible only in the SEM imaging. The shape of the cracks takes the lamellar form with the angle between the crack and the metal substrate of about 10 degrees.

In almost all cases, the type of damage was mixed, adhesive-cohesive. The next image (Figure 5(b)) shows the surface of the metal substrate after separation of the porcelain from the metal. The beginning of the cracking in the metal-ceramic joint was observed on the edge of the porcelain layer (maximum stresses). The presence of the thick residue ceramic layer at the ends of the layer area can be observed.

The hardness test showed that the highest values were found for the samples containing 50% of the recycled material (Table 5). The lowest hardness values characterized castings made entirely from the recycled material, but the differences between samples 0 and 100 were not statistically significant. The ANOVA analysis demonstrated that there were statistically significant differences between the hardness of samples 100 and 50 as well 50 and 0 (Table 5).

An explanation of different hardness and surface roughness is required due to the same parameters of the sandblasting process. Sandblasting is an abrasive process which effectiveness and results are dependent on the mechanical properties of the material. In turn, the mechanical properties are related to, except chemical composition, the microstructure. The microstructure determines the susceptibility for abrasion. Materials exhibiting ductility and high strength are less susceptible for abrasion. Materials which are hard and brittle exhibit higher susceptibility for the erosion process. Microstructure of prosthetic alloys consists of solid solution of alloying elements in nickel with precipitations of intermetallic phases [3]. The amount of precipitates, related to materials used and the manufacturing process, corresponds with the hardness. The presence of precipitates causes brittleness and as a result higher roughness. It is confirmed by the appearance of the surface obtained from AFM (Figure 4), where steep slopes with sharp edges are visible.

The results are consistent with the similar study for the Remanium CS+ alloy (from Dentarum Co.) obtained by Walczak [23, 24], but they are not consistent with the Ucar's et al. research [25] for the RemaniumCSe alloy (the same manufacturer). These two alloys (Remanium CS+ and

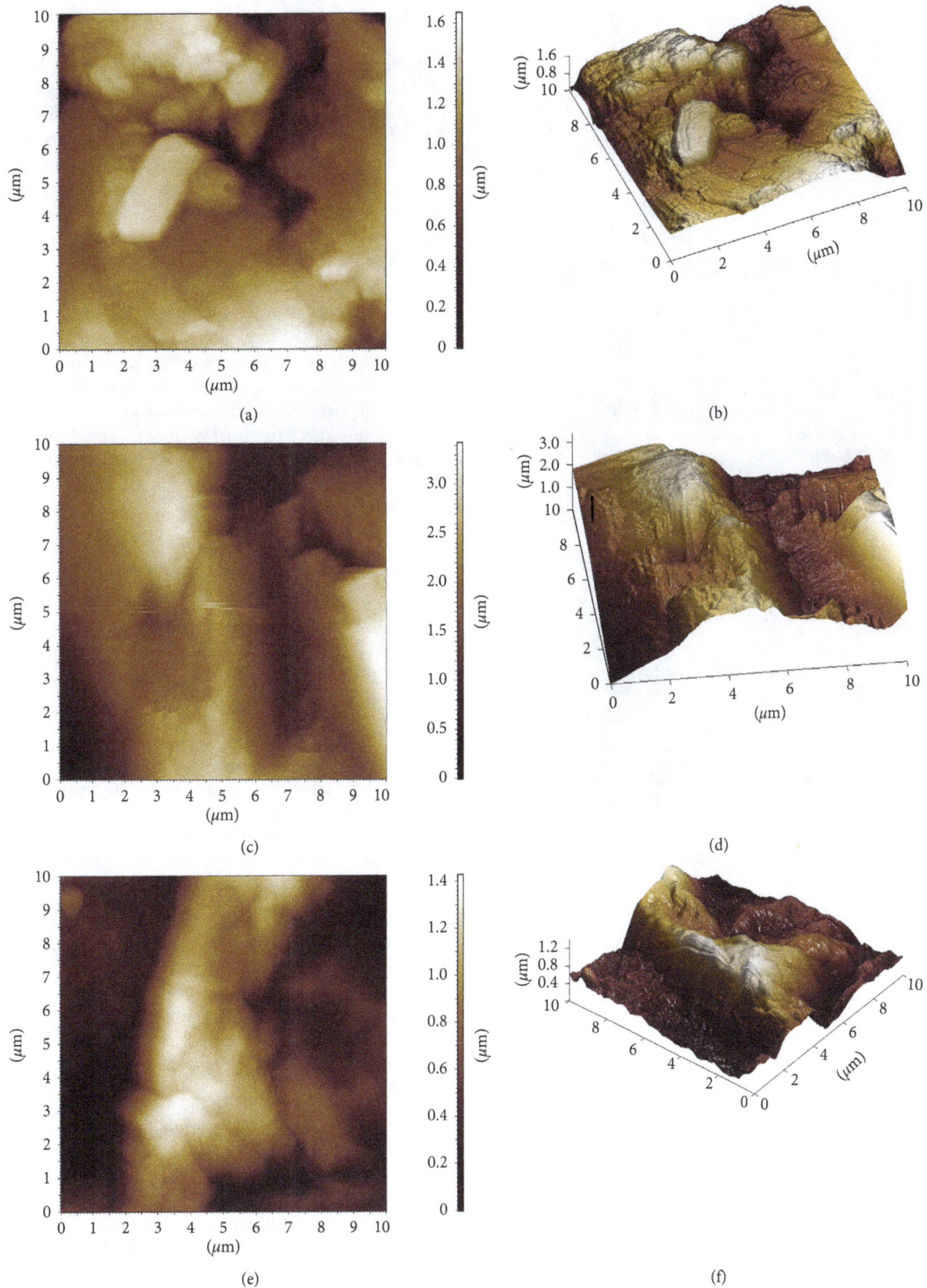

FIGURE 4: The AFM 2D (on the left) and 3D topography images (on the right) of specimen 100 (a, b), 50 (c, d), and 0 (e, f).

RemaniumCSe) have almost identical chemical composition. Ucar et al. [25] showed that the samples were evaluated visually (correctness of execution). However, in this study, the measurements of metal substrate hardness and the evaluation of surface roughness were not presented. Also, higher hardness values of metal framework cause greater

TABLE 4: The mean values of the τ_b index ($\pm$SD) calculated for each group of specimens.

Specimen	τ_b (MPa) $\pm$ SD
100	48.43 $\pm$ 3.59
50	49.21 $\pm$ 5.43
0	46.48 $\pm$ 6.58
Kruskal–Wallis H test	$H_{K\text{-}W} = 0.42$; $P = 0.8$

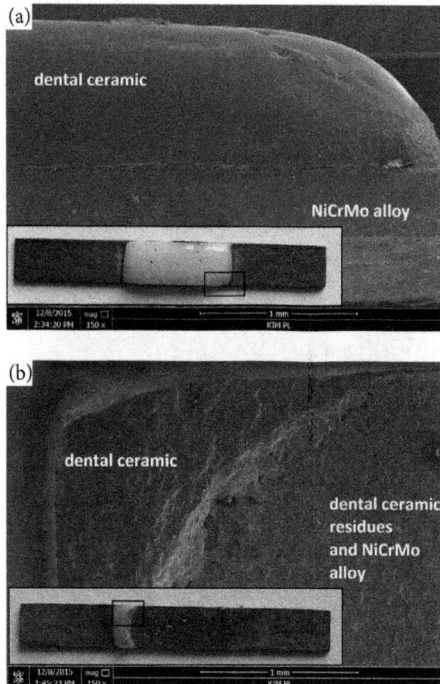

FIGURE 5: SEM images of the specimens after the three-point bending test: (a) specimen 100; (b) specimen 0.

TABLE 5: The mean values of hardness ($\pm$SD) for each group of specimens.

Specimen	HV 0.5 $\pm$ SD
100	253 $\pm$ 9.4[a]
50	271 $\pm$ 4.7[b]
0	262 $\pm$ 6.5[a]

The mean values with dissimilar indices are significantly different at $P < 0.05$. The letters a and b indicate the existence of significant differences between the mean values of hardness.

rigidity of the metal-ceramics system and result in quality differences of the metal-ceramic bond. A significant adverse effect of recasting NiCr alloys on the strength of the metal-ceramic bond and different points of view were presented by Madani et al. The authors reported the largest adhesion of dental porcelain on the castings made of the brand new material [26]. The study of multiple castings of nickel-chromium alloys was also conducted by Mirković [27]. He noticed the occurrence of inconsiderable linear decrease in the elastic modulus up to the sixth generation of castings.

4. Conclusions

The study showed very good metal-porcelain bonding for all kinds of investigated castings. The best quality of the metal–ceramic bond was recorded for the material containing 50% of the brand new alloy; however, the differences between the groups were not statistically significant. The SEM analysis conducted after the bending test confirmed good metal-ceramic connection and exhibited the adhesive-cohesive fracture. The results of the bond strength are also confirmed by the microanalyses of the surface (SEM). Analysis of the surface morphology (AFM) revealed that the highest values of calculated roughness parameters were obtained for the material containing 50% of the brand new alloy. The metal-ceramic bond does not deteriorate quality when the material from recasting is used. Any small differences are due to those in hardness of the metal substrate and/or roughness. They do not affect the final quality of bonding. Further researches are planned to obtain required strength and corrosion resistance of the recasted material and improve the metal–ceramic bond.

Conflicts of Interest

The authors declare that they have no conflicts of interest.

References

[1] C. M. Wylie, R. M. Shelton, G. J. P. Fleming, and A. J. Davenport, "Corrosion of nickel-based dental casting alloys," Dental Materials, vol. 22, no. 6, pp. 714–723, 2007.

[2] J. W. McLean, The Science and Art of Dental Ceramics. The Nature of Dental Ceramics and Their Clinical Use, Quintessence Publishing, vol. 1, Chicago, IL, USA, 1979.

[3] K. Beer, K. Pałka, B. Surowska, and M. Walczak, "A quality assessment of casting dental prosthesis elements," Eksploatacja i Niezawodnosc—Maintenance and Reliability, vol. 15, no. 3, pp. 230–236, 2013.

[4] J. Bauer, J. F. Costa, C. N. Carvalho, R. H. Grande, A. D. Loguercio, and A. Reis, "Characterization of two Ni-Cr dental alloys and the influence of casting mode on mechanical properties," Journal of Prosthodontic Research, vol. 56, no. 4, pp. 264–271, 2012.

[5] K. Beer-Lech and B. Surowska, "Research on resistance to corrosive wear of dental CoCrMo alloy containing postproduction scrap," Eksploatacja i Niezawodnosc—Maintenance and Reliability, vol. 17, no. 1, pp. 90–94, 2015.

[6] A. S. Vaillant-Corroy, P. Corne, P. De March, S. Fleutot, and F. Cleymand, "Influence of recasting on the quality of dental alloys: a systematic review," Journal of Prosthetic Dentistry, vol. 114, no. 2, pp. 205–211, 2015.

[7] A. Casucci, E. Osorio, R. Osorio et al., "Influence of different surface treatments on surface zirconia frameworks," Journal of Dentistry, vol. 37, no. 11, pp. 891–897, 2009.

[8] A. Kakaboura, M. Fragouli, C. Rahiotis, and N. Silikas, "Evaluation of surface characteristics of dental composites using profilometry, scanning electron, atomic force microscopy and gloss-meter," Journal of Materials Science: Materials in Medicine, vol. 18, no. 1, pp. 155–163, 2007.

[9] G. J. Lee, K. H. Park, Y. G. Park, and H. K. Park, "A quantitative AFM analysis of nano-scale surface roughness in various orthodontic brackets," *Micron*, vol. 41, no. 7, pp. 775–782, 2010.

[10] M. G. Subaşı and Ö. İnan, "Evaluation of the topographical surface changes and roughness of zirconia after different surface treatments," *Lasers in Medical Science*, vol. 27, no. 4, pp. 735–742, 2012.

[11] J. J. Roa, G. Oncins, J. Díaz, X. G. Capdevila, F. Sanz, and M. Segarra, "Study of the friction, adhesion and mechanical properties of single crystals, ceramics and ceramic coatings by AFM," *Journal of the European Ceramic Society*, vol. 31, no. 4, pp. 429–449, 2011.

[12] M. Vilotić, T. Lainović, D. Kakaš, L. Blažić, D. Marković, and A. Ivanišević, "Atomic force microscopy in metal forming and dental materials characterization," *Journal of Plastics Technology*, vol. 37, no. 2, pp. 173–187, 2012.

[13] I. Świetlicka, S. Muszyński, E. Tomaszewska et al., "Prenatally administered HMB modifies the enamel surface roughness in spiny mice offspring: an atomic force microscopy study," *Archives of Oral Biology*, vol. 70, pp. 24–31, 2016.

[14] I. M. Pelin, A. Piednoir, D. Machon, P. Farge, C. Pirat, and S. M. M. Ramos, "Adhesion forces between AFM tips and superficial dentin surfaces," *Journal of Colloid and Interface Science*, vol. 376, no. 1, pp. 262–268, 2012.

[15] G. Cortés-Sandoval, G. A. Martínez-Castañón, N. Patiño-Marín, P. R. Martínez-Rodríguez, and J. P. Loyola-Rodríguez, "Surface roughness and hardness evaluation of some base metal alloys and denture base acrylics used for oral rehabilitation," *Materials Letters*, vol. 144, pp. 100–105, 2015.

[16] March 2018, http://kulzer.com/int2/int/dental_technician/products_a_to_z/hera_1/NonPreciousMetalAlloysforCB.aspx.

[17] ISO 9693-1:2012, *Dentistry–Compatibility Testing–Part 1: Metal-Ceramic Systems*, ISO, Geneva, Switzerland, 2012.

[18] E. Alcamo, *The Microbiology Coloring Book*, Addison-Wesley, New York, NY, USA, 1995.

[19] B. Klaić, V. Svetličić, A. Čelebić et al., "Analysis of surface topography and surface roughness of CoCr alloy samples by atomic force microscopy," *Acta Stomatologica Croatica*, vol. 41, no. 4, pp. 306–314, 2007.

[20] S. Ţălu, S. Stach, B. Klaić, T. Mišić, J. Malina, and A. Čelebić, "Morphology of Co–Cr–Mo dental alloy surfaces polished by three different mechanical procedures," *Microscopy Research and Technique*, vol. 78, no. 9, pp. 831–839, 2015.

[21] K. Banaszek, W. Szymański, B. Pietrzyk, and L. Klimek, "Adhesion of *E. coli* bacteria cells to prosthodontic alloys surfaces modified by TiO_2 sol-gel coatings," *Advances in Materials Science and Engineering*, vol. 2013, Article ID 179241, 6 pages, 2013.

[22] R. G. Craig, *Restorative Dental Materials*, J. M. Powers and R. L. Sakaguchi, Eds., pp. 453–465, Elsevier Inc, New York, NY, USA, 12th edition, 2006.

[23] M. Walczak, *Influence of the Selected Technological Processes on Durability of Metal-Ceramic Systems Used in Dental Prosthetics*, Lublin University of Technology, Lublin Poland, 2014, in Polish.

[24] M. Walczak, J. Caban, and K. Gałuszko, "Studies of remelted nickel-based alloy in a simulated chemical environment," *Przemysl Chemiczny*, vol. 96, no. 6, pp. 1329–1332, 2017.

[25] Y. Ucar, Z. Aksahin, and C. Kurtoglu, "Metal ceramic bond after multiple castings of base metal alloy," *Journal of Prosthetic Dentistry*, vol. 102, no. 3, pp. 165–171, 2009.

[26] A. S. Madani, S. R. Rokni, A. Mohammadi, and M. Bahrami, "The effect of recasting on bond strength between porcelain and base–metal alloys," *Journal of Prosthodontics*, vol. 20, pp. 190–194, 2011.

[27] N. Mirković, "Effect of recasting on the elastic modulus of metal-ceramic systems from nickel-chromium and cobalt-chromium alloys," *Vojnosanitetski Pregled*, vol. 64, no. 7, pp. 469–473, 2007.

9

Transformation Mechanism of Fluormica to Fluoramphibole in Fluoramphibole Glass Ceramics

Wei Si,[1] Hua-Shen Xu,[1] Ming Sun,[1] Chao Ding,[2] and Wei-Yi Zhang[1]

[1]*School of Materials Science and Engineering, Dalian Jiaotong University, Dalian 116028, China*
[2]*Dalian Environmental Monitoring Center, Dalian 116023, China*

Correspondence should be addressed to Wei Si; siwei@djtu.edu.cn

Academic Editor: Santiago Garcia-Granda

During isothermal sintering at 820°C, the transformation mechanism of fluormica to fluoramphibole in powder compacts of fluormica and soda-lime glass was investigated using differential thermal analysis, infrared reflection spectrometry, X-ray diffraction, scanning electron microscopy, and so forth. Results show that an interaction between fluormica and glass occurred during isothermal heating; O^{2-}, Na^+, and Ca^{2+} ions were diffused from glass to fluormica. This diffusion facilitates the transformation of the sheet structures of fluormica crystals to double-chain structures by the breakage of bridge oxygen bonds in the sheet. Subsequently, the broken two parallel double chains were rearranged by relative displacement along the *c*-axis direction of the fluormica crystal and were linked by Na^+ and Ca^{2+} ions to form fluoramphibole. A crystallography model of fluormica-fluoramphibole transformation was established in this study.

1. Introduction

The world's first fluoramphibole glass ceramic was prepared by Beall by using a melting method [1]. He successfully prepared potassium fluorrichterite by using a two-step heat treatment based on K_2O-MgO-Al_2O_3-SiO_2-F fluormica glass ceramic formulation with CaO, NaO, Li_2O, BaO, and P_2O_5. Results showed that the transition phase tetrasilicic fluormica and diopside were first precipitated from the parent phase glass and then a reaction occurred with the parent phase glass and formed fluoramphibole during the high-temperature crystallization process [2, 3].

The traditional production model of glass ceramics has been a "one-to-one" process; that is, to obtain a desired glass ceramic, a batch of parent glass with the specific components should be melted [4–6]. The "one-to-multiple" preparation model of glass ceramics by the direct reactive crystallization of common soda-lime glass and the added crystal becomes a new research method. We have proposed a novel route called sintering-reactive crystallization to fabricate a series of different types of glass ceramics by using the waste soda-lime glass powder as the main raw material, adding different silicate crystals into the glass powder, and sintering. Fluoramphibole glass ceramics have wide applications in environmental protection building materials because they possess good machinability [7, 8]. This technique can be used for low-cost production without melting glass with the specific components and can significantly reduce sintering temperature and energy consumption.

The effect of promoter components, additions, and sintering temperatures on the crystallization, microstructures, and mechanical properties of fluoramphibole glass ceramics were studied in detail [9]. However, the mechanism to transform fluormica crystals into fluoramphibole crystals is rarely reported. The present study aims to establish a crystallography model of the polysomatic transformation of silicate crystals from fluormica to fluoramphibole.

2. Experimental

2.1. Preparation. Commercially available fluormica powder (size of 130 mesh) (Dashiqiao Chemical Plant, Liaoning, China) was used. The chemical composition of this fluormica was 6.33% K, 15.12% Mg, 5.01% Al, 19.69% Si, 43.48% O, and

10.36% F by weight, close to the stoichiometric composition of fluorphlogopite ($KMg_3AlSi_3O_{10}F_2$). The recycled window glass was used as the base glass. Its composition was 71.3% SiO_2, 9.8% CaO, 13.3% Na_2O, 4.3% MgO, 1.0% Al_2O_3, and 0.3% K_2O by weight, and its size was 100 mesh.

Two types of sample were prepared. (1) Mixed sample: the fluormica crystal (35 wt%) and recycled glass (65 wt%) powder mixtures were ball-milled for 4 h. Thereafter, 3 drops of 6 wt% PVA water solution was added as a binder. Finally, the mixtures were pressed in a hardened steel die to create cylindrical compacts with 15 mm diameter and 15 mm thickness. (2) Sandwich sample: in this sample, a thin fluormica powder layer (thickness of about 1 mm) was embedded within two layers of glass powders. The sample was pressed in a hardened steel die to create cylindrical compacts with 15 mm diameter and 15 mm thickness. The compacts were presintered at 400°C for 2 h to remove the binder and then sintered from room temperature to 820°C at different time durations to obtain the final glass ceramics.

2.2. Characterization. The crystallization reaction temperature of the glass and fluormica powders was identified by differential thermal analysis (DTA, STA449F3) at a heating rate of 10 K/min. The crystalline-precipitated type of sintered glass ceramics and the changes in the relative contents of fluormica during isothermal process were characterized by X-ray diffraction (XRD) (Rigaku D_{MAX}-RB) with Cu Kα radiation. The relative contents of fluormica were calculated according to the equation I_t/I_o [10], where I_o is the total strength of the five strongest peaks in fluormica XRD (d = 0.960, 0.330, 0.290, 0.199, and 0.167 nm) before sintering and I_t is the total strength of the above five strongest peaks in fluormica XRD at a certain sintered time during the isothermal process.

The changes in the chemical bonds of fluormica during the isothermal process were characterized by infrared reflection spectrometry (Nicolet-20DXB). The fractured surfaces of the sintered glass ceramics were coated with a thin film of gold and were observed by scanning electron microscopy (SEM) (JSM-6360LV). The fractured surfaces of the sandwich sample were polished and etched using 20% HF water solution for 30 s, washed with deionized water, and dried. The changes of element contents around the glass and fluormica interface in the sandwich sample were characterized by energy dispersive spectrometer (Oxford-INCA). The high-resolution transmission electron microscopy (HRTEM), selected area electron diffraction, and fast Fourier transform of the products were characterized by a JEM-2100F operated at an accelerating voltage of 200 kV.

3. Results and Discussion

The DTA curve of mixed samples is shown in Figure 1. The endothermic peak was observed at 600°C, thus showing the glass transition temperature. The smaller exothermic peak was observed at 823°C, which is attributed to the reaction crystallization between the glass and fluormica.

Figure 2 shows the XRD patterns of samples isothermally heated at 820°C. The main crystalline phase of the sample was

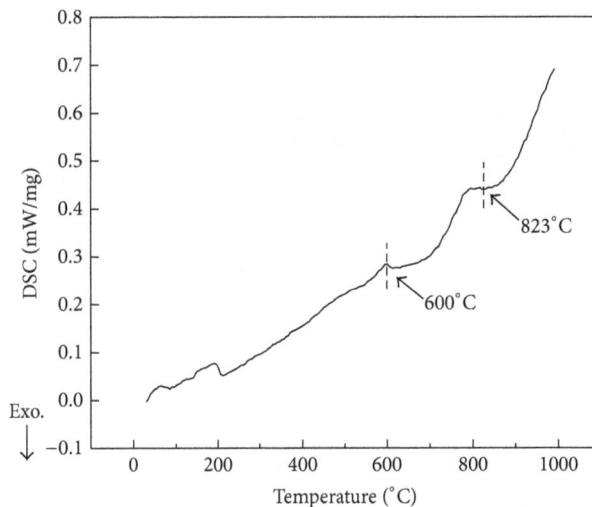

FIGURE 1: DTA curve of mixed samples.

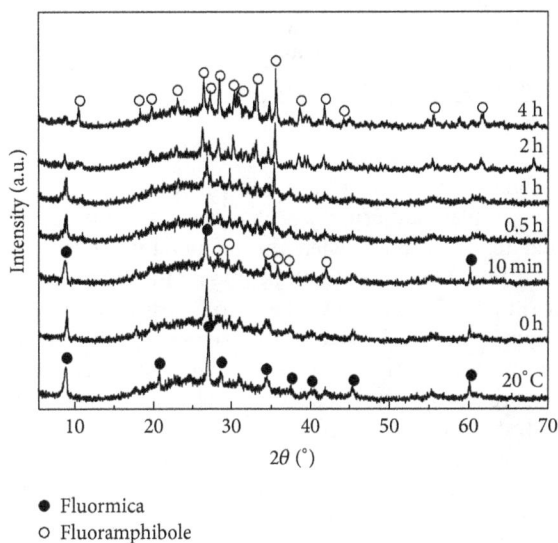

● Fluormica
○ Fluoramphibole

FIGURE 2: XRD patterns of samples isothermally heated at 820°C.

fluormica (JCPDS 16-0344) at room temperature (20°C, not sintered) and heated at 820°C, where no new crystalline phase formed. When isothermally heated at 820°C for 10 min, the main crystalline phase was still fluormica, but the intensity of the diffraction peak decreased. However, some diffraction peaks disappeared. The weak peak of the fluoramphibole (JCPDS 41-1429) appeared because of annealing. These new fluoramphibole phases were the reactive crystallization products between fluormica crystals and glass powder during the sintering process but only had few precipitation contents. With increasing isothermal time, the diffraction peak intensity of fluormica continued to decline. After isothermal treatment for 4 h, the diffraction peaks of fluormica almost completely disappeared and the main phase changed to fluoramphibole.

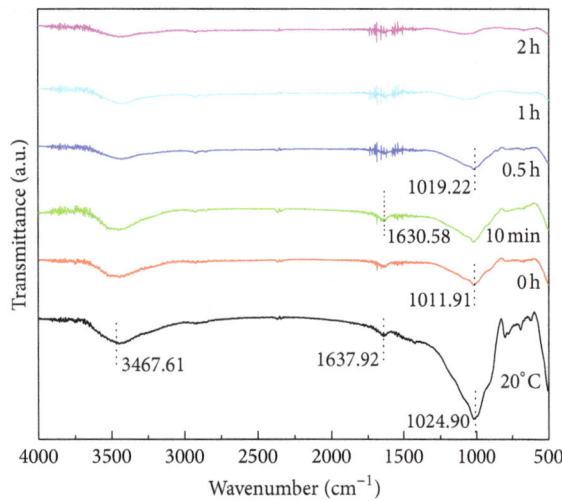

FIGURE 3: IR spectra of samples isothermally heated at 820°C.

TABLE 1: Relative contents of fluormica in mixed samples isothermally heated at 820°C.

Time	0 h	10 min	0.5 h	1 h	2 h	4 h
Relative content/%	91.8	58.8	56.7	46.5	39.3	5.36

TABLE 2: Composition change of sandwich samples isothermally heated at 820°C and 900°C (wt%).

| Element | 820°C | | | 900°C | | |
| | 0 h | | 2 h | 0 h | | 2 h |
	P_1	P_2	P_3	P_4	P_5	P_6
O	29.44	19.74	35.80	52.93	39.36	45.65
Si	42.35	31.83	35.20	25.25	25.37	30.57
Na	8.34	2.27	4.38	7.15	3.50	8.18
Ca	9.00	3.04	5.69	7.70	1.96	4.39
F	1.35	11.36	4.40	1.01	10.36	6.22
Mg	5.41	21.01	10.20	1.34	10.97	2.06
Al	1.72	5.59	1.67	1.52	4.40	1.41
K	2.39	5.16	2.66	3.10	4.08	1.52
Total	100.00	100.00	100.00	100.00	100.00	100.00

Table 1 shows the relative contents of fluormica in the mixed samples isothermally heated at 820°C. Few fluormica started to decompose when the temperature increased to 820°C. The fluormica decomposition rate increased with increasing isothermal time. The decomposition of fluormica was approximately 50% when the isothermal time was 1 h. Fluormica was nearly completely decomposed after 4 h.

Figure 3 shows the IR spectra of the samples isothermally heated at 820°C at different time points. At 20°C (before sintering), the wide band is due to the stretching of Si-O-Si; its maximum peak is at 1024.90 cm^{-1}. The band at 640.72 and 3467.61 cm^{-1} indicates the presence of O-H bond. All of the above peaks are characteristic peaks of fluormica.

The Si-O-Si bond weakened with increasing isothermal treatment time. This bond also has low strength when isothermally heated at 820°C for 2 h. The Si-O-Si bond was gradually broken, and fluormica was continually decomposed during the isothermal process. This result is consistent with the above measured results of XRD.

The crystalline microstructures of the glass ceramics isothermally heated at 820°C for different time points are shown in Figure 4. The microstructures of the sample contain the plate-like fluormica and glass phase when only heated at 820°C (Figure 4(a)). After isothermal heating at 820°C for 10 min, the sample was not sintered, whereas most of the glass particles were interconnected. The plate-like fluormica crystals were dispersed in the glass particles (Figure 4(b)).

After isothermal heating for 0.5 h (Figure 4(c)), a closed hole formed in the organization, and plate-like fluormica was surrounded by a glassy matrix. Among them, a little fine acicular fluoramphibole crystal was precipitated. After isothermal heating for 1 h, dense acicular fluoramphibole crystals

appeared in closed pores; they were exerted from the edge of the glassy matrix (Figure 4(d)). With increasing isothermal timing, plate-like fluormica crystals completely disappeared and the interlace of circular fluoramphibole crystals within the pores further became rod-shaped (Figures 4(e) and 4(f)).

Figure 5 shows the microstructures of the interface in sandwich samples isothermally heated at 820°C for 0 (a) and 2 h (b) and at 900°C for 0 (c) and 2 h (d). The obvious interface exists between the glass and fluormica when only heated at 820°C (indicated by the arrow in Figure 5(a)). After being isothermally heated at 820°C for 2 h, the glass phase gradually leaked into fluormica by viscous flow. Thereafter, the fluoramphibole crystals were precipitated at the interface of the glass and fluormica, as shown by the arrows in Figure 5(b). According to the results of DTA and XRD, the above acicular crystal was the reaction crystallization product between the glass and fluormica. At 900°C isothermal heating, the crystal was precipitated to rod-like fluoramphibole crystals (Figure 5(c)). After isothermal heating at 900°C for 2 h, the fluoramphibole crystals further increased in size and combined with the glass matrix (Figure 5(d)). Figure 5 shows the sizes of the rod-like crystals, and the amount of pores increased with the increasing sintering temperature and isothermal heating time.

Table 2 displays the composition change of microstructures near the interface isothermally heated at 820 and 900°C (for P_1, P_2, P_3, P_4, P_5, and P_6 in Figure 5). The raw materials showed that the glass does not contain any F elements and fluormica does not contain Na and Ca elements. When only heated to 820°C (Figure 5(a)) or 900°C (Figure 5(c)), a small amount of F element was observed in the glassy side of the interface (P_1, P_4), whose content was 1.35% and 1.01%, respectively. After isothermal heating for 2 h (P_3, P_6), the contents of the O, Si, Na, and Ca elements increased, whereas those of F, Mg, Al, and K decreased. Therefore, these elements were reciprocally diffused between the glass and fluormica. The O^{2-}, Si^{4+}, Na^+, and Ca^{2+} ions in the glass phase entered the fluormica phase through diffusion. The F^-, Mg^{2+}, Al^{3+}, and K^+ ions also entered into the glass phase through the diffusion, and fluoramphibole crystals were subsequently precipitated.

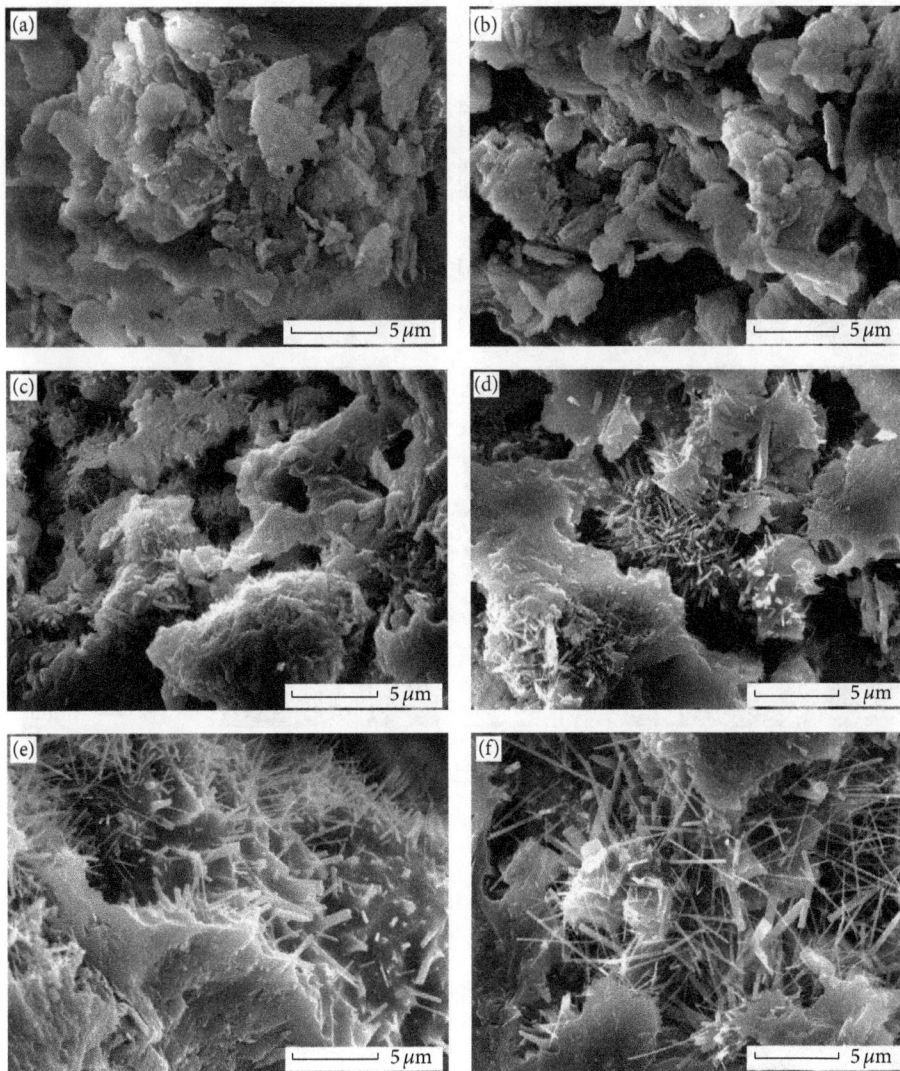

FIGURE 4: Microstructures of glass ceramic samples isothermally heated at 820°C with (a) 0 h, (b) 10 min, (c) 0.5 h, (d) 1 h, (e) 2 h, and (f) 4 h.

The above conclusion can be further illustrated by the SEM photographs of the sandwich samples after hydrofluoric acid etching. When using hydrofluoric acid etching, many pores formed in the glassy matrix because SiF_4 was formed via a chemical reaction of SiO_2 and HF; in this case, almost all rod-like fluoramphibole crystals were eluted (Figure 6(a), right illustration shows the enlarged area of the left red circle). The elution rod-like fluoramphibole crystals have smooth surfaces, and the length is above 40 μm (Figure 6(b)). This observation further proved that the fluoramphibole crystals were indirectly precipitated from the fluormica crystal and formed by the interdiffusion between fluormica crystals and glass powder.

The transmission electron microscope (TEM) and HRTEM analyses were conducted on the glass part of

the sandwich sample at 820°C isothermal heating for 2 h. Precipitated crystal has a long rod-like shape (Figure 7(a)). A good crystallinity crystal in the glassy matrix, which is a single crystal, can also be observed. The measured d value of the precipitated crystal is 0.267 nm (Figure 7(b)).

According to the XRD results, the d value of the crystal plane (151) with the strongest diffraction peak of the fluoramphibole crystals is 0.269 nm (JCPDS 41-1429). These two results obtained from XRD and HRTEM analyses are consistent with each other; they show that the precipitated crystal is a fluoramphibole crystal. Based on the experimental results, crystallization reaction occurred between the components of recycled glass powder and fluormica and subsequently formed the fluoramphibole crystals. This reaction can be written in the form of a chemical equation as follows:

$$KMg_3AlSi_3O_{10}F_2 + CaO + Na_2O + 5SiO_2 \longrightarrow KNa_2CaAlMg_3Si_8O_{22}F_2 \qquad (1)$$

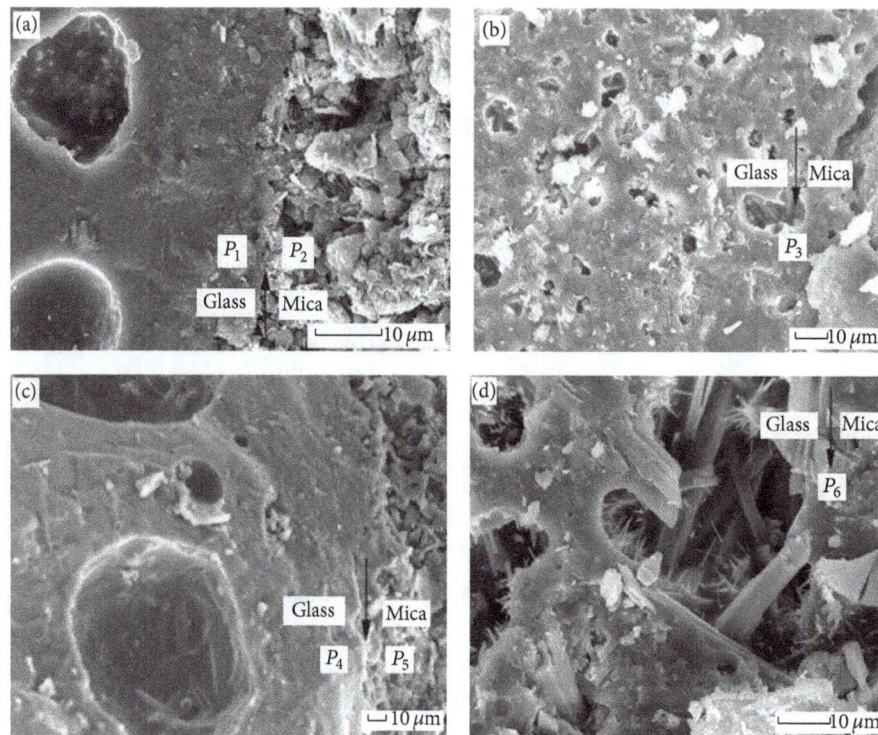

FIGURE 5: Microstructures of the interface in sandwich samples isothermally heated at 820°C for (a) 0 and (b) 2 h and at 900°C for (c) 0 and (d) 2 h.

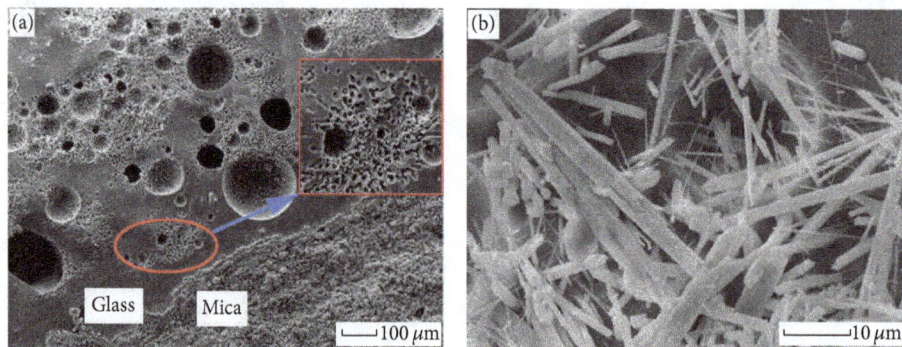

FIGURE 6: Microstructures of the interface in sandwich samples after hydrofluoric acid etching.

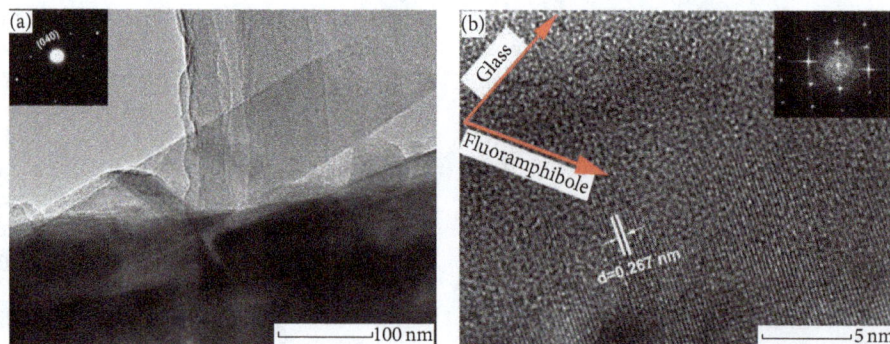

FIGURE 7: TEM (a, inset is the selected area electron diffraction) and HRTEM (b, inset is the fast Fourier transform) photographs of glass in sandwich samples.

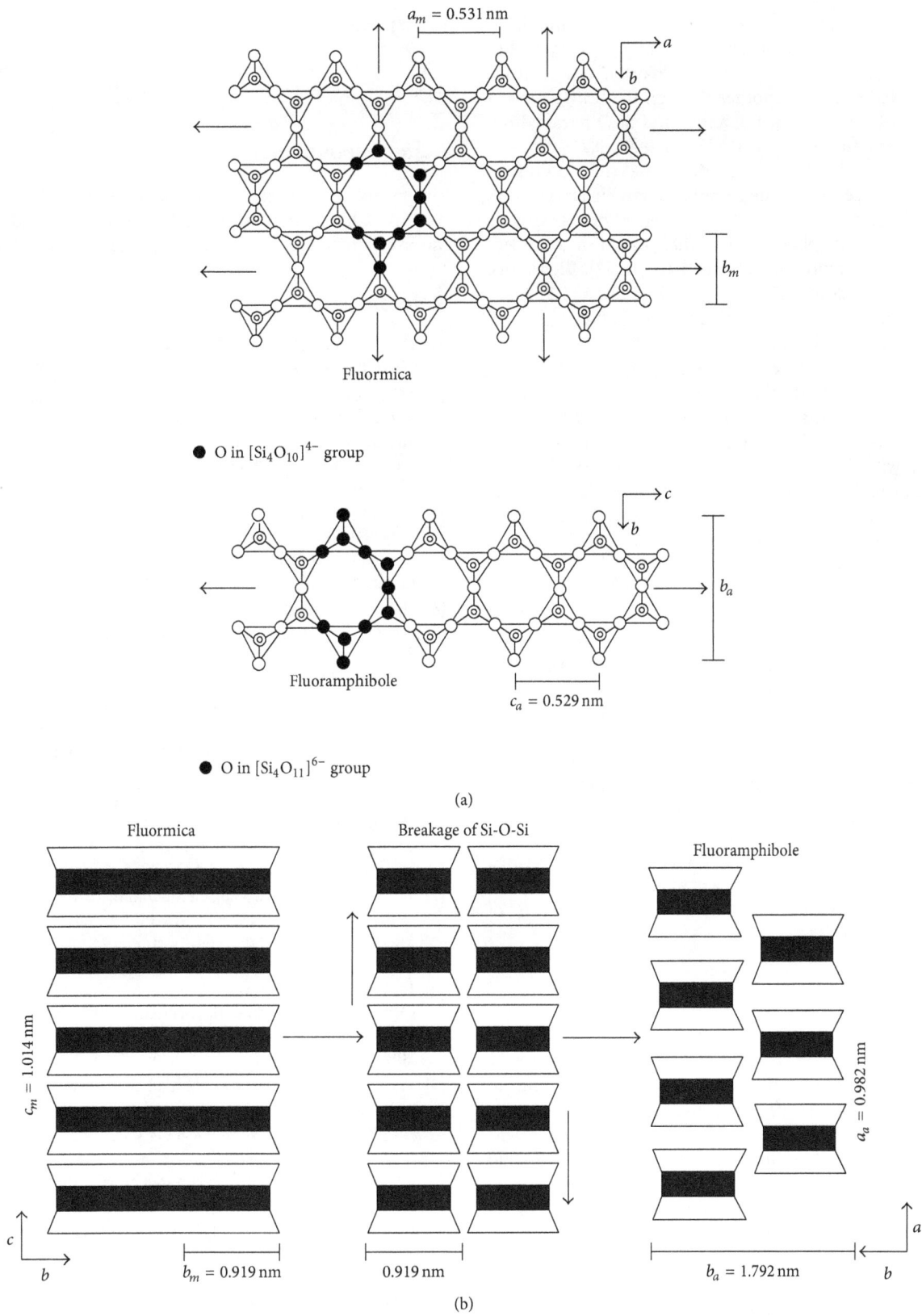

FIGURE 8: Schematic of transformation from fluormica to fluoramphibole. (a) The sheet and double-chain structures in fluormica and fluoramphibole. (b) Crystallography model of fluormica-fluoramphibole transformation.

A crystallographic model of fluormica–fluoramphibole transformation can be established. The fluormica crystal has a layered structure, and the silicon oxygen tetrahedrons are connected with one another through the bridge oxygen bonds, which form 2D plane network of Si-O tetrahedron; its lattice constants are $a_m = 0.531$ nm, $b_m = 0.919$ nm, and $c_m = 1.014$ nm. The fluoramphibole crystals have a double-chain structure; each fluoramphibole crystal cell has two rows of parallel double strands, but the two lines of double strands are not in the same plane. Double chain establishes connection through sodium and calcium ions [11, 12]. The lattice constants of the fluoramphibole crystals are $a_a = 0.982$ nm, $b_a = 1.792$ nm, and $c_a = 0.529$ nm; $a_m \approx c_a$, $2b_m \approx b_a$, and $c_m \approx a_a$ (Figure 8(a)).

In the isothermal process at 820°C, the O^{2-}, Na^+, and Ca^{2+} ions diffuse into the fluormica crystals. The results indicated that the mean silicon-bridge oxygen (Si-O-Si) in the plane network of Si-O tetrahedron was disconnected along the a-axis with the increase of oxygen contents [13, 14]. Thus, fluormica will decompose and finally transform to a double-chain structure (Table 1, Figure 3, and Table 2). Two rows of disconnected double strands along the c-axis direction of the fluormica crystals occurred at a relative displacement $1/4c$ and transformed into two rows of parallel double chain in fluoramphibole crystals [15]. The double-chain displacement was reconnected by the Na^+ and Ca^{2+} ions to form the ion bonds through diffusion into the fluormica crystals; subsequently, fluoramphibole crystals were formed.

Hence, the c-axis direction of fluormica crystals was transformed into the a-axis direction of fluoramphibole crystals, and the a-axis direction of fluormica crystals has been transformed into the c-axis direction of fluoramphibole crystals. Nevertheless, the lattice constant and b-axis direction of fluormica crystals were unchanged. However, one fluoramphibole crystal cell along b-axis direction was transformed from two juxtaposition fluormica crystal cells. Therefore, the fluoramphibole crystal lattice constant b_a is twice of the b_m of fluormica (Figure 8(b)).

4. Conclusions

(1) In the isothermal sintering process, fluormica crystals do not directly change into fluoramphibole crystals. First, these crystals are decomposed and precipitated; O^{2-}, Na^+, and Ca^{2+} ions diffuse from the glass into the fluormica crystals. The increase of oxygen contents causes the disconnection of the oxygen bridge bond along the a-axis from the plane network; consequently, the layered structure changes into a double-chain structure, and fluormica decomposition occurs.

(2) Two rows of disconnected double strands along the c-axis direction of the fluormica crystals occur at a relative displacement of $1/4c$ and transform into two rows of parallel double chain in the fluoramphibole crystals. The double-chain displacement is reconnected by the Na^+ and Ca^{2+} ions to form bonds through diffusion into the fluormica crystals; subsequently, fluoramphibole crystals are formed.

Competing Interests

The authors declare that there is no conflict of interests regarding the publication of this paper.

Acknowledgments

This project is supported by National Natural Science Foundation of China (Grant no. 51308086) and Program for Liaoning Excellent Talents in University (Grant no. LJQ2015020).

References

[1] G. H. Beall, "Chain silicate glass-ceramics," *Journal of Non-Crystalline Solids*, vol. 129, no. 1–3, pp. 163–173, 1991.

[2] M. Mirsaneh, I. M. Reaney, P. F. James, and P. V. Hatton, "Effect of CaF₂ and CaO substituted for MgO on the phase evolution and mechanical properties of K-fluorrichterite glass ceramics," *Journal of the American Ceramic Society*, vol. 89, no. 2, pp. 587–595, 2006.

[3] I. L. Denry and J. A. Holloway, "Effect of sodium content on the crystallization behavior of fluoramphibole glass-ceramics," *Journal of Biomedical Materials Research Part A*, vol. 63, no. 1, pp. 48–52, 2002.

[4] G.-H. Chen and X.-Y. Liu, "Sintering, crystallization and properties of MgO-Al₂O₃-SiO₂ system glass-ceramics containing ZnO," *Journal of Alloys and Compounds*, vol. 431, no. 1-2, pp. 282–286, 2007.

[5] S. R. Bragança and C. P. Bergmann, "Traditional and glass powder porcelain: technical and microstructure analysis," *Journal of the European Ceramic Society*, vol. 24, no. 8, pp. 2383–2388, 2004.

[6] E. Bernardo, M. Varrasso, F. Cadamuro, and S. Hreglich, "Vitrification of wastes and preparation of chemically stable sintered glass-ceramic products," *Journal of Non-Crystalline Solids*, vol. 352, no. 38-39, pp. 4017–4023, 2006.

[7] W. Y. Zhang and H. Gao, "Preparation of machinable fluoramphibole glass-ceramics from soda-lime glass and fluormic," *International Journal of Applied Ceramic Technology*, vol. 5, no. 4, pp. 412–418, 2008.

[8] W. Y. Zhang, H. Gao, and Y. Xu, "Sintering and reactive crystal growth of diopside-albite glass-ceramics from waste glass," *Journal of the European Ceramic Society*, vol. 31, no. 9, pp. 1669–1675, 2011.

[9] W.-Y. Zhang, H. Gao, B.-Y. Li, and Q.-B. Jiao, "A novel route for fabrication of machinable fluoramphibole glass-ceramics," *Scripta Materialia*, vol. 55, no. 3, pp. 275–278, 2006.

[10] J. Rocherullé and P. Bénard-Rocherullé, "The devitrification of a LAS glass matrix studied by X-ray powder diffraction," *Solid State Sciences*, vol. 4, no. 7, pp. 999–1004, 2002.

[11] D. Y. Pushcharovskiĭ, Y. S. Lebedeva, I. V. Pekov, G. Ferraris, A. A. Novakova, and G. Ivaldi, "Crystal structure of magnesioferrikatophorite," *Crystallography Reports*, vol. 48, no. 1, pp. 16–23, 2003.

[12] F. C. Hawthorne and R. Oberti, "Amphiboles: crystal chemistry," *Reviews in Mineralogy and Geochemistry*, vol. 67, no. 1, pp. 1–54, 2007.

[13] J. Najorka and M. Gottschalk, "Crystal chemistry of tremolite-tschermakite solid solutions," *Physics and Chemistry of Minerals*, vol. 30, no. 2, pp. 108–124, 2003.

[14] D. M. Jenkins, K. N. Bozhilov, and K. Ishida, "Infrared and TEM characterization of amphiboles synthesized near the tremolite-pargasite join in the ternary system tremolite-pargasite-cummingtonite," *American Mineralogist*, vol. 88, no. 7, pp. 1104–1114, 2003.

[15] D. R. Vnnren and P. R. Buseck, "Microstructures and reaction mechanisms in biopyriboles," *American Mineralogist*, vol. 65, no. 7-8, pp. 599–623, 1980.

10

The Effect of Commercial Rice Husk Ash Additives on the Porosity, Mechanical Properties, and Microstructure of Alumina Ceramics

Mohammed Sabah Ali,[1,2] **M. A. Azmah Hanim,**[1,3] **S. M. Tahir,**[1] **C. N. A. Jaafar,**[1]
Norkhairunnisa Mazlan,[3,4] **and Khamirul Amin Matori**[5]

[1]*Department of Mechanical and Manufacturing Engineering, Faculty of Engineering,*
 Universiti Putra Malaysia, 43400 Serdang, Selangor, Malaysia
[2]*Department of Agriculture Machinery & Equipment Engineering Techniques, Technical College, Al-Mussaib, Iraq*
[3]*Laboratory of Biocomposite Technology, Institute of Tropical Forestry and Forest Products,*
 Universiti Putra Malaysia, 43400 Serdang, Selangor, Malaysia
[4]*Department of Aerospace Engineering, Faculty of Engineering, Universiti Putra Malaysia, 43400 Serdang, Selangor, Malaysia*
[5]*Department of Physics, Faculty of Science, Universiti Putra Malaysia, 43400 Serdang, Selangor, Malaysia*

Correspondence should be addressed to Mohammed Sabah Ali; mohammed.sabah94@yahoo.com

Academic Editor: Amit Bandyopadhyay

A porous ceramic is made from composite materials which consist of alumina and commercial rice husk ash. This type of ceramics is obtained by mixing the commercial rice husk ash as a source of silica (SiO_2) and a pore forming agent with alumina (Al_2O_3) powder. To obtain this type of ceramic, a solid-state technique is used with sintering at high temperature. This study also investigated the effects of the rice husk ash ratios on the mechanical properties, porosity, and microstructure. The results showed that, by increasing the content of the rice husk ash from 10 to 50 wt%, there is an increase in the porosity from 42.92% to 49.04%, while the mechanical properties decreased initially followed by an increase at 30 wt% and 50 wt%; the hardness at 20 wt% of the ash content was recorded at 101.90 HV_1. When the ash content was increased to 30 wt% and 50 wt%, the hardness was raised to 150.92 HV_1 and 158.93 HV_1, respectively. The findings also revealed that the tensile and compressive strengths experienced a decrease at 10 wt% of the ash content and after that increase at 30 wt% and 50 wt% of rice husk ash. XRD analysis found multiple phases of ceramic formation after sintering for the different rice husk ash content.

1. Introduction

Porous ceramic material is commonly used in the filtration of liquids and gases, absorption of shock, catalyst support, molten-metal filtration, thermal insulators, high temperature applications, and environmental protection [1–4]. This is due to its excellent properties, which include resistance to high temperature and chemical corrosion and low thermal conductivity as well as high surface area and low density [1–3]. The mullite ceramic phase is important in the Al_2O_3-SiO_2 system. The mullite phase has good mechanical and physical properties [5] such as high oxidation resistance, excellent

chemical stability, good creep resistance, and low thermal conductivity [1, 6, 7]. In addition, porous mullite ceramic has many important uses, for example, in insulating materials, catalyst support insulation, and membrane filters for gases [8].

One of the most common and effective ways for the manufacturing of porous ceramic is the pore-forming method [4, 9]. Recently, there has been renewed interest in using agricultural waste materials in porous ceramics. Sengphet et al. [10] used wastes from kenaf powder for the fabrication of porous clay ceramics, while Njeumen Nkayem et al. [11] used corn cob for the preparation of porous ceramic brick. As for

Sooksaen et al. [12], they used wood dust as a pore agent in producing porous glass ceramics. Unfortunately, most of the pore agents that have been used to produce porous ceramics have resulted in a decrease in the mechanical properties due to porosity. This paper recommends the use of commercial rice husk ash from agricultural waste as a source of silica (SiO_2) and as a pore forming agent for producing porous mullite and cristobalite-corundum ceramics. Commercial rice husk ash is directly mixed with different ratios in the alumina matrix using commercial sucrose (sugar) as a binder and sintered at high temperature to produce porous mullite and cristobalite-corundum ceramics. Previously, several researchers used the Al_2O_3-SiO_2 system to produce porous ceramics with good mechanical properties. Dong et al. [13] used a mixture of fly ash from industrial waste and bauxite to fabricate bulk porous mullite supports for ceramic membranes using dry pressing and sintering. Serra et al. [8] used rice husk ash to produce mullite ceramics through a sintering reaction of alumina and rice husk ash using different sintering temperatures. Cao et al. [14] prepared low-cost porous mullite supports for ceramic membranes using recycled coal fly ash with natural bauxite. While Hua et al. [1] fabricated anorthite-mullite-corundum porous ceramics using the waste from construction sites and alumina powder (Al_2O_3). Therefore, the aim of this study is to produce mullite and cristobalite-corundum ceramic composite materials with high porosity and mechanical properties but low cost through the direct mixing of commercial rice husk ash and alumina powder, followed by sintering at high temperature.

2. Experimental Procedure

2.1. Materials and Specimen Preparation. The matrix materials used included a commercial aluminum oxide Al_2O_3 powder ($\rho = 3.94 \, g/cm^3$) with high level of purity (99.99%) and a particle size of 0.5 μm. Commercial sucrose (sugar) used as the binder (10%) was added to the ceramic mixture according to the maximum solubility of sugar in distilled water. For the use of the binder in this study, the concentration was fixed at 60 wt% [15]. The binder was mixed manually with ceramic powder for 3–5 minutes using an agate mortar. Then, the rice husk ash powder was added to the ceramic slurry at ratios of 10, 20, 30, 40, and 50 wt%. After that, all the batches were mixed in a mortar for around 5–10 minutes followed by ball milling for 3 hrs in a plastic container homogenously mixed at a ratio of weight 3 : 1, which is 3 parts alumina balls to one part weight of powder. Later, the dry mixtures were pressed uniaxially in a circular steel die (diameter = 20 mm and thickness = 5 mm) using an Instron hydraulic press at a pressure of 90 MPa. Following this, the green compacts were dried in the oven at 110°C for 24 hrs. The organic burnout of the dried samples was carried out in an ambient atmosphere in an electrically heated, programmable furnace. The rate of heating was set at 1.5°C/min for each increment in the temperature. According to the TGA (thermogravimetric analysis) of sucrose and rice husk ash, the samples were sintered at 200°C, 300°C, 500°C, and 900°C in an electric furnace for a duration of 1 hr of soaking time. The rate of

heating and cooling was fixed at 1.5°C for the removal of the carbon and organic materials in the rice husk ash as well as the sucrose. After that, the ceramic specimens were sintered at 1600°C for a duration of 2 hrs of soaking time while the rate of heating and cooling was fixed at 5°C/min. The findings revealed that all the samples with different ratios of rice husk ash and (binder) sucrose showed uniform dimensions without cracks after undergoing sintering.

2.2. Ceramic Samples Characterisation. The microstructure and the chemical composition of the porous ceramics were examined using a field emission scanning electron microscope (FESEM) and an EDX technique (SE-440). The phase compositions of the porous ceramics were determined using an X-ray diffractometer (PANAlytical (Philips) X'pert ProPW3050/60) with Cu radiation (wave length = 1.54060 Å) that had been set at 40 mA and 40 Kv using X-pert software.

The pore size distribution was determined from the FESEM images that had been taken using the image analysis software (image-J) [16, 17]. The compressive strength of the samples with dimensions of 20 mm in diameter and 24 mm in height were measured using an Instron machine. The measurements were in accordance with the ASTM-1424-10 standard [18] at a crosshead rate of 0.5 mm/min. The maximum mechanical load and cross-sectional area were used to calculate the compressive strength of the samples [19]. The hardness values of the samples were measured. A micro Vickers hardness machine was used to measure the value of the hardness of the samples whereby the applied force was 9.81 N for 15 sec at full load [20]. All the samples were ground and polished using polishing media and then thermally etched. A Brazilian test was performed using samples of dimensions 20 mm in diameter and 5 mm in thickness by means of an Instron machine. An average of five samples was used to obtain the average value for the strength and hardness for each composition. The calculation of the diametrical tensile strengths of the porous alumina ceramics was based on the tensile strength of the Brazilian test.

$$\sigma \text{ (tensile MPa)} = \frac{(2 * P)}{(\pi * t * d)}, \tag{1}$$

where σ (compressive) is in MPa, P is the applied force (N), d is the diameter of the samples (mm), and t is the thickness of the sample (mm) [21].

A water immersion method based on Archimedes' principle, as specified in ASTM C20-00, was used to determine the overall porosity, density, and open porosity of the sintered samples using the following equations:

$$P_{\text{open}} = \frac{M_{\text{wet}} - M_{\text{dry}}}{M_{\text{wet}} - M_{\text{suspended}} + M_{\text{wire}}}$$

$$P_{\text{overall}} = \left(1 - \frac{\rho}{\rho_{\text{theoritical}}}\right) \times 100 \tag{2}$$

$$\rho = \frac{M_{\text{dry}} \times \rho_{\text{water}}}{M_{\text{wet}} - M_{\text{suspended}} + M_{\text{wire}}},$$

TABLE 1: The chemical compositions of rice husk ash powder and sucrose.

Materials	Elements	C	O_2	Si	P	S	K	AL	Mg	Ca
Rice husk ash	Weight%	6.78	56.40	35.74	—	—	—	—	0.43	0.66
Sucrose (sugar)	Weight%	59.13	40.88	—		—		—	—	—

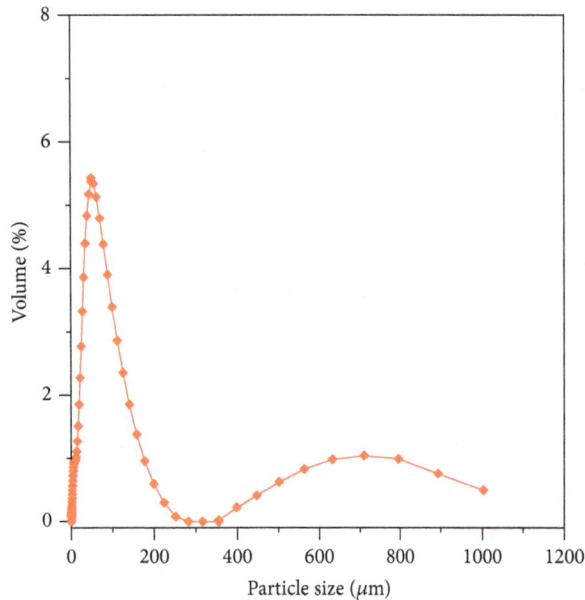

FIGURE 1: Distribution of particle size of the rice husk ash powder.

where M_{dry} is the dry mass [22] of the sample, $M_{suspended}$ is the mass of the sample suspended in distilled water, M_{wet} is the mass of the sample after soaking in water, M_{wire} is the mass of the suspending system, $P_{overall}$ is the volume fraction of the overall porosity (vol.%) of the sample, and P_{open} is the volume fraction of the open porosity of the sample [23, 24]. The theoretical density (true) of alumina (Al_2O_3), recorded at 3.94 g/cm^3, has been used as a reference. This measurement was obtained using the Accupyc II 1340. The relative density of the porous alumina ceramic samples was determined from the ratio of the measured bulk density to the theoretical density [25].

3. Results and Discussion

3.1. Characterisation of the Rice Husk Ash and Binder. The commercial rice husk ash was ground in an electrical grinder (type RT-02A, 3000 rpm) for 1 min. Then, it was sieved in an electrical sieve (type Retsch, As 200) to obtain a particle size of 250 μm.

The actual measurement of the density of rice husk ash powder was obtained using a gas pycnometer (Accupyc II 1340). Meanwhile, a particle size analyser (Malvern, master size 2000) was used to analyse the distribution of particle size for the rice husk ash powder as shown in Figure 1.

The actual measurement of the density for rice husk ash is 2.17 g/cm^3 as indicated by the Accupyc II 1340. The thermogravimetric (TGA) tests for rice husk ash and sucrose

(sugar) powder were conducted under nitrogen gas at a temperature rate of 10°C/min. It is necessary to perform TGA analysis of the rice husk ash in order to determine the thermal decomposition range.

Figures 2(b) and 2(c) are the FESEM images of the rice husk ash delivered from the factory and the sucrose. The average length of the rice husk ash particles was 355.3 μm. The FESEM image of the sucrose shows the shape of the sucrose particles.

Figure 2(a) presents the TGA results of commercial rice husk ash and sucrose. The weight loss of the rice husk ash is around 6.48%. The removal of the unburnt elements and organic materials, such as carbon, started from around 200°C to 650°C. The maximum weight loss occurs at approximately 300°C.

Based on the weight loss of the rice husk ash, the pyrolysis of the rice husk ash was incomplete. This incomplete pyrolysis of the rice husk ash was attributed to the presence of ceramic oxide (SiO_2). Meanwhile, the decomposition and burning of the organic materials in the sucrose take place between 180°C and 500°C based on the sucrose TGA. Based on the TGA data of rice husk ash and sucrose, the sintering temperature of porous alumina ceramics was selected in order to perform the sintering process [26, 27]. It is necessary to achieve a controlled burnout process for the rice husk ash and sucrose in order to obtain defect free samples [28].

The chemical compositions of rice husk ash and sucrose (sugar) were measured using an EDX machine. Table 1 shows the chemical compositions of the powder obtained from rice husk ash and sucrose.

Figure 3 shows the results obtained from the XRD and EDX tests for rice husk ash. The findings reveal the presence of amorphous silica (SiO_2) in the powder of the rice husk ash. This agrees with the chemical compositions tabulated in Table 1.

3.2. The Porosity, Density, Mechanical Properties, and Pore Size Distribution of Porous Ceramics. Generally, rice husk plays a role in the formation of pores due to the nature of its composition. Rice husk consists of ash (17–23%), fixed carbon (10–15%), and volatile matter (60–65%). Rice husk comprises around 20% silica, 30% group of lignin, and 40% cellulose [29]. Therefore, after burning the materials inside the ceramic during sintering, materials such as carbon, cellulose, and a group of lignin will turn into another form such as CO_2 and evaporate, which leads to the formation of pores in the ceramic body, and around 92–97% silica with other oxides in a minor percentage will remain in the matrix. The silica (SiO_2) in the rice husk or rice husk ash is found in either an amorphous or crystalline form [30]. In addition, the rice husk has an important effect on the sinterability of the ceramic body. As compared to other

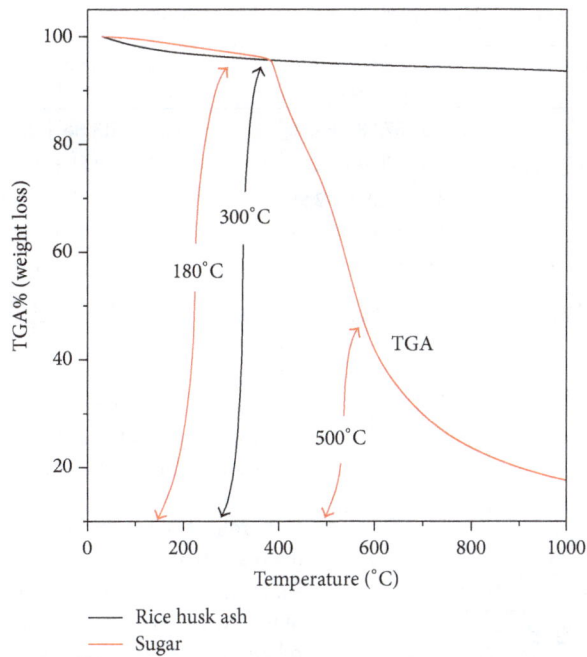

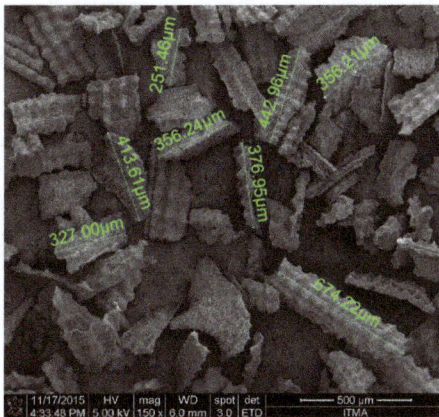

(b) Rice husk ash

(c) Sugar

Figure 2: (a) TGA, (b) FESEM images of the rice husk ash, and (c) sucrose.

Table 2: Different ratios of rice husk ash with alumina used to fabricate porous ceramics, porosity characterization, and density.

Rice husk ash content (wt%)	Al$_2$O$_3$ ratio (wt%)	Overall porosity (vol%)	Open porosity (vol%)	Green density (g/cm^3)	Sintered bulk density (g/cm^3)	Relative density (%)
0	100	18.45	10.76	1.92	2.99	75.89
10	90	42.97	24.29	1.86	2.25	57.11
20	80	48.09	27.38	1.76	2.04	51.78
30	70	47.63	23.94	1.67	2.06	52.28
40	60	49.44	24.00	1.58	1.99	50.51
50	50	49.04	20.15	1.53	2.01	51.02

factors, such as the sintering temperature which was constant (1600°C) in this research study, the ratio of rice husk ash has more effect on the sintering parameters such as the density, average pore size, porosity, and the formation of the phases [8].

Table 2 shows the ratios for the rice husk ash, the bulk density for the green and sintered samples, the overall porosity, and the open porosity.

The bulk density and porosity were characterised quantitatively by determining the values of the overall and open

TABLE 3: Mechanical properties of porous ceramics.

Rice husk ash content (wt%)	Porosity (vol%)	Hardness (HV$_1$)	Compressive strength (MPa)	Tensile strength (MPa)
0	18.45	372.83 ± 7.50	218.66 ± 2.52	27.39 ± 0.92
10	42.97	110.00 ± 2.11	59.81 ± 2.27	20.17 ± 1.19
20	48.09	101.90 ± 2.99	59.33 ± 3.96	18.99 ± 1.26
30	47.64	150.92 ± 1.88	69.72 ± 2.87	24.12 ± 0.93
40	49.45	114.92 ± 3.97	53.87 ± 1.55	21.88 ± 1.03
50	49.04	158.93 ± 3.96	60.14 ± 2.01	21.98 ± 0.84

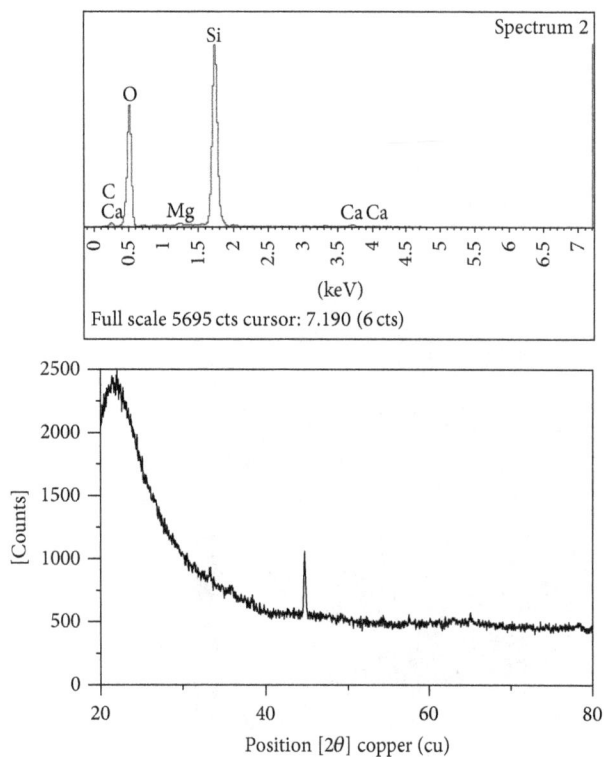

FIGURE 3: EDX and XRD for rice husk ash.

high weight materials, whilst open pores are beneficial for separation and filtration [27].

Table 3 shows the variations in the mechanical properties for the different ratios of rice husk ash content and porosity. Generally, the mechanical properties of the porous materials decrease with an increase in the porosity [33]. According to Rice's formula ($\sigma = \sigma_{\circ} e^{(-bp)}$), where $\sigma_{\circ}$ and σ are the strengths of the nonporous and porous materials, b is the constant value related to the pore characteristics and p is the porosity of the porous materials [34–36].

Figure 5 presents the data in Table 3 in graph form. Initially, the hardness of the porous alumina ceramics decreases at 20% after which it increases at 30% and 50%, due to the variations of porosity and the formation of the ceramic phases, namely, mullite and cristobalite, which have a high value of hardness as shown in Figure 5(a) [37–39]. The compressive and tensile strengths of porous alumina ceramics reached a peak at 30% rice husk ash. This improvement in the mechanical properties was attributed to the formation of the phases of the ceramic, that is, mullite, cristobalite, and corundum. Figure 5(b) shows the variation of the porosities.

Figure 6 shows the distribution of the pore size (open pore) of the samples. The distribution of the pore size (open pore) was strongly affected by the ratios of rice husk ash. All the samples displayed a unimodal pore size distribution for all the ratios. The average pore sizes recorded were 18.28 μm for 10 wt% rice husk ash, 28.35 μm for 30 wt% rice husk ash, and 50.02 μm for 50 wt% rice husk ash. It can be concluded that the shape and pore size distribution in porous ceramics corresponds to the required functions of certain applications [27]. Generally, during sintering at high sintering temperature with different ratios of rice husk ash, the relative density of the porous alumina ceramics has various values. With sintering at 1600°C, the decrease in the relative density was in the range from 18.78% (relative density 57.11%) to 24.84% (relative density 51.02%) at 10 wt% and 50 wt% rice husk ash due to the porosity, which affected the pore size distribution. With a decrease in the relative density, the porosity increased, leading to pore formation. In addition, it can be seen that there is not much difference in the open pore size distribution above 20 wt% rice husk ash due to the presence of the glassy phase, which leads to an increase in the relative density of 52.28% at 30 wt% of rice husk ash and a decrease in the porosity to 47.63%, resulting in little difference in the pore size of the ceramic body [40, 41].

porosities of the samples of porous alumina. Figure 4 shows the variations of open and total porosities with different ratios of rice husk ash. An increase in the contents of rice husk ash showed an increase in the total and open porosities, followed by a decrease due to the formation of glassy phases [31]. Meanwhile, the green and bulk densities decreased from 1.92 to 1.53 g/cm^3 and from 2.99 to 2.01 g/cm^3, respectively. This is attributed to the burnout of the carbon, the organic materials, and the binder after sintering at high temperature [32]. The decrease in the green densities was attributed to the low density of rice husk ash (2.17 g/cm^2) compared with the density of the alumina matrix. The maximum porosity is at 49.45% while the lowest porosity is recorded at 42.97%. Generally, the characteristics of the pores were categorised into pore size, pore morphology, and porosity. The pore morphology can be classified into closed and open pores. Closed pores are important for heat insulating materials and

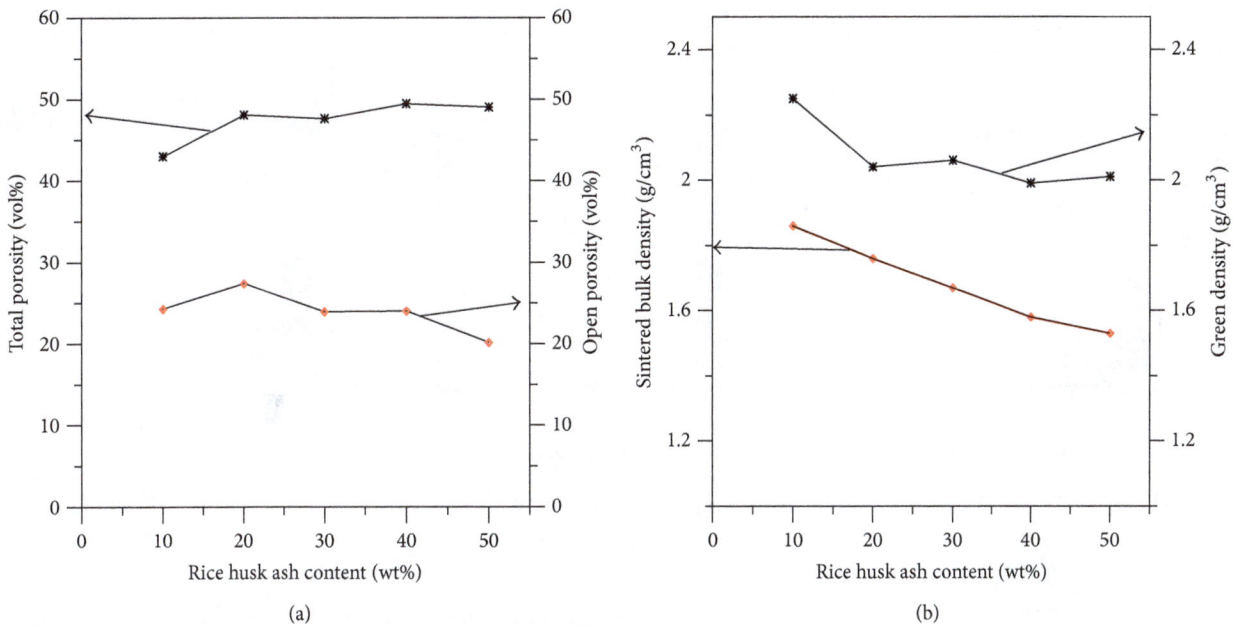

FIGURE 4: (a) Trend of total porosity and open porosity with increases in the content of rice husk ash; (b) trend of sintered bulk density and green density with increases in the content of rice husk ash. All samples were sintered at 1600°C for 2 hrs.

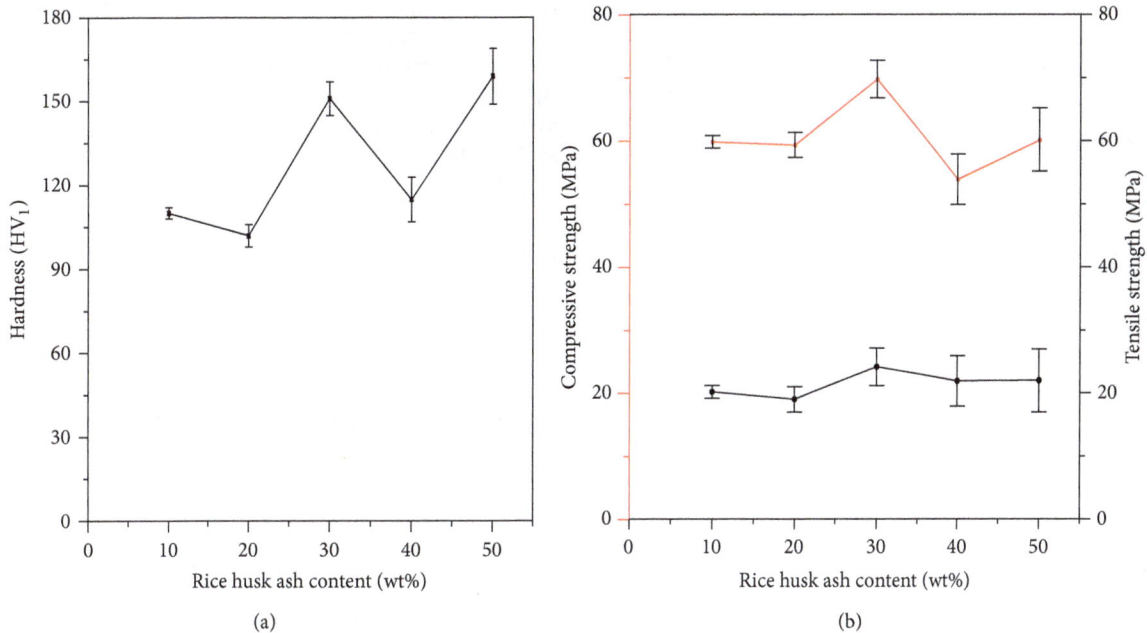

FIGURE 5: The variations in the mechanical properties of the content of rice husk ash of the samples of porous alumina ceramic sintered at 1600°C for 2 hrs; (a) hardness and (b) compressive strength and tensile strength.

3.3. XRD Analysis, Al_2O_3-SiO_2 Reaction, and the Microstructure of Porous Alumina Ceramics. The objective of the XRD analysis in this research study is to observe the phase formation in the porous alumina ceramics and how it affects the mechanical properties. Figure 7 indicates that the XRD patterns of porous alumina ceramics samples sintered at 1600°C for 2 hrs with different ratios of rice husk ash have different peaks, which refer to some of the ceramic phases including corundum (Al_2O_3) (JCPDS file number 01-075-0785), cristobalite (SiO_2) (JCPDS file number 00-001-0424), mullite ($3Al_2O_3 \cdot 2SiO_2$) (JCPDS file number 00-006-0259), and sillimanite (Al_2SiO_5) (JCPDS file number 01-088-0890). It was found that, with increasing ratios of rice husk ash and sintering at high temperature, the phases detected in the samples of the porous alumina with rice husk ash included corundum (Al_2O_3) at 10 wt% rice husk ash, cristobalite (SiO_2)

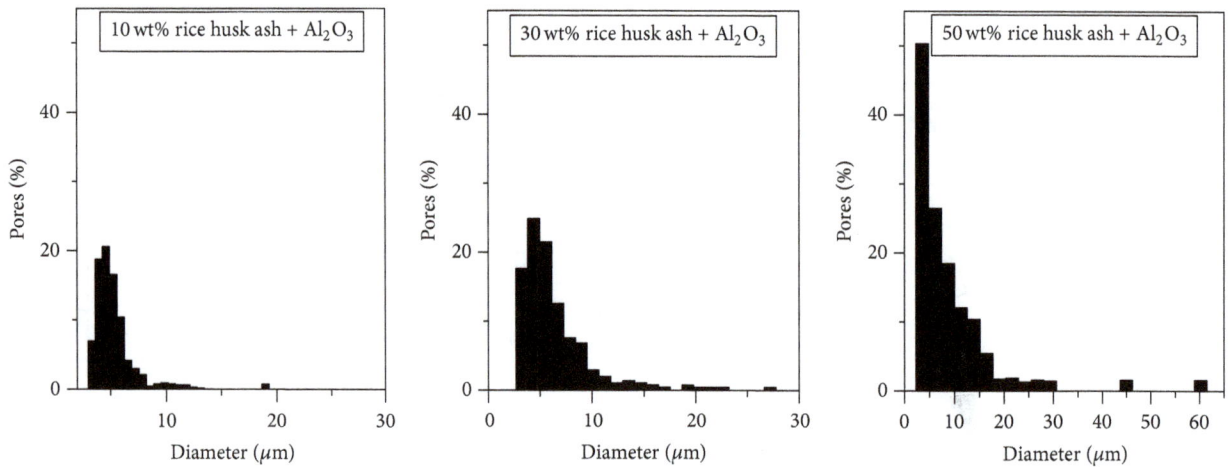

FIGURE 6: Distribution of the open pore size of the samples of porous alumina ceramic sintered at 1600°C for 2 hrs with different ratios of rice husk ash.

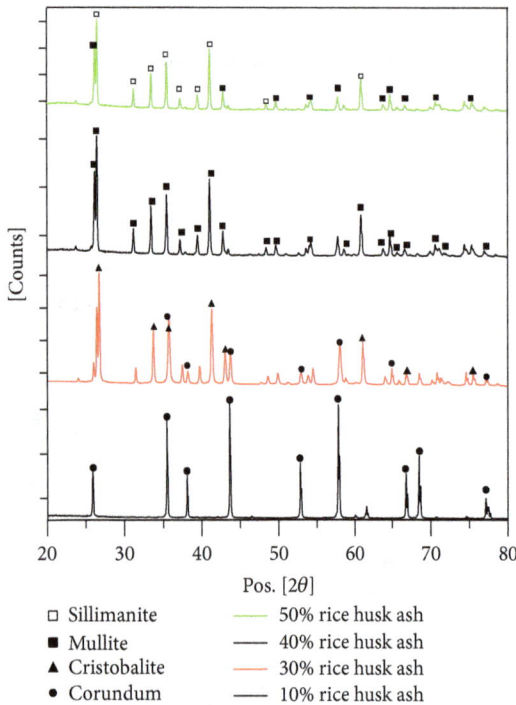

FIGURE 7: XRD pattern for the porous alumina samples sintered at 1600°C for 2 hrs with rice husk ash.

and corundum (Al_2O_3) at 30 wt% rice husk ash, and mullite ($3Al_2O_3 \cdot 2SiO_2$) at 40 wt% rice husk ash while sillimanite (Al_2SiO_5) and mullite ($3Al_2O_3 \cdot 2SiO_2$) existed at 50 wt% rice husk ash. All these phases were matched with their JCPDS file number.

For the ratio of 10 wt% rice husk ash, only the corundum (A_2O_3) phase was detected because the ratio of SiO_2 in the rice husk ash was not sufficient to react with Al_2O_3 for production or transformation into another ceramic phase like cristobalite (SiO_2) and mullite ($3Al_2O_3 \cdot 2SiO_2$); this is evident in Figure 7. Based on the XRD pattern, there is no sign of

the silica (SiO_2) peak. Meanwhile, for the ratio of 30 wt% rice husk ash, the cristobalite (SiO_2) and corundum (Al_2O_3) phases were detected, which lead to improvements in the mechanical properties. The big peaks of cristobalite (SiO_2) can be seen due to the increase in the rice husk ash ratio. After sintering at high temperature, there is no sign of the silica peak for the 40 wt% rice husk ash sample. This is evident in Figure 7 based on the XRD pattern. This proves that the silica (SiO_2) reacted with the surrounding alumina (Al_2O_3) to produce a stable mullite ($3Al_2O_3 \cdot 2SiO_2$) phase, according to the following equation [15]:

$$3Al_2O_3 + 2SiO_2 \longrightarrow 3Al_2O_3 \cdot 2SiO_2 \qquad (3)$$

By increasing the percentage ratio of mullite ($3Al_2O_3 \cdot 2SiO_2$) at 40 wt%, the densification rate was retarded and increased the porosity, which lead to a reduction in the mechanical properties of porous ceramics [42]. The silica peak was also unclear from the 50 wt% rice husk ash sample. It is believed that the alumina had dissolved into the residual glassy phase of the silica (SiO_2) to form the mullite ($3Al_2O_3 \cdot 2SiO_2$) and sillimanite (Al_2SiO_5) or ($Al_2O_3 \cdot SiO_2$) phases [43, 44] according to the following reaction:

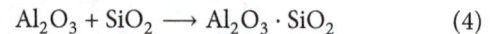

$$Al_2O_3 + SiO_2 \longrightarrow Al_2O_3 \cdot SiO_2 \qquad (4)$$

Thus, it can be concluded that all these phases of the ceramics formed have an effect on the mechanical properties of porous ceramics [15]. In this case, the ratios of the rice husk ash controlled the sintering parameters such as the density, the average pore size, porosity, and the formation of the phases.

Figure 8 illustrates the FESEM images of the samples of porous alumina ceramic with different ratios of rice husk ash (10 wt%, 30 wt%, and 50 wt% rice husk ash; the sintering temperature was set at 1600°C for 2 hrs). An increase in the ratio of the rice husk ash not only causes an increase in the porosity but also brings changes to the microstructure. Figures 8(a), 8(b), and 8(c) show the longitudinal shape of the pores in the samples of the porous alumina ceramics, which were attributed to the shape and size of the pore agent

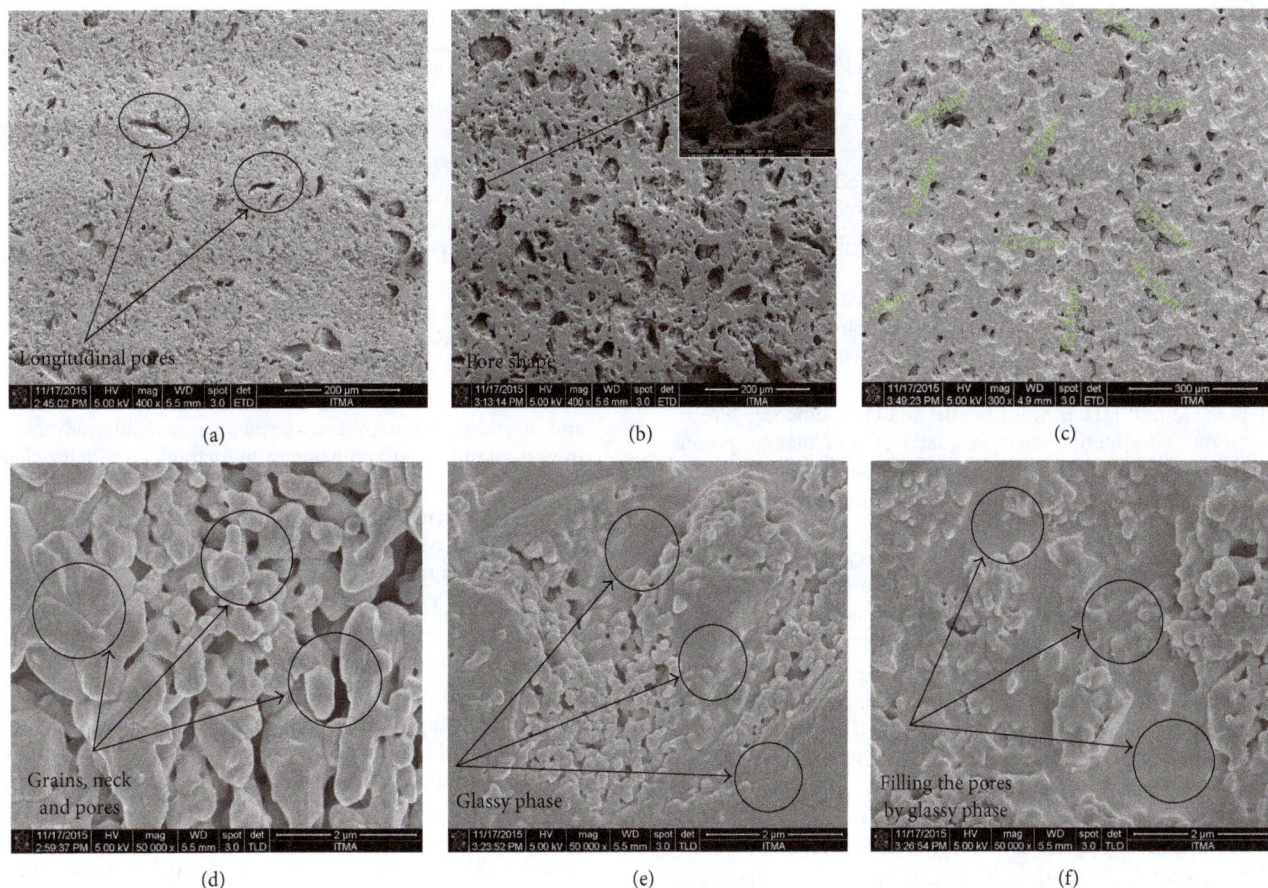

FIGURE 8: FESEM images of the samples of porous ceramic sintered at 1600°C for 2 hrs with different ratios of rice husk ash; (a, d) 10 wt% rice husk ash, (b, e) 30 wt% rice husk ash, and (c, f) 50 wt% rice husk ash.

particles as shown in Figure 2. The pores, grains, glassy phase, and neck shapes of the samples of the porous alumina can be clearly seen in Figures 8(d), 8(e), and 8(f). The growth of the grains and necks was a result of sintering at high temperature [45], while each gap between the particles and the neck becomes a pore in addition to the effects of the pore agent [46]. In addition, the different sizes of the pores resulted in an increase in the ratio of the rice husk ash and sucrose burnout [15]. The linking of the open pores resulted in fine and big pores.

Figures 8(e) and 8(f) present the glassy phase of 30 wt% and 50 wt% rice husk ash during sintering at 1600°C for 2 hrs. The presence of the glassy phase for 30 wt% and 50 wt% rice husk ash was to fill the porous structure, which leads to a decrease in the porosity from 48.09% to 47.64% and from 49.45% to 49.04% for 30 wt% and 50 wt% rice husk ash, respectively [31]. Hence, decreasing the porosity, through the formations of the glassy phase ceramic phases, has improved the mechanical properties.

4. Conclusion

The use of alumina and commercial rice husk ash as a source of silica (SiO_2) and as a pore agent has successfully resulted in the preparation of porous mullite and cristobalite-corundum ceramic composite materials. The porosity increased with an increase in the ratios of the rice husk ash from 10 wt% to 50 wt%. However, the tensile and compressive strengths showed an initial decrease, followed by an increase. The the improvement in the mechanical properties was attributed to the formation of the phases of the ceramic (sillimanite, mullite, corundum, and cristobalite) at high sintering temperature and higher rice husk ash ratios. Finally, the distributions of the pore size corresponded to the ratio of the rice husk ash.

Competing Interests

The authors declare that they have no competing interests.

Acknowledgments

The financial support provided by Putra Grant 2013 (GPIBT/ 2013/9410600) is much appreciated. The authors would like to thank and appreciate the Iraqi Government/Ministry of Higher Education and Scientific Research for its financial support.

References

[1] K. Hua, A. Shui, L. Xu, K. Zhao, Q. Zhou, and X. Xi, "Fabrication and characterization of anorthite–mullite–corundum porous ceramics from construction waste," *Ceramics International*, vol. 42, no. 5, pp. 6080–6087, 2016.

[2] G. Xu, Y. Ma, H. Cui, G. Ruan, Z. Zhang, and H. Zhao, "Preparation of porous mullite-corundum ceramics with controlled pore size using bioactive yeast as pore-forming agent," *Materials Letters*, vol. 116, pp. 349–352, 2014.

[3] F. Nyongesa, N. Rahbar, S. Obwoya, J. Zimba, B. O. Aduda, and W. Soboyejo, "An investigation of thermal shock in porous clay ceramics," *ISRN Mechanical Engineering*, vol. 2011, Article ID 816853, 9 pages, 2011.

[4] E. S. Gómez, J. D. J. A. F. Cuautle, and O. T. Jiménez, "Preparation and sintering effect in quartz-barium titanate porous ceramics and permeability modulation using an implanted electrode," *Advances in Materials Science and Engineering*, vol. 2014, Article ID 721245, 5 pages, 2014.

[5] A. Tomba, M. A. Camerucci, G. Urretavizcaya, A. L. Cavalieri, M. A. Sainz, and A. Caballero, "Elongated mullite crystals obtained from high temperature transformation of sillimanite," *Ceramics International*, vol. 25, no. 3, pp. 245–252, 1999.

[6] H. S. Tripathi, B. Mukherjee, S. K. Das, A. Ghosh, and G. Banerjee, "Effect of sillimanite beach sand composition on mullitization and properties of Al_2O_3-SiO_2 system," *Bulletin of Materials Science*, vol. 26, no. 2, pp. 217–220, 2003.

[7] Y. M. Park, T. Y. Yang, S. Y. Yoon, R. Stevens, and H. C. Park, "Mullite whiskers derived from coal fly ash," *Materials Science and Engineering A*, vol. 454-455, pp. 518–522, 2007.

[8] M. Serra, M. Conconi, M. Gauna, G. Suárez, E. Aglietti, and N. Rendtorff, "Mullite ($3Al_2O_3 \cdot 2SiO_2$) ceramics obtained by reaction sintering of rice husk ash and alumina, phase evolution, sintering and microstructure," *Journal of Asian Ceramic Societies*, vol. 4, no. 1, pp. 61–67, 2016.

[9] J.-M. Wu, X.-Y. Zhang, J. Xu et al., "Preparation of porous Si_3N_4 ceramics via tailoring solid loading of Si_3N_4 slurry and Si_3N_4 poly-hollow microsphere content," *Journal of Advanced Ceramics*, vol. 4, pp. 260–266, 2015.

[10] K. Sengphet, K. Pasomsouk, T. Sato, M. A. Fauzi, O. Radzali, and S. P. Pinang, "Fabrication of porous clay ceramics using kenaf powder waste," *International Journal of Scientific and Research Publications*, vol. 3, no. 8, 2013.

[11] D. E. Njeumen Nkayem, J. A. Mbey, B. B. Kenne Diffo, and D. Njopwouo, "Preliminary study on the use of corn cob as pore forming agent in lightweight clay bricks: physical and mechanical features," *Journal of Building Engineering*, vol. 5, pp. 254–259, 2016.

[12] P. Sooksaen, S. Suttiruengwong, K. Oniem, K. Ngamlamiad, and J. Atireklapwarodom, "Fabrication of porous bioactive Glass-Ceramics via decomposition of natural fibres," *Journal of Metals, Materials and Minerals*, vol. 18, pp. 85–91, 2008.

[13] Y. Dong, J.-E. Zhou, B. Lin et al., "Reaction-sintered porous mineral-based mullite ceramic membrane supports made from recycled materials," *Journal of Hazardous Materials*, vol. 172, no. 1, pp. 180–186, 2009.

[14] J. Cao, X. Dong, L. Li, Y. Dong, and S. Hampshire, "Recycling of waste fly ash for production of porous mullite ceramic membrane supports with increased porosity," *Journal of the European Ceramic Society*, vol. 34, no. 13, pp. 3181–3194, 2014.

[15] K. Mohanta, A. Kumar, O. Parkash, and D. Kumar, "Processing and properties of low cost macroporous alumina ceramics with tailored porosity and pore size fabricated using rice husk and sucrose," *Journal of the European Ceramic Society*, vol. 34, no. 10, pp. 2401–2412, 2014.

[16] G. L. Re, F. Lopresti, G. Petrucci, and R. Scaffaro, "A facile method to determine pore size distribution in porous scaffold by using image processing," *Micron*, vol. 76, pp. 37–45, 2015.

[17] M. Gan and J. Wang, *Applications of Image Processing Technique in Porous Material Characterization*, InTech Open Access, 2012.

[18] S. R. Elsen and T. Ramesh, "Optimization to develop multiple response hardness and compressive strength of zirconia reinforced alumina by using RSM and GRA," *International Journal of Refractory Metals and Hard Materials*, vol. 52, pp. 159–164, 2015.

[19] W. C. Oliver and G. M. Pharr, "Measurement of hardness and elastic modulus by instrumented indentation: advances in understanding and refinements to methodology," *Journal of Materials Research*, vol. 19, no. 1, pp. 3–20, 2004.

[20] A. s. ASTM standard C 1327-03, "ASTM standard C 1327-03 Standard Test Method for Vickers Indentation Hardness of Advanced Ceramics," 2004.

[21] B. S. M. Seeber, U. T. Gonzenbach, and L. J. Gauckler, "Mechanical properties of highly porous alumina foams," *Journal of Materials Research*, vol. 28, no. 17, pp. 2281–2287, 2013.

[22] J. A. Junkes, B. Dermeik, B. Gutbroda, D. Hotzab, P. Greila, and N. Travitzky, "Influence of coatings on microstructure and mechanical properties of preceramic paper-derived porous alumina substrates," *Journal of Materials Processing Technology*, vol. 213, no. 2, pp. 308–313, 2013.

[23] L. Hu, R. Benitez, S. Basu, I. Karaman, and M. Radovic, "Processing and characterization of porous Ti_2AlC with controlled porosity and pore size," *Acta Materialia*, vol. 60, no. 18, pp. 6266–6277, 2012.

[24] R. L. Menchavez and L.-A. S. Intong, "Red clay-based porous ceramic with pores created by yeast-based foaming technique," *Journal of Materials Science*, vol. 45, no. 23, pp. 6511–6520, 2010.

[25] L. F. Hu and C.-A. Wang, "Effect of sintering temperature on compressive strength of porous yttria-stabilized zirconia ceramics," *Ceramics International*, vol. 36, no. 5, pp. 1697–1701, 2010.

[26] J. Dittmann and N. Willenbacher, "Micro structural investigations and mechanical properties of macro porous ceramic materials from capillary suspensions," *Journal of the American Ceramic Society*, vol. 97, no. 12, pp. 3787–3792, 2014.

[27] J.-H. Eom, Y.-W. Kim, and S. Raju, "Processing and properties of macroporous silicon carbide ceramics: a review," *Journal of Asian Ceramic Societies*, vol. 1, no. 3, pp. 220–242, 2013.

[28] O. L. Ighodaro, O. I. Okoli, M. Zhang, and B. Wang, "Ceramic preforms with 2D regular channels for fabrication of metal/ceramic-reinforced composites," *International Journal of Applied Ceramic Technology*, vol. 9, no. 2, pp. 421–430, 2012.

[29] G. Görhan and O. Şimşek, "Porous clay bricks manufactured with rice husks," *Construction and Building Materials*, vol. 40, pp. 390–396, 2013.

[30] K. A. Matori, M. Haslinawati, Z. Wahab, H. Sidek, T. Ban, and W. Ghani, "Producing amorphous white silica from rice husk," *MASAUM Journal of Basic and Applied Sciences*, vol. 1, no. 3, pp. 512–515, 2009.

[31] K. Hua, A. Shui, L. Xu, K. Zhao, Q. Zhou, and X. Xi, "Fabrication and characterization of anorthite-mullite-corundum porous ceramics from construction waste," *Ceramics International*, vol. 42, no. 5, pp. 6080–6087, 2016.

[32] W. Guo, H. Lu, and C. Feng, "Influence of La_2O_3 on preparation and performance of porous cordierite from rice husk," *Journal of Rare Earths*, vol. 28, no. 4, pp. 614–617, 2010.

[33] Z. Negahdari, M. Willert-Porada, and C. Pfeiffer, "Mechanical properties of dense to porous alumina/lanthanum hexaaluminate composite ceramics," *Materials Science and Engineering A*, vol. 527, no. 12, pp. 3005–3009, 2010.

[34] G. P. Kennedy, K.-Y. Lim, Y.-W. Kim, I.-H. Song, and H.-D. Kim, "Effect of SiC particle size on flexural strength of porous self-bonded SiC ceramics," *Metals and Materials International*, vol. 17, no. 4, pp. 599–605, 2011.

[35] D.-M. Liu, "Influence of porosity and pore size on the compressive strength of porous hydroxyapatite ceramic," *Ceramics International*, vol. 23, no. 2, pp. 135–139, 1997.

[36] L. Yin, X. Zhou, J. Yu, and H. Wang, "Preparation of high porous silicon nitride foams with ultra-thin walls and excellent mechanical performance for heat exchanger application by using a protein foaming method," *Ceramics International*, vol. 42, no. 1, pp. 1713–1719, 2016.

[37] N. Kayal, A. Dey, and O. Chakrabarti, "Synthesis of mullite bonded porous SiC ceramics by a liquid precursor infiltration method: effect of sintering temperature on material and mechanical properties," *Materials Science and Engineering A*, vol. 556, pp. 789–795, 2012.

[38] W. Yan, N. Li, and B. Han, "Effects of sintering temperature on pore characterisation and strength of porous corundum-mullite ceramics," *Journal of Ceramic Processing Research*, vol. 11, no. 3, pp. 388–391, 2010.

[39] B. V. Manoj Kumar, J.-H. Eom, Y.-W. Kim, I.-H. Song, and H.-D. Kim, "Effect of aluminum hydroxide content on porosity and strength of porous mullite-bonded silicon carbide ceramics," *Journal of the Ceramic Society of Japan*, vol. 119, no. 1389, pp. 367–370, 2011.

[40] P. Lemes-Rachadel, H. Birol, A. P. N. Oliveira, and D. Hotza, "Development of alternative glass ceramic seal for a planar solid oxide fuel cell," *Advances in Materials Science and Engineering*, vol. 2012, Article ID 346280, 6 pages, 2012.

[41] J. Cao, J. Lu, L. Jiang, and Z. Wang, "Sinterability, microstructure and compressive strength of porous glass-ceramics from metallurgical silicon slag and waste glass," *Ceramics International*, vol. 42, no. 8, pp. 10079–10084, 2016.

[42] A. Sedaghat, E. Taheri-Nassaj, G. D. Soraru, and T. Ebadzadeh, "Microstructure development and phase evolution of alumina-mullite nanocomposite," *Science of Sintering*, vol. 45, no. 3, pp. 293–303, 2013.

[43] L. Zhu, Y. Dong, S. Hampshire, S. Cerneaux, and L. Winnubst, "Waste-to-resource preparation of a porous ceramic membrane support featuring elongated mullite whiskers with enhanced porosity and permeance," *Journal of the European Ceramic Society*, vol. 35, no. 2, pp. 711–721, 2015.

[44] R. C. Bradt, "The sillimanite minerals: andalusite, kyanite, and sillimanite," in *Ceramic and Glass Materials*, pp. 41–48, Springer, Berlin, Germany, 2008.

[45] S. Ding, S. Zhu, Y.-P. Zeng, and D. Jiang, "Fabrication of mullite-bonded porous silicon carbide ceramics by in situ reaction bonding," *Journal of the European Ceramic Society*, vol. 27, no. 4, pp. 2095–2102, 2007.

[46] W. D. Callister Jr. and D. G. Rethwisch, *Materials Science and Engineering*, chapter 12, John Wiley & Sons, Inc, New York, NY, USA, 8th edition, 2010.

11

Experimental Investigation on the Material Removal of the Ultrasonic Vibration Assisted Abrasive Water Jet Machining Ceramics

Tao Wang, Rongguo Hou, and Zhe Lv

School of Mechanical Engineering, Shandong University of Technology, Zibo, China

Correspondence should be addressed to Rongguo Hou; hourongguo212@163.com

Academic Editor: Charles C. Sorrell

The ultrasonic vibration activated in the abrasive water jet nozzle is used to enhance the capability of the abrasive water jet machinery. The experiment devices of the ultrasonic vibration assisted abrasive water jet are established; they are composed of the ultrasonic vibration producing device, the abrasive supplying device, the abrasive water jet nozzle, the water jet intensifier pump, and so on. And the effect of process parameters such as the vibration amplitude, the system working pressure, the stand-off, and the abrasive diameter on the ceramics material removal is studied. The experimental result indicates that the depth and the volume removal are increased when the ultrasonic vibration is added on abrasive water jet. With the increase of vibration amplitude, the depth and the volume of material removal are also increased. The other parameters of the ultrasonic vibration assisted abrasive water jet also have an important role in the improvement of ceramic material erosion efficiency.

1. Introduction

Abrasive water jet (AWJ) technology, as one of the fastest growing nonconventional machining processes, has been applied in many engineering fields such as aeronautics and mechanical manufacture [1, 2]. The abrasive water jet has so many merits such as no heat deformation, high machining efficiency, high accuracy, and broad machining range. In order to improve the capability of the water jet machining, the pulsed water jet, which could cause a greater removal of target material by generating water-hammer effect, has been a research focus [3–5]. With an aim to produce high speed pulsed water jets, the mechanical methods including rotation, reciprocating, and wobbling have been successfully applied. However, the devices using these methods require tedious mechanical maintenance and their durability and reliability are in harsh working environments. Ultrasonic vibration has been used widely in abrasive water jet technology; for example, it is used to determine the vibration emission frequency when the abrasive particles impact the materials [6, 7]. It is also used to produce the pulsed water jet; for example,

Vijay et al. [8] and Foldyna et al. [9, 10] have introduced a kind of high frequency pulsed water jet with the use of ultrasonic vibration activated in the nozzle; the jet's energy is pretty low. Lehocka et al. [11] have investigated the surface topography, morphology, and anisotropy of copper alloys created by pulsating water jet, and the results indicate that this new way of metal eroding can be used in the automotive and engineering industries in the future. Hloch et al. [12] have used the selective property of ultrasonic pulsating water jet for the disintegration of the interface created by bone cement as a potential technique for revision arthroplasty, and the results positively support an assumption that pulsating water jet has a potential to be a suitable technique for the quick and safe disintegration of bone cement during revision arthroplasty. Zelenak et al. [13] have tested the applicability of the shadowgraph technique combined with PIV processing algorithms to visualize water jet structure and analyze flow velocity field, and its results focus on the visualization of pulsating and continuous water jets, while there is little literature about experiment investigation on the material removal of the ultrasonic vibration assisted abrasive water jet

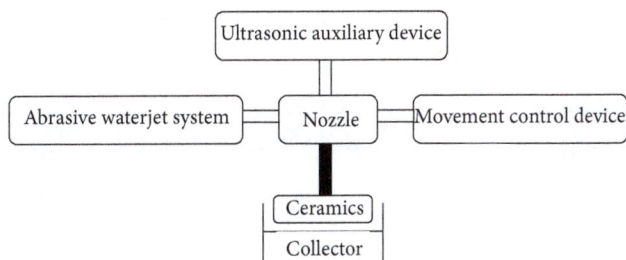

FIGURE 1: The sketch of experiment scheme.

(a) The pneumohydraulic intensifier

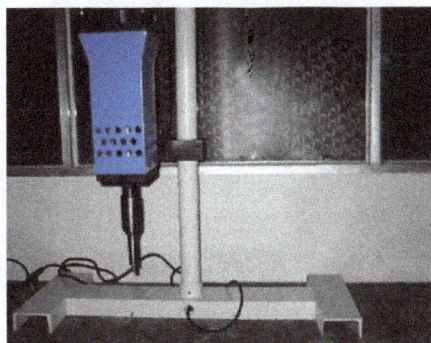

(b) The device of ultrasonic aided water jet

FIGURE 2: The device of ultrasonic vibration aided abrasive water jet.

impacting ceramics. With an aim to investigate the effect of the ultrasonic variations on the efficiency of abrasive water jet impacting the ceramics, a serial of experiments are carried out, and the effect of the vibration amplitude, the system working pressure, the distance of stand-off, and the abrasive diameter on the ceramics material removal is studied.

2. Experimental Set-Up

The experiment device of ultrasonic vibration assisted abrasive water jet, as Figure 1 shows, is composed of the ultrasonic auxiliary device, the abrasive water jet system including the abrasive water jet nozzle, the movement control device, the work piece (ceramics), the collector, and so on.

Among them, the ultrasonic auxiliary device, as is shown in Figure 2(b), can provide the vibration with a frequency of 20 KHz, and the amplitude is about 20 um, the vibration amplifier is used to change vibration aptitude, and it is connected with the abrasive nozzle by stud bolts; the abrasive water jet system, as is shown in Figure 2(a), can provide the water jet with a pressure within 0–70 Mpa, the working system pressure of water jet in this study is between 5 and 25 MPa, the value is pretty low because these experiments aim to study the micromachine the ceramics, and the ultrasonic vibration also helps to decrease the working pressure. The water jet nozzle diameter is 1 mm. The stand-off is within 3–7 mm and the abrasive supply rate is 2.5 mg/s.

The erosion surface morphology of experimental sample is observed and analyzed with the help of an instrument of 3D Laser Shape Measurement.

The alumina ceramics produced by National Engineering Laboratory of Ceramics are used as the work piece material and the purity of these alumina ceramics is 96%; the mechanical property is shown in Table 1.

The green silicon carbide is used as abrasive, which is produced by a corundum manufacturing company named HeXing in ZhengZhou, and the mechanical property is shown in Table 2.

3. Experimental Procedure

The way of fixed-point erosion is used in this experiment, and each processing time is 8s. In this experiment, by changing the amplitude, the water jet pressure, the stand-off, and the abrasive grain, the instrument of 3D Laser Shape Measurement will be used to observe and analyze the change law of area, shape, and depth in the erosion field; each sample point will suffer erosion for three times, and the average value will be used. The experiment parameters are shown in Table 3.

According to the method of single factor experiment, the designed experiment scheme is shown in Table 4.

4. Experimental Result and Its Discussion

Figure 3 shows the erosion surface morphology of alumina ceramics with and without the ultrasonic vibration assisted abrasive water jet. The erosion surface morphology of alumina ceramics is under the experimental condition that the size of abrasive grain is 320#, the stand-off is 5 mm, the supply rate of abrasive is 0.25 g/s, and the processing time is 8 s.

TABLE 1: The mechanical property of alumina ceramic [14].

Material	Density (g/cm^3)	Vickers hardness (Gpa)	Bending strength (Mpa)	Fracture toughness (Mpa·m$^{1/2}$)
96% alumina	3.7	20	320	3

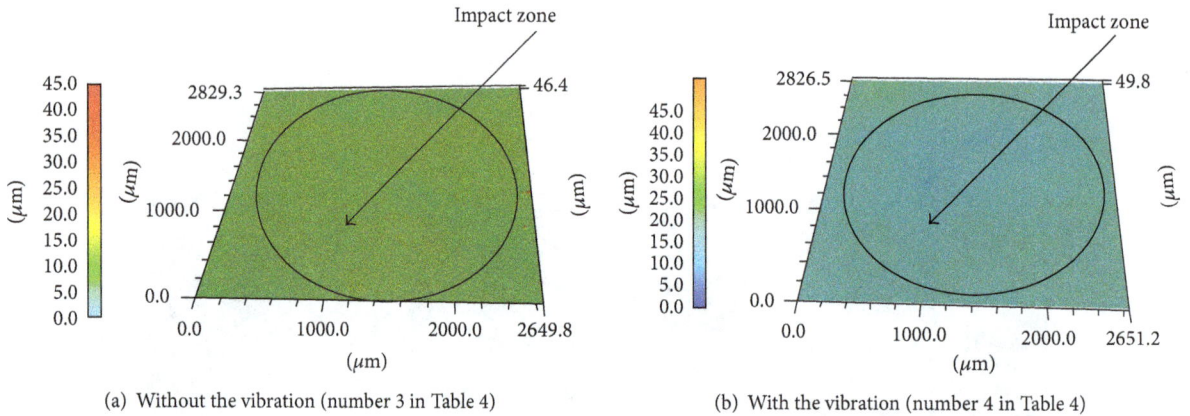

(a) Without the vibration (number 3 in Table 4)

(b) With the vibration (number 4 in Table 4)

FIGURE 3: The erosion surface morphology of alumina ceramics.

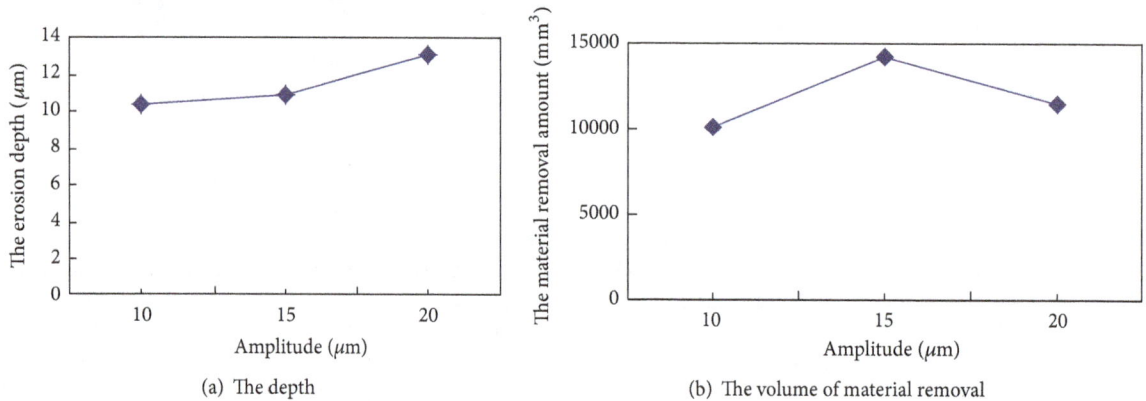

(a) The depth

(b) The volume of material removal

FIGURE 4: The effect of amplitude on the depth and the material removal amount.

TABLE 2: The mechanical property of green silicon carbide [14].

Material	Density (g/cm^3)	Mosh hardness scale
Green silicon carbide (SiC)	3.2	9.2

As is shown in Figure 3 obviously, there are sporadic erosive pits in the erosion field without the ultrasonic vibration assisted abrasive water jet, and, for these pits, the depth and the volume of material removal are very small. In this experiment, the amount of erosive pits, the maximum depth of erosive pits and the volume of material removal increase simultaneously with the ultrasonic vibration assisted abrasive water jet. And the erosive area increases slightly with the ultrasonic vibration assisted abrasive water jet.

Figure 4 shows the effect of vibration amplitude on the depth of erosive pits and the volume of material removal. As is shown obviously, with the increase of amplitude, the depth of erosive pits and the volume of the material removal will increase firstly but decrease subsequently.

Figure 5 shows the effect of water jet pressure on the depth of erosive pits and the volume of material removal with different values of pressure, such as 5 MPa, 10 MPa, 15 MPa, 20 MPa, and 25 MPa. As is shown obviously, when the workpiece surface suffers vertical impact by abrasive water jet, the maximum depth will increase with the increase of water jet pressure, and that is because the lower the water jet pressure is, the smaller the normal component of the impact load becomes. And the surface roughness of initial workpiece is very large, and the volume of material removal is very small. However, the normal component of impact load, the depth, and the volume of material removal will increase simultaneously with the increase of water jet pressure.

Figure 6 shows the effect of stand-off on the depth and the volume of material removal. As is shown obviously, the stand-off has a small effect on the depth and the volume of material removal.

TABLE 3: The experiment parameters.

Abrasive (mesh/#)	Pressure (Mpa)	Amplitude (mm)	Stand-off (mm)	Material
280, 320, 400	5, 10, 15, 20, 25	22, 17, 10	3, 5, 7	Alumina ceramic, glass

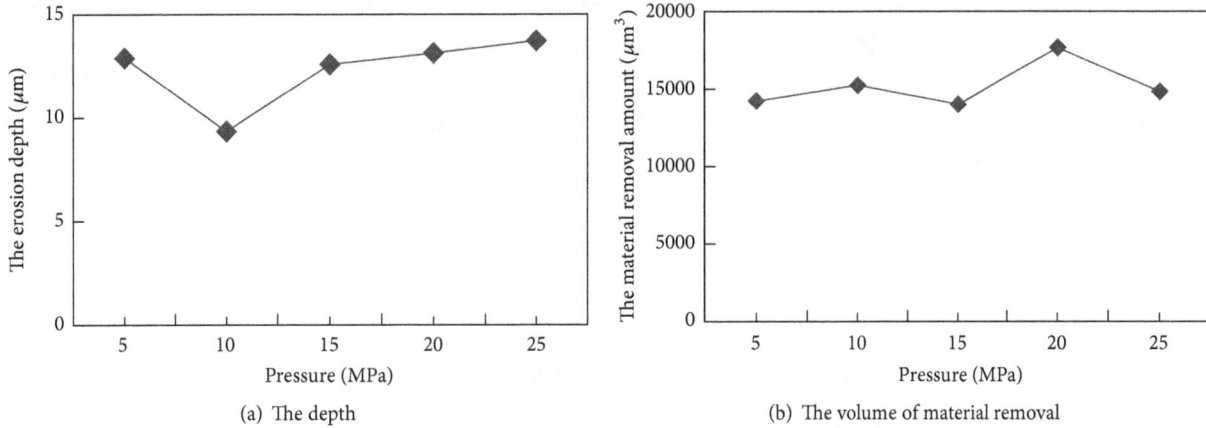

(a) The depth

(b) The volume of material removal

FIGURE 5: The effect of water jet pressure on the depth and the volume of material removal.

TABLE 4: The experiment scheme.

Serial number	System pressure (Mpa)	Stand-off (mm)	Amplitude (mm)	Particle size (mesh/#)
1	20	3	0	320
2	20	3	22	320
3	20	5	0	320
4	20	5	22	320
5	20	7	0	320
6	20	7	22	320
7	20	5	0	280
8	20	5	22	280
9	20	5	0	400
10	20	5	22	400
11	25	5	0	320
12	25	5	22	320
13	15	5	0	320
14	15	5	22	320
15	10	5	0	320
16	10	5	22	320
17	5	5	0	320
18	5	5	22	320
19	20	3	17	320
20	20	5	17	320
21	20	7	17	320
22	20	5	17	280
23	20	5	17	400
24	25	5	17	320
25	15	5	17	320
26	10	5	17	320
27	5	5	17	320

TABLE 4: Continued.

Serial number	System pressure (Mpa)	Stand-off (mm)	Amplitude (mm)	Particle size (mesh/#)
28	20	3	10	320
29	20	5	10	320
30	20	7	10	320
31	20	5	10	280
32	20	5	10	400
33	25	5	10	320
34	15	5	10	320
35	10	5	10	320

Figure 7 shows the effect of abrasive grain on the erosion surface morphology with different values of abrasive, such as 280#, 320#, and 400#. As is shown obviously, the depth and the volume of material removal will increase firstly but decrease subsequently with the increase of abrasive. It indicates that, for the alumina ceramics, the abrasive has a small effect on the depth and the volume of material removal.

5. Establishment of Experimental Model

By analyzing the experimental result, it is easy to obtain the experimental model of depth and material removal amount for the ceramics which is machined by the ultrasonic vibration assisted abrasive water jet. The symbol of water jet pressure is P_s, the symbol of stand-off is B, the symbol of amplitude is a, and the symbol of abrasive is M; they all have a great effect on the surface morphology. Therefore, the above processing parameters which play an important role on the depth and the material removal amount will satisfy the relation, and the predicted model of depth and material

TABLE 5: The coefficients of regression equation.

Project	Coefficient				
	w_0	w_1	w_2	w_3	w_4
h_1	8.894	0.115	0.491	0.049	−0.008
Q_1	8.457	0.239	−0.586	0.072	0.0077

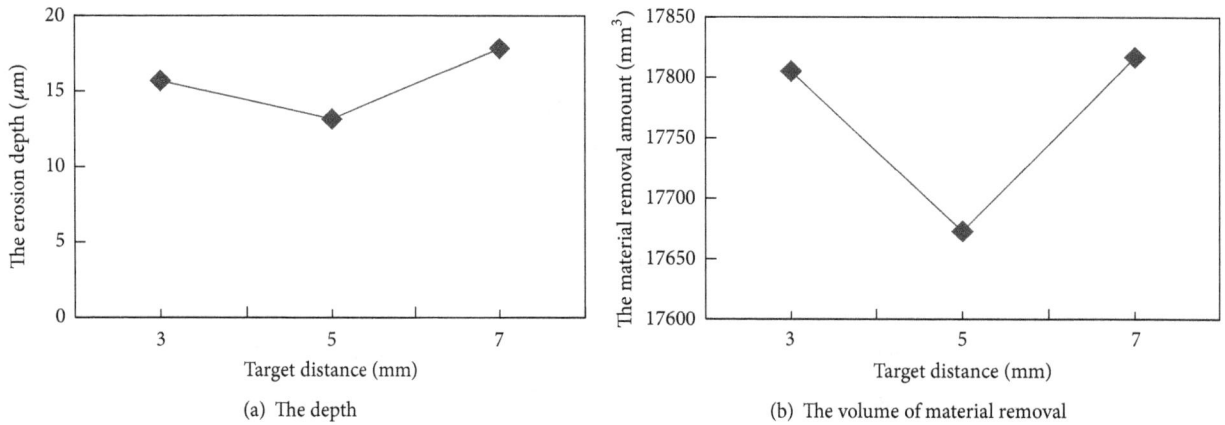

(a) The depth

(b) The volume of material removal

FIGURE 6: The effect of stand-off on the depth and the volume of material removal.

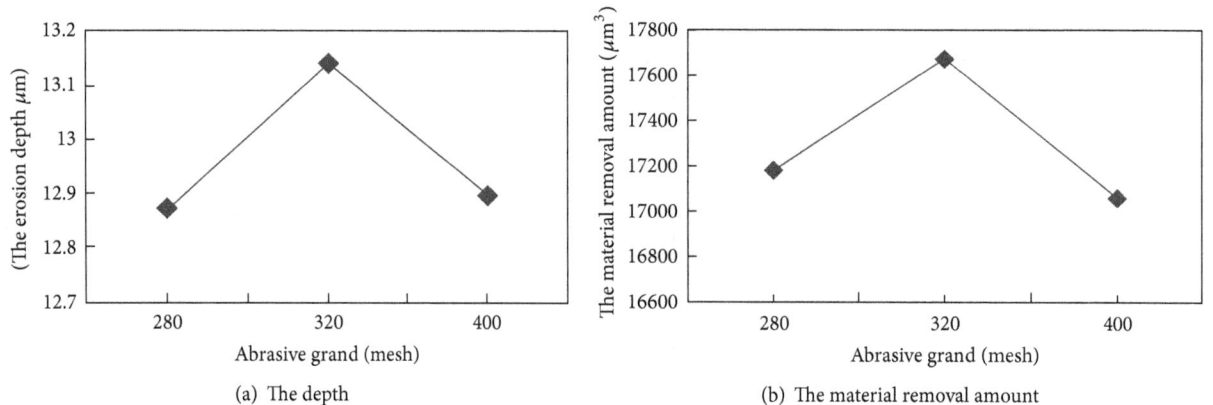

(a) The depth

(b) The material removal amount

FIGURE 7: The influence of abrasive on the depth and the material removal amount.

removal amount for the alumina ceramics machined by the ultrasonic vibration assisted abrasive water jet is obtained:

$$h = CP_s^{w_1} B^{w_2} a^{w_3} M^{w_4}$$
$$Q_1 = CP_s^{w_1} B^{w_2} a^{w_3} M^{w_4}, \tag{1}$$

where h is depth (mm), Q_1 is material removal amount (mm^3), C is coefficient which is related to the abrasive, and w_1, w_2, w_3, and w_4 are undetermined coefficients.

According to the experimental date, the MATLAB language is used to do the regression analysis [15, 16]. The coefficients of the above regression equation are shown in Table 5.

Figure 8 shows the regression residual analysis. As is shown obviously, the regression coefficients of depth and material removal amount are within the confidence interval

(95%), and it indicates that the above regression coefficients are reliable.

According to the comparison between all effects caused by different parameters, for the depth and the material removal amount, the stand-off and the water jet pressure have the biggest effect, the amplitude has a great effect, and the abrasive grain has a small effect.

6. Conclusions

This experimental study focuses on the ultrasonic vibration assisted abrasive water jet impacting the alumina ceramics. By measuring the surface morphology of alumina ceramics, the results indicate that the ultrasonic vibration improves the processing capability of abrasive water jet. Meanwhile, it also shows that both the depth and the material removal amount will increase when the ultrasonic vibration is applied.

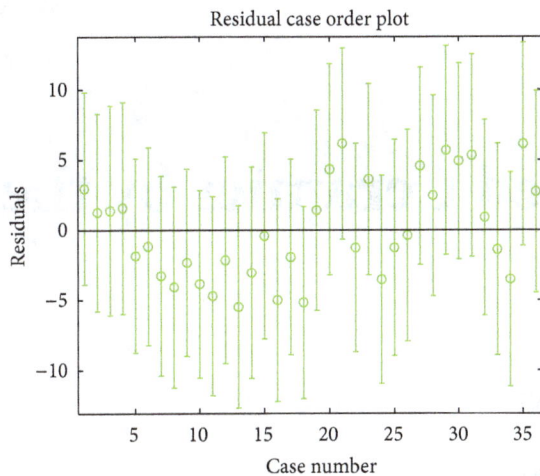

FIGURE 8: The analysis of residual.

A predicted model of depth and material removal amount is built. The results indicate that, for the depth and the material removal amount, the water jet pressure and the stand-off have the biggest effect, the amplitude has a great effect, and the abrasive grain has a small effect.

Conflicts of Interest

The authors declare that they have no conflicts of interest.

Acknowledgments

This work is financially supported by the National Natural Science Foundation of China (51405274) and Program for the Young Development of Shandong University of Technology.

References

[1] M. Hashish, "Pressure effects in Abrasive-WaterJet (AWJ) machining," *Journal of Engineering Materials and Technology*, vol. 111, no. 3, pp. 221–228, 1989.

[2] J. Wang, "Particle velocity models for ultra-high pressure abrasive waterjets," *Journal of Materials Processing Technology*, vol. 209, no. 9, pp. 4573–4577, 2009.

[3] G. L. Chahine, A. Kapahi, J.-K. Choi, and C.-T. Hsiao, "Modeling of surface cleaning by cavitation bubble dynamics and collapse," *Ultrasonics Sonochemistry*, vol. 29, pp. 528–549, 2016.

[4] G. L. Chahine, K. M. Kalumuck, and G. S. Frederick, "The use of self-resonating cavitating water jets for rock cutting," in *Proceedings of the 8th American Water Jet Conference*, 1995.

[5] S. Ghadi, K. Esmailpour, S. M. Hosseinalipour, and A. Mujumdar, "Experimental study of formation and development of coherent vortical structures in pulsed turbulent impinging jet," *Experimental Thermal and Fluid Science*, vol. 74, pp. 382–389, 2016.

[6] P. Hreha, A. Radvanská, S. Hloch, V. Peržel, G. Królczyk, and K. Monková, "Determination of vibration frequency depending on abrasive mass flow rate during abrasive water jet cutting," *The International Journal of Advanced Manufacturing Technology*, vol. 77, no. 1, pp. 763–774, 2015.

[7] V. Peržel, P. Hreha, S. Hloch, H. Tozan, and J. Valíček, "Vibration emission as a potential source of information for abrasive waterjet quality process control," *The International Journal of Advanced Manufacturing Technology*, vol. 61, no. 1, pp. 285–294, 2012.

[8] M. M. Vijay, J. Foldyna, and J. Remisz, "Ultrasonic modulation of high-speed waterjets," in *Proceedings of the International Conference on Geomechanics*, pp. 327–332, Netherlands, 1993.

[9] J. Foldyna, L. Sitek, J. Ščučka, P. Martinec, J. Valíček, and K. Páleníková, "Effects of pulsating water jet impact on aluminium surface," *Journal of Materials Processing Technology*, vol. 209, no. 20, pp. 6174–6180, 2009.

[10] J. Foldyna, L. Sitek, B. Švehla, and Š. Švehla, "Utilization of ultrasound to enhance high-speed water jet effects," *Ultrasonics Sonochemistry*, vol. 11, no. 3-4, pp. 131–137, 2004.

[11] D. Lehocka, J. Klich, J. Foldyna et al., "Copper alloys disintegration using pulsating water jet," *Measurement*, vol. 82, no. 3, pp. 375–383, 2016.

[12] S. Hloch, J. Foldyna, F. Pude et al., "Experimental in-vitro bone cements disintegration with ultrasonic pulsating water jet for revision arthroplasty," *Journal of Tehnicki Vjesnik—Technical Gazette*, vol. 22, no. 6, pp. 1609–1615, 2015.

[13] M. Zelenak, J. Foldyna, J. Scucka, S. Hloch, and Z. Riha, "Visualisation and measurement of high-speed pulsating and continuous water jets," *Measurement*, vol. 72, pp. 1–8, 2015.

[14] Z. W. Liu, *Study on Removal Mechanism of Hard Brittle Materials and Premix Micro Abrasive Waterjet Polishing Technology*, Shandong University, 2011.

[15] X. D. Yang, "Application of regression analysis method in ultra-precision machining surface roughness prediction," *Value Engineering*, vol. 33, no. 24, pp. 34-35, 2014.

[16] B. Zhao, Y. Wu, C. S. Liu, A. M. Gao, and X. S. Zhu, "The study on ductile removal mechanisms of ultrasonic vibration grinding nano-ZrO_2 ceramics," *Key Engineering Materials*, vol. 304-305, pp. 171–175, 2006.

12

Manufacture of Fine-Pored Ceramics by the Gelcasting Method

Bronisław Psiuk,[1] Anna Gerle,[1] Małgorzata Osadnik,[2] and Andrzej Śliwa[1]

[1]*Refractory Materials Division, Institute of Ceramics and Building Materials, Toszecka 99, 44-100 Gliwice, Poland*
[2]*Institute of Non-Ferrous Metals, Sowińskiego 5, 44-100 Gliwice, Poland*

Correspondence should be addressed to Bronisław Psiuk; b.psiuk@icimb.pl

Academic Editor: Fernando Lusquiños

The fine-pored materials represent a wide range of applications and searches are being continued to develop methods of their manufacturing. In the article, based on measurements on fine-grained powders of Al_2O_3, TiO_2, and SiO_2, it has been demonstrated that gelcasting can be relatively simple method of obtaining of nanoporous materials with high values of both specific surface area and open porosity. The powders were dispersed in silica sol, and the gelling initiator was NH_4Cl. The usefulness of experiment design theory for developing of fine-pored materials with high porosity and specific surface area was also shown.

1. Introduction

The classification proposed by the International Union of Pure and Applied Chemistry (IUPAC) divides pores into 3 groups, depending on their size: micropores (up to 2 nm), mesopores (2–50 nm), and macropores (above 50 nm) [1]. A commonly used term is also that of microporous materials, when the pores are in the order of micrometres (i.e., [2]). Moreover, the informal notion of nanopores conventionally includes pores having a diameter of up to 100 nm [3]. Therefore, in this article, mesoporous, microporous, or nanoporous materials, regardless of the classification, will be referred to as fine-pored materials.

Such materials represent a wide range of applications. An example of fine-pored products applications include the following: biomaterials used as medical implants [4], carriers of permeating membranes or catalysts [5], molecular sieves [6], thermoinsulating materials [7], diffusion layers in fuel cells [8], composite materials matrixes [9], and porous ceramics infiltrated with metals or polymers [10].

A good method of obtaining porous materials with a set shape is gelcasting [11–22]. Gelcasting is a combination of the traditional method of forming ceramic materials from casting slips with the polymerization reaction. The basic idea of this relatively new but intensively developing method involves making a ceramic powder suspension with an addition of a substance which is subject to gelation when influenced by an appropriate initiator [23–25]. Proportions between the suspension components and the time of suspension homogenization are so selected that they enable casting the suspension into the mould before the gelation process and, at the same time, ensure that gelation does not occur too late due to sedimentation processes (if they are not desired) [26]. One of the simplest gelling systems is that based on silica sol (dispersing agent) and NH_4Cl (gelling initiator) [27].

The aim of this article is to prove that the gelcasting method can be applied in the manufacture of fine-pored materials, using relatively simple measures. In the presented examples of obtaining fine-pored ceramic materials by the gelcasting method, the suspensions were made with silica sol as a dispersing agent, whereas dispersing powders in the suspension were compounds commonly used in the ceramic industry. Investigations were conducted for the group of powders: Al_2O_3–TiO_2–SiO_2. Since the radius of pore diameter is a fraction of the diameter of grains in a particular monofraction [28], the investigations were carried out for dusty fractions.

The proportions of powders in subsequent sets of samples were selected according to the experiment design theory for investigations into the composition-properties correlation.

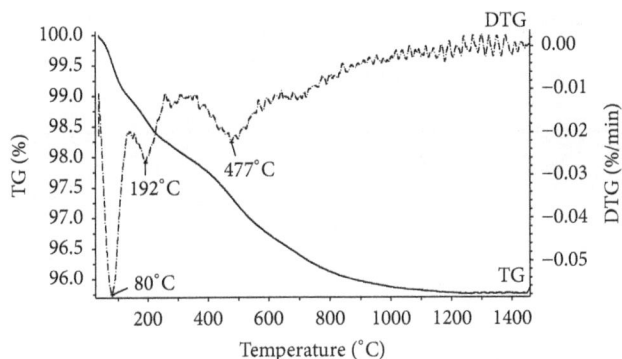

FIGURE 1: Thermogravimetric analysis of the investigated powders suspension in silica sol (gelled under NH_4Cl solution). 154.5 mg of sample was heated in the air with heating rate 3.5°C/min.

The so-called simplex-network Scheffe test for a three-component system was applied [29].

2. Manufacture of Materials

The experiments were done on a gelling set based on silica sol, in which the gelling initiator was NH_4Cl. The following ceramic powders from Evonic were used: SiO_2 Aerosil P90 (having a developed surface area of 90 m²/g), Al_2O_3 Aeroxide Alu C (100 m²/g), and TiO_2 Aeroxide P90 (90 m²/g).

Each time the sample preparation procedure was very similar: after adding a set amount of powder to the silica sol (the proportions will be given in tables in the further part of the article) the suspensions were homogenized during an appropriate time (usually half an hour) in a magnetic stirrer. Next, an appropriate amount of gelling initiator was added so that the time of sample gelling would range from several to several dozen minutes. To avoid uneven gelling (formation of lumps), the silica sol-based sets gelling initiator—NH_4Cl—was added to the suspension in the form of a water solution. A 23% solution was prepared. As the different suspensions require the use of a solution in various concentrations, the solution prepared was most frequently diluted. Table 2 does not contain the concentrations calculated each time but the proportions of additional diluting (this manner of providing information was considered better as in reality, apart from the additive concentration, also the content of water in the solution was changed). The gelled suspensions were dried slowly (for 2-3 days the temperature was gradually increased to 195°C). After drying all the samples were fired at an appropriate temperature (600°C, 1000°C, 1400°C), with a hold time of 4 h. Temperature progression was 2°C/min. The samples fired at 600°C and 1000°C were additionally held at 250°C (60 min) and 400°C (60 min). Temperatures at which the samples were additionally held were aimed at preventing the samples' cracking due to a release of a large amount of gases—they were determined on the basis of thermogravimetric investigations (Figure 1). For the samples fired at 1400°C slow heating rate was only applied.

3. Measurement Techniques

The open porosity, apparent density, and water absorption of the prepared samples were determined by the method of hydrostatic weighing in water. The crystal structure of manufactured powders was investigated by X-ray Powder Diffraction (XRD), using an X'Pert PRO MPD (CuKα) diffractometer produced by PANalytical. The specific surface area (BET method) was determined by means of a Gemini 2360 analyzer produced by Micromeritics using nitrogen as an adsorbed gas. Moreover, for selected samples measurements of pore distribution were taken with AutoPore IV9500 Mercury Porosimeter produced by Micromeritics. Photos on the investigating samples were taken with the use of high-resolution scanning electron microscope (SEM) Mira III from Tescan.

4. Results and Their Analysis

As mentioned before, the experiments were based on Scheffe test designs. Seven samples of the appropriate proportions (Table 1) were prepared in a way which enabled calculating the coefficients of the third-degree polynomial (in this case) for the three-component mixture.

For the design in question one can determine the result of composition correlation from the following formula:

$$Y = b_1 x_1 + b_2 x_2 + b_3 x_3 + b_{12} x_1 x_2 + b_{13} x_1 x_3 + b_{23} x_2 x_3 + b_{123} x_1 x_2 x_3, \tag{1}$$

where relevant coefficients are calculated from the following dependence:

$$\begin{aligned} b_1 &= y_1 \quad (\text{analogously } b_2 = y_2, \ b_3 = y_3) \\ b_{12} &= 4 y_{12} - 2 y_1 - 2 y_2 \text{ etc.} \\ b_{123} &= 27 y_{123} - 12 \left(y_{12} + y_{13} + y_{23} \right) \\ &\quad + 3 \left(y_1 + y_2 + y_3 \right). \end{aligned} \tag{2}$$

Based on initial investigations, it was found that in the case of mixture homogenized with the magnetic stirrer (for this type of devices the stirring force is relatively small) the proportion for the abovementioned oxide materials would be 9 g of oxides mixture powder per 60 g of silica sol. The first batch of samples (obtained by firing at 600°C) were based on suspensions having proportions given in Table 2.

It can be visible (Table 3) that results of apparent density, open porosity, and water absorption obtained on the basis of hydrostatic method are coherent with values of specific surface area from BET method. The highest values of porosity and specific surface area have samples with presence of Alu C powder. The selection of powders based on the experiment design theory allows observing certain general tendencies concerning the selected properties of the samples obtained, depending on the substrates' proportions. In this paper the parameter analyzed by the three-component Scheffe estimation is open porosity. The determined function of

TABLE 1: Design matrix for the third-degree polynomial for the three-component system.

Measurement number	Components and their contents in relation to the maximum amount			Result
	x_1	x_2	x_3	Y
1	1	0	0	y_1
2	0	1	0	y_2
3	0	0	1	y_3
4	0.5	0.5	0	y_{12}
5	0.5	0	0.5	y_{13}
6	0	0.5	0.5	y_{23}
7	0.33	0.33	0.33	y_{123}

TABLE 2: Compositions of suspensions for silica sol-based samples (proportion: 60 ml silica sol per 9 g of oxides mixture powder).

Sample number	Al_2O_3 Alu C [g]	SiO_2 P90 [g]	TiO_2 P90 [g]	Silica sol [ml]	23% water solution of NH_4Cl + water [ml]
1	9.0	0.0	0.0	60	1.0 + 5.0
2	0.0	9.0	0.0	60	2.0 + 4.0
3	0.0	0.0	9.0	60	6.0 + 0.0
4	4.5	4.5	0.0	60	1.5 + 4.5
5	4.5	0.0	4.5	60	1.5 + 4.5
6	0.0	4.5	4.5	60	3.5 + 1.5
7	3.0	3.0	3.0	60	2.0 + 4.0

TABLE 3: Results of selected physical properties obtained for samples having compositions contained in Table 2 (obtained by firing at 600°C).

Sample number	Apparent density [g/cm³]	Open porosity [%]	Water absorption [%]	Specific surface area [m²/g]
1	1.12	55.8	49.9	114.54
2	1.63	25.5	15.7	0.73
3	1.88	23.2	12.3	0.44
4	1.04	56.1	54.2	110.22
5	1.17	53.3	45.4	106.24
6	1.88	17.4	9.2	1.47
7	1.11	54.2	48.9	106.92

open porosity changes (using formulas (1) and (2)) has the following form:

$$Y = 55{,}8x_1 + 25{,}5x_2 + 23{,}2x_3 + 61{,}8x_1x_2 + 55{,}2x_1x_3$$
$$- 27{,}8x_2x_3 + 255{,}3x_1x_2x_3. \tag{3}$$

Transition from the coded scale of coefficients x_1, x_2, and x_3 to the natural scale, where the role of the independent variable is played by the masses of particular powders (please compare the proportion in Tables 1 and 2) and the results calculated by substituting a series of mutual proportions of the examined oxides into formula (3) allowed drawing an isoline of the samples' open porosity depending on their initial composition. An approximate course of these lines is presented in Figure 2. The estimation indicates that, for the suspension containing 9 g of the examined oxide powders, the highest porosity values in 60 ml of silica sol should be expected to be found in the following components: 50–80% of Al_2O_3, 10–40% of SiO_2, and 0–30% of TiO_2. However, the interpretation of the diagram should be taken into consideration more qualitatively. The isolines course show, primarily, that the porosity value depends mainly on Alu C content in the mixture of powders. When the content of Alu C reduces in the range from 50% to 0% the estimated values of the samples porosity decrease rapidly. On the other hand, when the content of the alumina is between 50% and 100% the estimated values of porosity become high (close to 60%) and are only slightly sensitive to the changes of powders

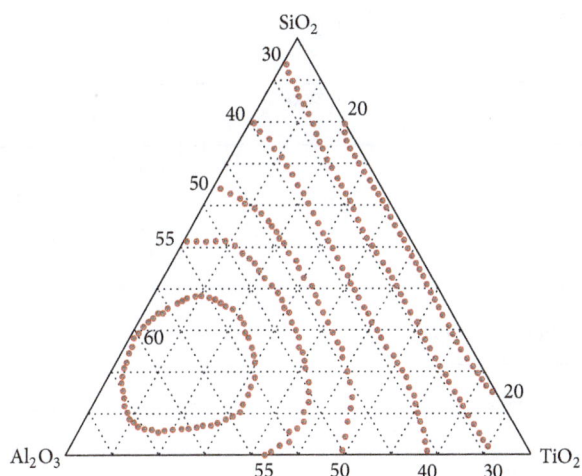

FIGURE 2: Open porosity isolines (circles) for the system from Table 2, drawn on the basis of calculations conducted based on formula (3). The dotted lines define the specified powder portion in the mixture (100% at the indicated apex of the triangle and 0% at an opposed side).

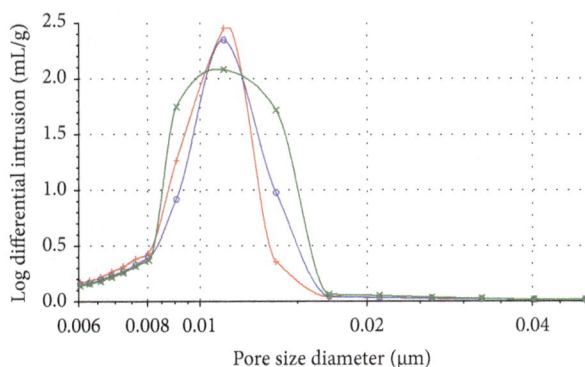

FIGURE 3: Intrusion of mercury as a function of pore diameter for samples 1 (+), 4 (×), and 5 (∘).

stoichiometry. The estimation indicates that the highest value of porosity is obtained for the systems where content of Alu C is at least 50%.

Thus it was shown that the gelcasting method allows manufacturing the materials with relatively high values of open porosity.

For samples 1, 4, and 5 pore size distribution was additionally determined and presented at Figure 3.

The results show that manufactured samples are formed as fine-pored material. The distribution of pores is similar for all three presented samples. However it should be noted that with higher content of Alu C powder the pore distribution in the sample has the most sharpness. The dominant pore size diameter is close to 10 nm.

The above results are confirmed by SEM investigations. The micrographs presented at Figure 4 show that the materials of all the samples are formed by highly packed grains with a diameter close to 100 nm. Some of the grains exceed 100 nm but there are numerous specimens with sizes of the

order of tens of nanometers. It can be visible that the range of the grains diameter in sample 4 is slightly wider than that in samples 1 and 5. Moreover the particles shape in samples 1 and 5 looks more isometric than in sample 4. These observations are in correlation with the mentioned results of the pore distributions in the investigated samples.

For the compositional range with the high porosity additionally the influence of firing temperature (600°C, 1000°C, and 1400°C) was determined. The contents of powders in suspensions and the determined results have been given in Tables 4 and 5.

As can be seen, firing at 600°C and 1000°C does not cause significant differences in the parameters of the samples obtained. They are dominated by amorphous or metastable (delta Al_2O_3) substrates; samples with TiO_2 also contain anatase and, to a lesser extent, rutile. The slightly higher sintering process is visible at 1000°C. On the other hand, the temperature of 1400°C clearly results in the material sintering, crystallization of silica phases, and mullitization of the material. In samples containing TiO_2 also the transformation of anatase into rutile is clearly visible. Thus it can be visible that temperature of sintering should be carefully selected for the system based on silica sol and the fine powders.

It should be also added that the described gelled suspensions were characterized by high drying shrinkage, reaching 25%, which makes it difficult to manufacture large-sized components, as they could crack in the drying process. The obtained small samples (order of cm) are characterized by sufficient operational strength, allowing their free movement and arrangement during potential subsequent processes. For example, owing to the porosity parameters and high surface area development, these materials can be applied as cores of porous matrixes infiltrated with polymers (including the ones maintained by capillary forces) or metals (macroscopically, smooth surfaces enable thin-walled casting and the fine-pored matrix removes the process gases).

5. Summary

The aim of measurements described in this work was to obtain materials characterized by high open porosity and small pore size by the gelcasting method.

The investigations were conducted using fine-grained powders of Al_2O_3, TiO_2, and SiO_2 with a specific surface area reaching ca 100 m^2/g. The powders were dispersed in silica sol, and the gelling initiator was NH_4Cl. Using relatively simple methods, we were able to obtain ceramics characterized by high porosity, small pore diameters, and high surface area development. The relatively high drying shrinkage causes that it is hard to obtain large-sized components from the examined systems. The materials obtained are characterized by sufficient operational strength, allowing their free movement and arrangement. The usefulness of experiment design theory for developing of fine-pored materials with high porosity and specific surface area was also shown. The simplex-network Scheffe test allows observing certain general tendencies concerning the selected properties of the samples obtained, depending on the substrates' proportions.

TABLE 4: Composition of suspensions based mainly on Al_2O_3, TiO_2, and SiO_2 for silica sol-based samples (proportion: 60 ml silica sol per 9 g of oxides mixture powder).

Sample number	Al_2O_3 Alu C [g]	TiO_2 P90 [g]	SiO_2 P90 [g]	Silica sol [ml]
A	5	0	4	60
B	5	4	0	60
C	5	2	2	60
D	9	0	0	60

TABLE 5: Results of selected physical properties for samples with compositions given in Table 4.

Sample number (firing temperature [°C])	Apparent density [g/cm^3]	Open porosity [%]	Phase composition
A 600	1.01	55.8	Amorphous phase, δ-Al_2O_3
B 600	1.09	54.5	Amorphous phase, anatase, δ-Al_2O_3, traces of rutile?
C 600	1.06	54.7	Amorphous phase, anatase, δ-Al_2O_3, traces of rutile?
D 600	1.04	56.2	Amorphous phase, δ-Al_2O_3
A 1000	1.28	45.6	Amorphous phase, δ-Al_2O_3
B 1000	1.4	44.2	Amorphous phase, anatase, rutile, δ-Al_2O_3
C 1000	1.40	42.9	Amorphous phase, anatase, rutile, δ-Al_2O_3
D 1000	1.20	52.4	Amorphous phase, δ-Al_2O_3
A 1400	2.38	2.3	Cristobalite, mullite, quartz, tridymite
B 1400	2.52	1.9	Cristobalite, mullite, rutile, tridymite
C 1400	2.45	1.6	Cristobalite, mullite, rutile, tridymite
D 1400	2.48	1.9	Cristobalite, mullite, quartz, tridymite

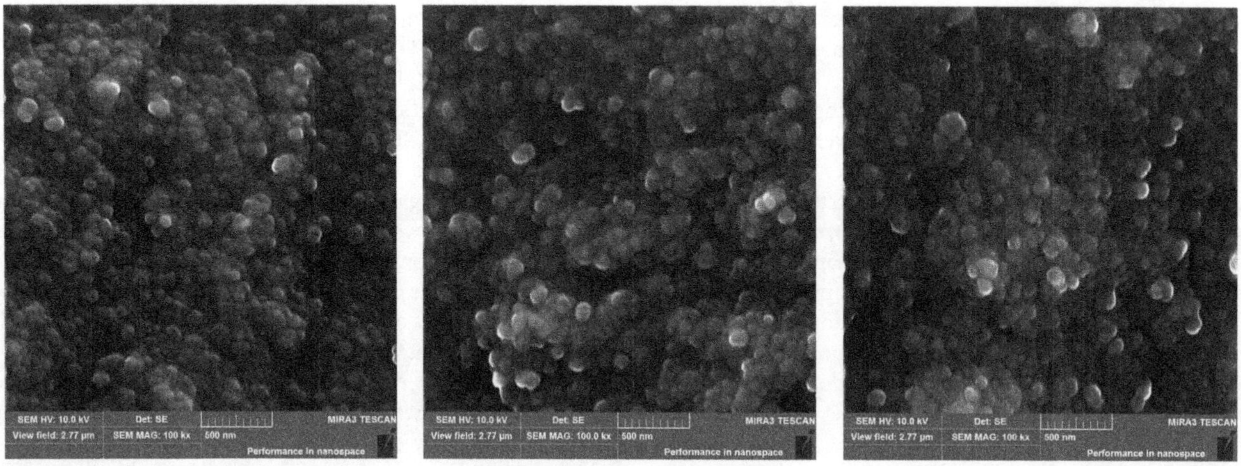

FIGURE 4: SEM micrographs for samples 1 (a), 4 (b), and 5 (c).

Conflicts of Interest

The authors declare that they have no conflicts of interest.

References

[1] J. Rouquerol, D. Avnir, C. W. Fairbridge et al., "Recommendation for the characterization of porous solids," *Pure and Applied Chemistry*, vol. 66, no. 8, pp. 1739–1758, 1994.

[2] Glossary of Soil Science Terms, 2008, https://www.soils.org/files/publications/soils-glossary/table-2.pdf.

[3] Z. Sarbak, "Materiały mikro- i nanoporowate (Micro- and nano-porous materials)," *LAB Laboratoria, Aparatura, Badania*, vol. 9, pp. 6–12, 2004 (Polish).

[4] A. V. Shevchenko, E. V. Dudnik, A. K. Ruban, Z. A. Zaitseva, and L. M. Lopato, "Functional graded materials based on ZrO_2 and Al_2O_3. Production methods," *Powder Metallurgy and Metal Ceramics*, vol. 42, no. 3-4, pp. 145–153, 2003.

[5] W. Kotowski, "Catalytic membrane reactors," *Ecological Chemistry and Engineering S*, vol. 15, pp. 43–60, 2008 (Polish).

[6] P. Kozyra, "Badanie materiałów mikroporowatych metodą spektroskopii IR," 2017 http://www2.chemia.uj.edu.pl/~kozyra/dydaktyka/inzynier/pory.pdf.

[7] D. R. Clarke and S. R. Phillpot, "Thermal barrier coating materials," *Materials Today*, vol. 8, no. 6, pp. 22–29, 2005.

[8] R. Włodarczyk, A. Dudek, R. Kobyłecki, and S. Bis, "Characteristic of fuel cells in aspect of theirs productions and applications" (Polish), 2009, http://wis.pol.lublin.pl/kongres3/tom2/30.pdf.

[9] M. Szafran, G. Rokicki, E. Bobryk, and A. Lamenta, "Ceramics-polymer composites based on porous ceramic material with porosity gradient," *Kompozyty*, vol. 4, pp. 231–236, 2004 (Polish).

[10] M. Szafran, G. Rokicki, W. Lipiec, K. Konopka, and K. Kurzydłowski, "Porous ceramic infiltrated by metals and polymers," *Kompozyty*, vol. 2, pp. 313–317, 2002 (Polish).

[11] G. Meng, H. Wang, W. Zheng, and X. Liu, "Preparation of porous ceramics by gelcasting approach," *Materials Letters*, vol. 45, no. 3, pp. 224–227, 2000.

[12] P. Wiecinska and M. Bachonko, "Processing of porous ceramics from highly concentrated suspensions by foaming, in situ polymerization and burn-out of polylactide fibers," *Ceramics International*, vol. 42, no. 13, pp. 15057–15057, 2016.

[13] C. Bartuli, E. Bemporad, J. M. Tulliani, J. Tirillò, G. Pulci, and M. Sebastiani, "Mechanical properties of cellular ceramics obtained by gel casting: characterization and modeling," *Journal of the European Ceramic Society*, vol. 29, no. 14, pp. 2979–2989, 2009.

[14] M. Potoczek, A. Zima, Z. Paszkiewicz, and A. Ślósarczyk, "Manufacturing of highly porous calcium phosphate bioceramics via gel-casting using agarose," *Ceramics International*, vol. 35, no. 6, pp. 2249–2254, 2009.

[15] A. G. A. Coombes and J. D. Heckman, "Gel casting of resorbable polymers. 1. Processing and applications," *Biomaterials*, vol. 13, no. 4, pp. 217–224, 1992.

[16] H. Varma, S. P. Vijayan, and S. S. Babu, "Transparent hydroxyapatite ceramics through gelcasting and low-temperature sintering," *Journal of the American Ceramic Society*, vol. 85, no. 2, pp. 493–495, 2002.

[17] M. Takahashi, R. L. Menchavez, M. Fuji, and H. Takegami, "Opportunities of porous ceramics fabricated by gelcasting in mitigating environmental issues," *Journal of the European Ceramic Society*, vol. 29, no. 5, pp. 823–828, 2009.

[18] Y.-F. Liu, X.-Q. Liu, H. Wei, and G.-Y. Meng, "Porous mullite ceramics from national clay produced by gelcasting," *Ceramics International*, vol. 27, no. 1, pp. 1–7, 2001.

[19] H. T. Wang, X. Q. Liu, and G. Y. Meng, "Porous α-Al2O3 ceramics prepared by gelcasting," *Materials Research Bulletin*, vol. 32, no. 12, pp. 1705–1712, 1997.

[20] F.-Z. Zhang, T. Kato, M. Fuji, and M. Takahashi, "Gelcasting fabrication of porous ceramics using a continuous process," *Journal of the European Ceramic Society*, vol. 26, no. 4-5, pp. 667–671, 2006.

[21] Y. Gu, X. Liu, G. Meng, and D. Peng, "Porous YSZ ceramics by water-based gelcasting," *Ceramics International*, vol. 25, no. 8, pp. 705–709, 1999.

[22] F.-H. Liu, "Manufacturing porous multi-channel ceramics by laser gelling," *Ceramics International*, vol. 37, no. 7, pp. 2789–2794, 2011.

[23] O. O. Omatete, M. A. Janney, and R. A. Strehlow, "Gelcasting a new ceramic forming process," *American Ceramic Society Bulletin*, vol. 70, no. 10, pp. 1641–1649, 1991.

[24] M. Szafran, P. Bednarek, and D. Jach, "Moulding of ceramic materials by the gelcasting method," *Materiały Ceramiczne/Ceramic Materials*, vol. 59, no. 1, pp. 17–25, 2007 (Polish).

[25] J. Yang, J. Yu, and Y. Huang, "Recent developments in gelcasting of ceramics," *Journal of The European Ceramic Society*, vol. 31, no. 14, pp. 2569–2591, 2011.

[26] B. Psiuk, P. Wiecinska, B. Lipowska, E. Pietrzak, and J. Podwórny, "Impulse excitation Technique IET as a non-destructive method for determining changes during the gelcasting process," *Ceramics International*, vol. 42, no. 3, pp. 3989–3996, 2016.

[27] D. Kong, H. Yang, S. Wei, D. Li, and J. Wang, "Gel-casting without de-airing process using silica sol as a binder," *Ceramics International*, vol. 33, no. 2, pp. 133–139, 2007.

[28] M. Szafran and P. Wiśniewski, "Effect of the bonding ceramic material on the size of pores in porous ceramic materials," *Colloids and Surfaces A: Physicochemical and Engineering Aspects*, vol. 179, no. 1-3, pp. 201–208, 2001.

[29] S. Ł. Achnazarowa and W. W. Kafarow, *Optymalizacja eksperymentu w chemii i technologii chemicznej (Optimizing Experiment in Chemistry And Chemical Technology)*, Wydawnictwa Naukowo-Techniczne, Warszawa, Poland, 1982.

13

Molding of Polymeric Composite Reinforced with Glass Fiber and Ceramic Inserts: Mathematical Modeling and Simulation

Túlio R. N. Porto ⓘ, Wanderley F. A. Júnior, Antonio G. B. De Lima ⓘ, Wanderson M. P. B. De Lima, and Hallyson G. G. M. Lima

Department of Mechanical Engineering, Federal University of Campina Grande, Campina Grande 58429-900, Brazil

Correspondence should be addressed to Túlio R. N. Porto; trnporto@gmail.com

Academic Editor: Giorgio Pia

This work provides a numerical study of a polymer composite manufacturing by using liquid composite material molding. Simulation of resin flow into a porous media comprising fiber perform (reinforcement) inserted in a mold with preallocated ceramic inserts has been performed, using the Ansys FLUENT® software. Results of resin volumetric fraction, stream lines and pressure distribution inside the mold, and mass flow rate (inlet and outlet gates) of the resin, as a function of filling time, have been presented and discussed. Results show that the number of inserts affects the filling time whereas the distance between them has no influence in a process.

1. Introduction

Composite is a material originating from the joining of two or more different component materials, exhibiting specific properties that are not observed in the constituent phases, acting separately [1].

Most of composites are described as having a matrix phase and a dispersed phase (reinforcement). The matrix is the continuous phase, which is responsible to transfer the stresses exerted on the part. Polymeric matrices are the most used in the composite material production.

Polymers are called thermosets when, after cure reaction, they have a molecular structure that does not allow process reversibility and thermoplastics, when the molecular structure, presented before cure, can be achieved as the polymer is remelted. The epoxy resin is a thermosetting polymer that has better thermal, electrical, and mechanical properties than the other polymer matrices, working in the range of −60 to 180°C [2].

The reinforcement of the composite may consist of continuously disposed, discontinuous, aligned or random fibers, particles, with different sizes and structures, either laminated or in sandwich-panels [1, 3].

Due to the ability of the composites to merge different properties in a single material, they have various applications. In the aeronautical and naval sector, there is a great demand for materials that present lightness associated with high mechanical resistance. Thus, the application of polymer composites to structural components of aircraft and vessels is constantly increasing. Currently, internal, external, wing ribs, landing gear doors, flaps, structural parts, and aircraft leading edges are being made of composite materials consisting of continuous fibers in a thermoset polymer matrix [4]. Vessels are able to associate low weight and maintenance cost with high wear resistance when they are manufactured by composites reinforced. In the scope of the armored structures, composites are processed from the union of ceramic inserts and reinforcing fibers, imbedded in a polymer matrix. This composition promotes to the armored equipment, both structural properties, sufficient to support high loads, as well as protection against ballistic attacks and reduction on the equipment weight [5–8].

The fiber-reinforced composites constitute a porous medium. Interconnected voids between the fibers are distributed along the preform through which the resin flows during the filling mold. The pore geometry and its

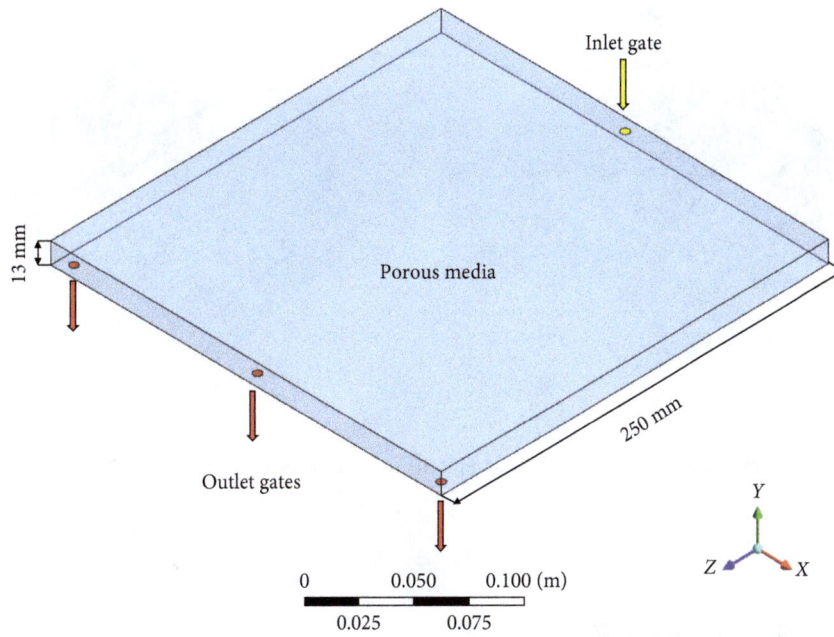

FIGURE 1: The geometry of the physical problem studied.

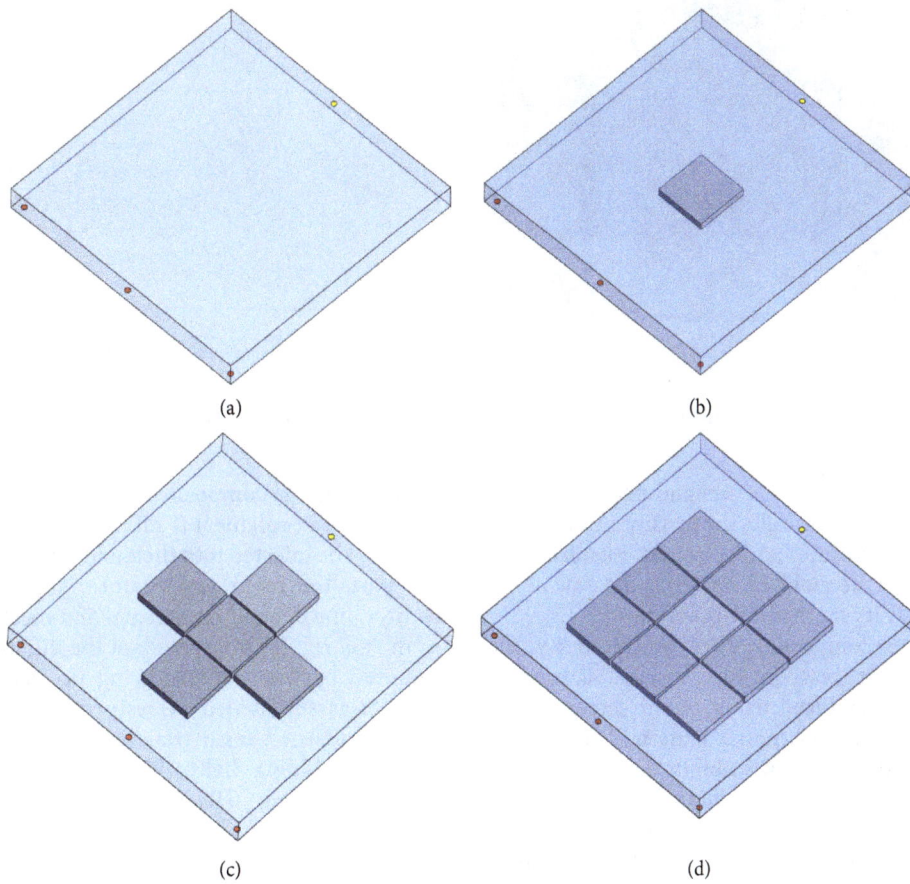

(a)

(b)

(c)

(d)

FIGURE 2: Continued.

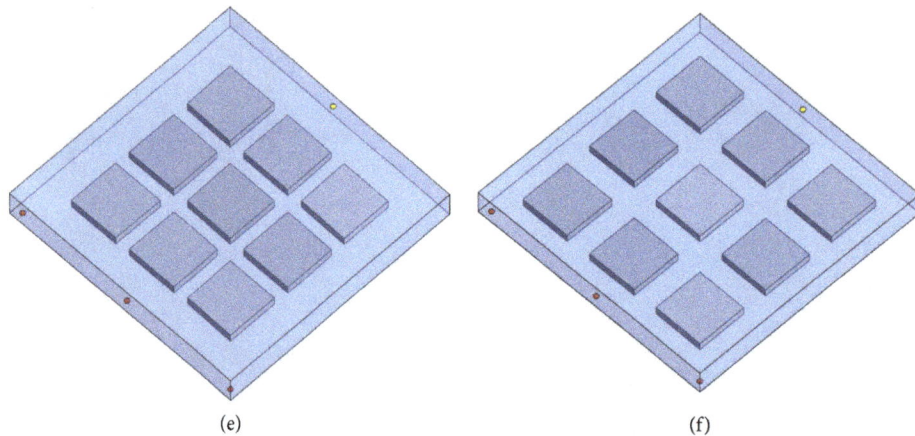

(e)   (f)

FIGURE 2: Reinforcement configuration with 0 (a), 1 (b), 5 (c), and 9 (d) ceramic inserts 2 mm apart and in the case of 9 inserts, with 15 mm apart (e) and 25 mm apart (f).

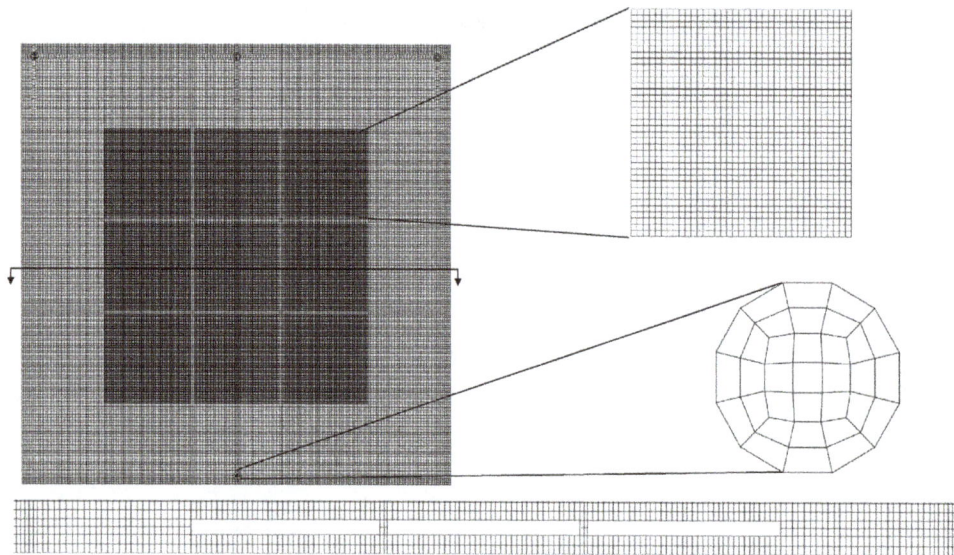

FIGURE 3: Numerical mesh and details of inserts and outlet gate.

distribution, described by the porosity and permeability of the medium, measured empirically, the density and viscosity of the resin, and the pressures and velocities, established at the inlet and outlet of the mold, characterize the flow in the porous medium and its mechanical properties [9–13].

Historically, the composites were produced by the manual lamination process. In this process, laminated layers are manually produced using reinforcements previously impregnated with the matrix material [14]. Due to the high costs and the operator's ability dependence, associated with the manual lamination process, developments in manufacturing with the aim of promoting reductions in process costs and in the number of failures are increasing. For this, the technique of liquid composite molding (MLC) was developed. In this technique, liquid resin is injected into a closed mold, where the reinforcement is preallocated, under specific conditions of pressure, velocity, and temperature. As the resin fills the mold cavity, impregnating the preform and cure process

are finished, the composite is produced [15]. In general, the liquid phase (polymer) is mixed with a chemical hardener before to be injected into the mold. In addition, the shape of the part, the mold temperature, and the maximum injection time depend on thermal and mechanical properties of the matrix. In order to adapt the different specifications, required by the manufacturing projects, the MLC technique was subdivided in resin transfer molding (RTM), vacuum-assisted resin transfer molding (VARTM), resin transfer molding light (RTML), and compressed resin transfer molding (CRTM) [16–19].

During the molding process, a multiphase flow is observed along the mold. A resin-air interface develops inside the mold as the liquid resin is injected into the mold and air, which previously occupied the entire porous volume is repelled through the strategically projected outlets. In the course of the process, the interface extension decreases as air is removed from the mold and phase mixture regions are observed to promote dispersed air bubbles in the polymer matrix. After the curing process, these bubbles correspond to voids in the solidified part, which give rise to cracks in the composite, drastically reducing the composite mechanical strength. Thus, the transient control of the fluid flow and the pressure distributions along the molded part are directly associated with the quality of the composite. These fluid dynamics parameters are dependent on the composite geometry, the reinforcement and the resin physical properties, the distribution of injectors and air outlets, and their relative pressures and velocity that potentiate the flow [16].

Experiments have shown that undesired results, such as voids inside the mold, high injection times, mold deformations, displacement and deformation of the preform, in locations near the injection points, are influenced by the flow behaviour. In this way, an accurate knowledge of the transport phenomena associated with the flow type to currently control the process is necessary in order to have a structure with desired mechanical properties.

In the search for optimize industrial processes, numerical simulation is characterized as a fundamental tool. From the discretized physical conservation equations, the fluid dynamics aspects of processes such as composite molding can be described and analyzed in order to predict the best operating conditions and the physical implications of the mold and reinforcement geometries used. The prior knowledge of how the flow occurs significantly reduces the logistical venture and costs that would have been used to obtain experimental results. As the computational tool is validated, numerical results are enabling to guiding the industrial processes.

In this sense, the present work carries out a computational study of the resin transfer molding process during the manufacturing of a composite composed by an epoxy resin polymer matrix, reinforced with glass fiber and ceramic inserts. As contribution in this research area, the description of the multiphase flow fronts, rate relative results of the resin and air volume fractions over the time, the resin mass flow at the mold inlet, and the pressures and velocities distributions inside the mold are numerically obtained. In addition, the numerical study allowed to evaluate the influence of the number of inserts and the distance between them in the mold filling process.

2. Mathematical Modeling

2.1. The Physical Problem and the Geometry. As shown in Figure 1, the physical problem consists of the filling, by injection of resin, of a square closed mold, with 250 mm of side and 13 mm thickness; the mold is composed by three air outlets with 5 mm diameter, distributed symmetrically on one side of the mold lower surface, 112.5 mm apart; one inlet on the opposite side, on the mold upper surface, with 5 mm diameter. A preform of glass fibers is allocated in the mold cavity, forming a porous medium, through which the resin passes during filling process.

The influence of the square-base prismatic ceramic inserts, with 50 mm of side and 4 mm of thickness, placed between the fibers and centered on the mold thickness, on the multiphase flow behaviour, will be analyzed. Figure 2

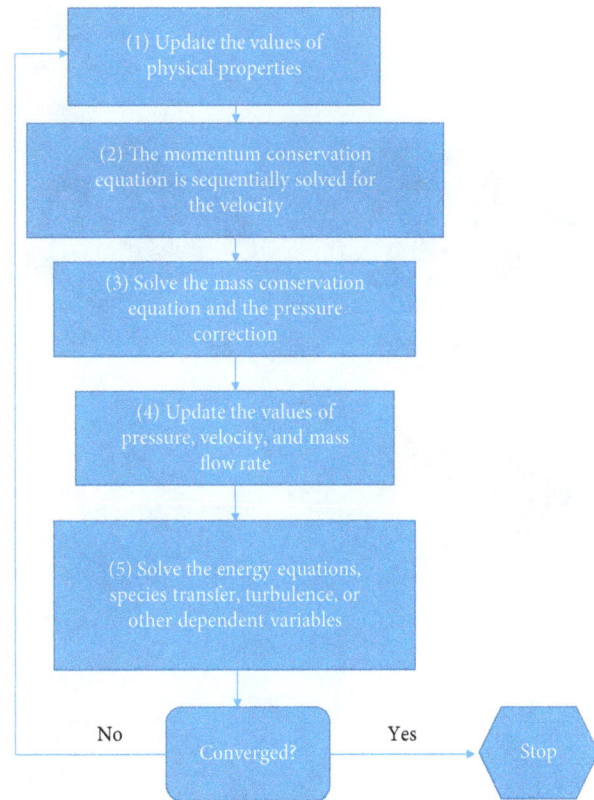

FIGURE 4: SIMPLE pressure-based solution method.

TABLE 1: Simulated cases.

Case	Number of inserts	Distances between the inserts (mm)
1	9	25
2	0	—
3	1	2
4	5	2
5	9	2
6	9	15

TABLE 2: The two physical properties of the fluid phases at 25°C and 1 atm.

Properties	Viscosity (Pa·s)	Density (kg/m³)
Air	$1.7984e^{-05}$	1.2257
Resin	0.35	1200

illustrates different configurations of composite reinforcements, relative to the number of inserts and the distances between them. The cases referring to the 0, 1, 5, and 9 inserts, with a distance of 2 mm between them, are presented in Figures 2(a)–2(d), respectively. The cases referring to the variations in the distance between the inserts of 15 and 25 mm for 9 inserts are shown, respectively, in Figures 2(e) and 2(f).

2.2. The Mathematical Model. Among the mathematical models used to describe the multiphase flow, the volume of

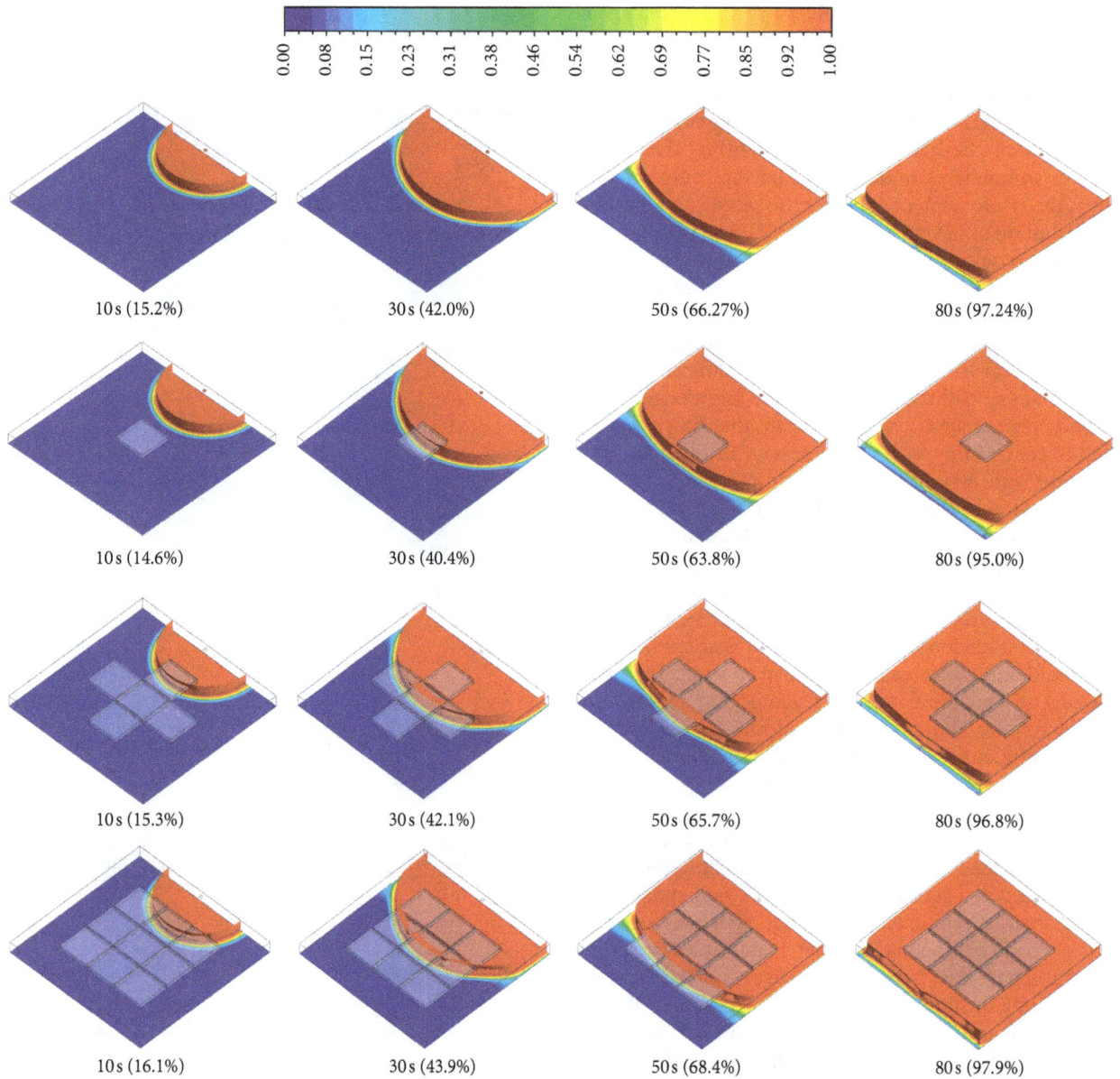

FIGURE 5: Distribution of the resin volumetric fraction inside the mold containing 0, 1, 5 and 9 insert spaced 2 mm apart, at different instants of process.

Fluid (VOF) model is suitable for composite molding processes. This model, through the solution of the conservation equations of mass and momentum, is able to trace the interfaces of a flow composed of two or more immiscible fluids with great accuracy. As the mold is filled with liquid resin, this model is able to specify the air and resin flow rates and the interface location between these fluids. In this way, it is possible to identify the resin front in the mold and the regions with air bubbles during the injection process.

2.2.1. Mass Conservation. In the mass conservation equation (Equation (1)), the transient, convective, and source terms are related to the volumetric fraction of the secondary phase "q." Making the calculation for the secondary phases present

in the flow, by volumetric completion, the conservative values for the primary phase "p" are obtained:

$$\frac{1}{\rho_q}\left[\frac{\partial\left(\alpha_q\rho_q\right)}{\partial t} + \nabla\cdot\left(\alpha_q\rho_q\,\vec{v_q}\right) = S_{\alpha q} + \sum_{p=1}^{n}\left(\dot{m}_{pq} - \dot{m}_{qp}\right)\right],$$

(1)

where t is the time variable, $S_{\alpha q}$ is the source term relative to "q" phase and its respective volumetric fraction α, and this term is related to the generation or mass sink of phase q. The terms $\dot{m}_{pq}$ and $\dot{m}_{qp}$ are related to mass transfer from phase "p" to phase "q" and from phase "q" to phase "p," respectively, which occurs when there is phase transformation associated to this physical problem.

2.2.2. Momentum Conservation.

From the momentum equation solution, the velocity and pressure fields, described along the flow, are obtained, which depend on the instant of analysis, the interactions of the fluid with the geometric structure, as well as the external and internal surface and field forces, to the control volume.

$$\frac{\partial}{\partial t}(\rho \vec{v}) + \nabla \cdot (\rho \vec{v} \vec{v}) = -\nabla p$$

$$+ \nabla \cdot \left[\mu\left(\nabla \vec{v} + \nabla \vec{v}^{T}\right) + \rho \vec{g} + \vec{F}\right]. \tag{2}$$

In Equation (2), $\vec{F}$ is the external force vector, $\vec{g}$ is the gravity acceleration vector, and p is the pressure distributed on the volume control surface. The physical properties inserted in Equation (2) correspond to the mixture of phases, in each control volume. These are measured by a weight average between the constituent phase properties of the flow, described in Equation (3), for example, to density:

$$\rho = \alpha_q \rho_q + (1 - \alpha_q)\rho_p. \tag{3}$$

In this way, the properties and appropriate variables are weighted in each region of the multiphase flow.

2.2.3. The Porous Media Flow.

The term of momentum conservation equation, relative to the pressure variation in a physical domain, for fluid flowing through an isotropic porous medium is described by Darcy's empirical law (Equation (4)). In this analysis, the Reynolds number (Equation (5)), a dimensionless value, describing the relationships between the inertial forces in relation to the viscous forces, is calculated as a function of the pore size or the particle diameter (d_p) that constitutes the porous medium, and consequently, their value is very small [12].

$$\nabla P = -\frac{\mu \vec{v}_s}{K}, \tag{4}$$

$$Re = \frac{\rho v_s d_p}{\mu}. \tag{5}$$

In the Equations (4) and (5), μ is the viscosity of the fluid, $\vec{v}_s$ is the fluid superficial velocity vector, which is determined by considering the porous medium as continuous and neglecting the effects of the geometric details of the porous medium structure, and K is the porous media permeability and can be calculated through Equation (6). This parameter is dependent on the porosity ϕ and the parameter "a" which is related to porous geometric microstructure:

$$K = \frac{\phi^3 d_p^2}{a(1 - \phi)^2}. \tag{6}$$

The approach applied in the Ansys Fluent 15.0 software does not treat the porous medium through its geometric variations, but as described in Equation (7), considering the resistance to the flow (σ) that the porous medium represents

$$\sigma = \frac{1}{K}. \tag{7}$$

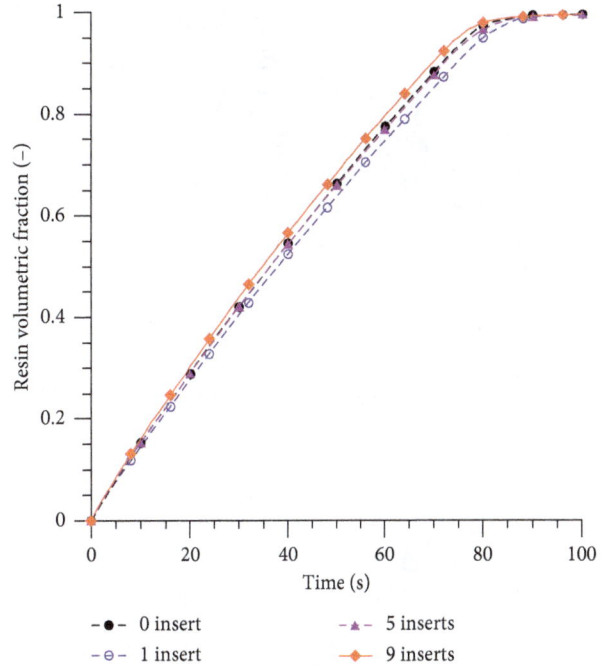

FIGURE 6: Total volumetric fraction of resin as a function of the filling time (inserts spaced apart with 2 mm).

2.3. Numerical Solution

2.3.1. Numerical Mesh.

The molds were described in different meshes, the most refined containing 297,942 elements. The number of elements was enough to describe the process, with considerable precision and physical coherence, presented in the results. Figure 3 illustrates one grid used is this work, produced by the ANSYS ICEM® CDF 15.0 software, with particular emphasis for the inserts surfaces and resin inlet, which was the same used to and air outlets. To reduce the number of elements and consequently the simulation time, the mesh was developed considering only the surfaces of the inserts. No element was treated within its volumes.

2.3.2. Spatial Discretization.

In order to obtain the numerical solution of the conservation equations presented, discretization of the governing equations is necessary, that work within differential limits. Therefore, we transform the partial differential equations in algebraic equations, defined for the finite three-dimensional limits of the numerical mesh. Taking Φ as a representative of the transport variables, velocity or pressure, referring to conservation equations of mass and momentum, we have the discretization of the transport general equation, as follows:

$$\frac{\partial(\rho \Phi)}{\partial t}V + \sum_{f=1}^{N_f} \rho_f \Phi_f \vec{v}_f \vec{A}_f = \sum_{f=1}^{N_f} \Gamma_{\Phi_f} \nabla \Phi_f \cdot \vec{A}_f + S_\Phi V, \tag{8}$$

where Γ_Φ is the general term relative to the characteristic physical properties of each conservation equation, $\vec{A}_f$ is the area vector corresponding to the faces (f) of the control volume, S_Φ is the source term per unit of its volume (V), and N_f

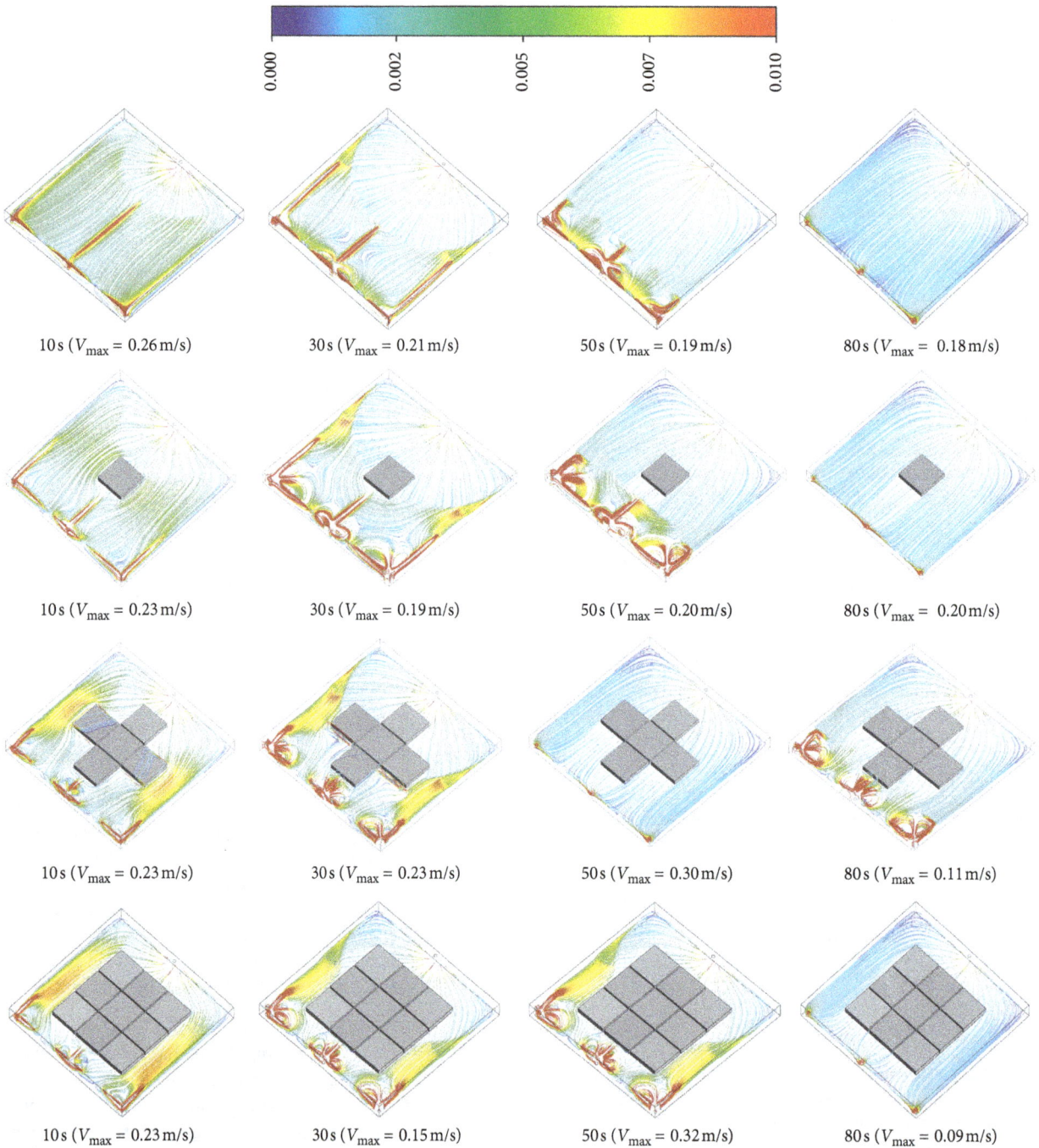

FIGURE 7: Streamlines with velocities (m/s) inside the molds containing 0, 1, 5, and 9 inserts spaced apart with 2 mm, at different instants of process.

is the number of faces in which the conservative equation is analyzed. Knowing the value of the variable Φ in the centroids $(c_0, c_1, \ldots, c_n)$ of the cells and their values (Φ_f) on each cell faces, solutions of conservation equations can be obtained along the physical space. Herein, we use the least squares cell-based method [20] to determine the gradient $\nabla\Phi$, the quadratic upwind implicit differential convective kinematics (QUICK) method [21] for discretization of the volumetric fraction (continuity), the second-order upwind method [22] for discretization of the continuity equation and the PRESTO [23] for pressure numerical model discretization.

2.3.3. Temporal Discretization. Taking the temporal differential of the general transport equation

$$\frac{\partial \Phi}{\partial t} = F(\Phi), \tag{9}$$

where $F(\Phi)$ incorporates all discretized spatial variables.

Using the implicit method [23] for temporal discretization, Equation (9) can be rewritten as follows:

$$\frac{\Phi^{n+1} - \Phi^n}{\Delta t} = F\left(\Phi^{n+1}\right), \tag{10}$$

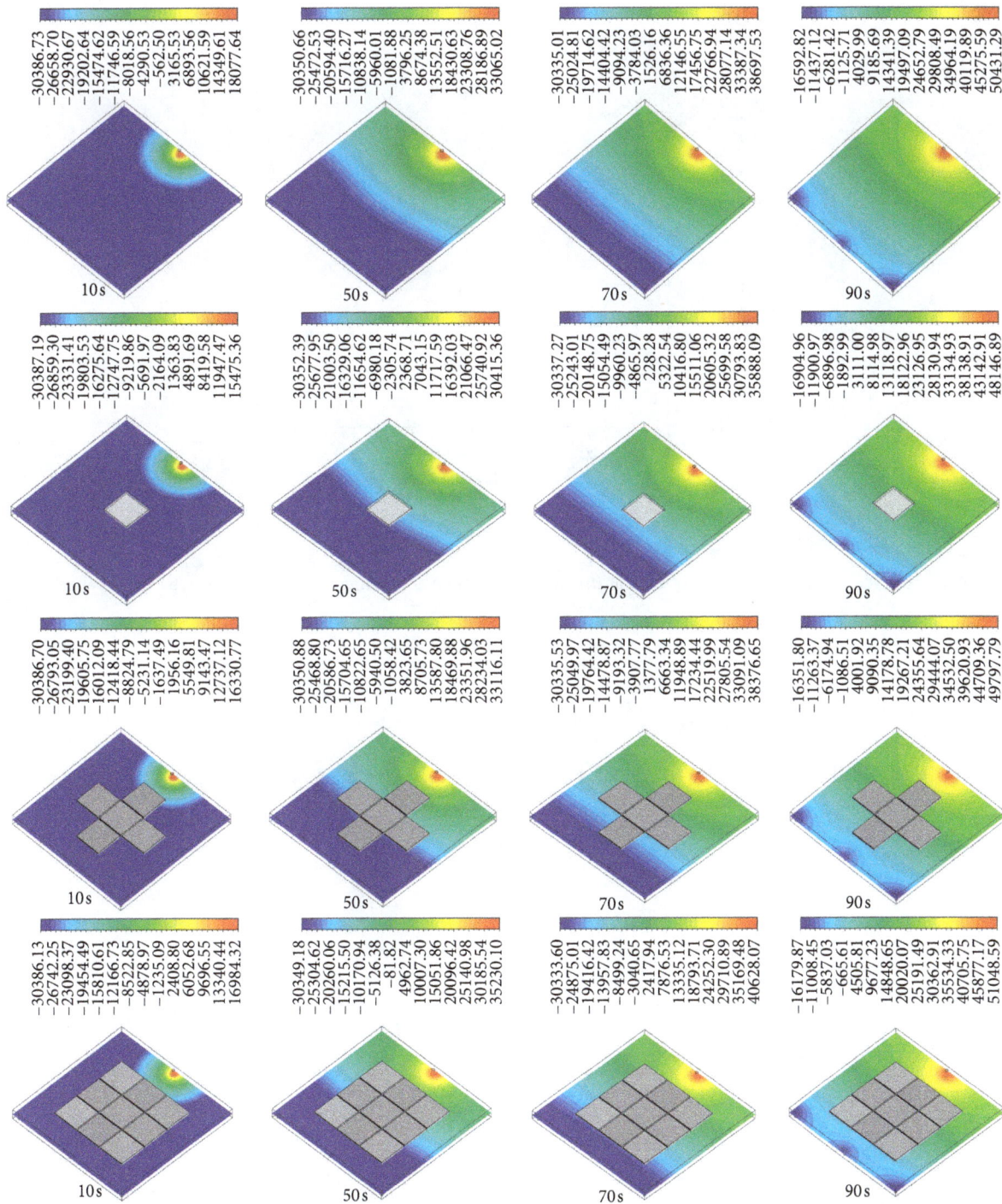

FIGURE 8: Pressure fields, measured (Pa), in the plane $y = 6.5\,\text{mm}$ for different filling times of the molds containing 0, 1, 5, and 9 inserts, spaced apart with 2 mm, for different instants of time.

where Φ^{n+1} refers to the value of the variable Φ, in the central mesh position of the cell, in the later time step and Φ^{n}, in the current time step. The discretized variables in relation to space are treated in future or later time, $F(\Phi^{n+1})$. Thus, in conjunction of specified initial and boundary conditions, numerical iterations are performed at each time step, and the transient behaviour of conservation equations is obtained.

2.3.4. Solution Procedure. Concerning the fluid flow problem treated here, a pressure-based solution method SIMPLE (semi-implicit method for pressure linked equation) by [24], which is traditionally used in incompressible flow simulations that develop at low velocities, is applied. This method is described by the following solution steps presented in Figure 4.

From the solution of conservation equations, it is possible to describe how the flow occurs during the molding

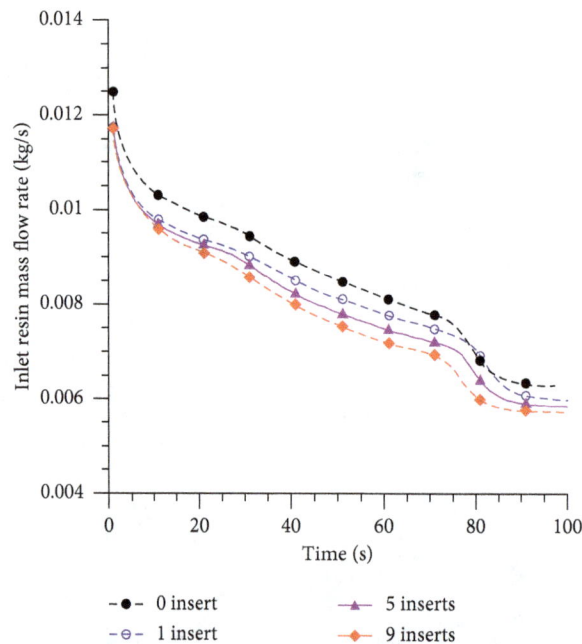

FIGURE 9: Resin mass flow rate as a function of time at the inlet of the mold containing 0, 1, 5, and 9 inserts, spaced apart with 2 mm.

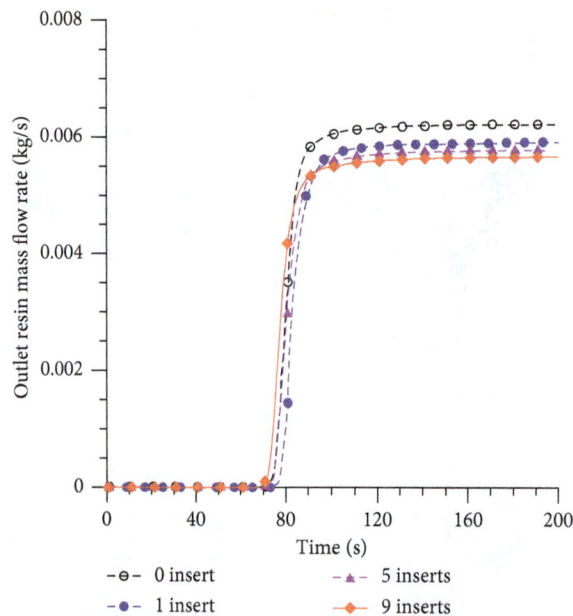

FIGURE 10: Resin mass flow rates in the outputs of the mold containing 0, 1, 5, and 9 inserts, spaced apart by 2 mm.

process of a fiber-reinforced polymer composite. With this procedure, the pressure fields, velocities, and volume fractions are obtained in each time step time.

2.3.5. Simulated Cases. The 6 cases, described in Table 1, were simulated in a commercial computational program (Ansys FLUENT® 15.0), and the results are presented in the next section. In all situations, the mesh with 297942 elements, a time step of 0.05 seconds with the maximum number of 100 iterations per time step, and convergence criterion for all variables as 10^{-5} were used.

2.3.6. Initial and Boundary Conditions and Fluid Properties. For the solution of the cases, the following initial and boundary conditions were applied:

(a) Prescribed pressure at the resin inlet: 101325 Pa (normal to inlet).

(b) Vacuum pressure at the air outlet: -30397 Pa (normal to outlets).

(c) Resin volumetric fraction in the inlet: 1.

(d) Resin volumetric fraction in the outlet: 0.

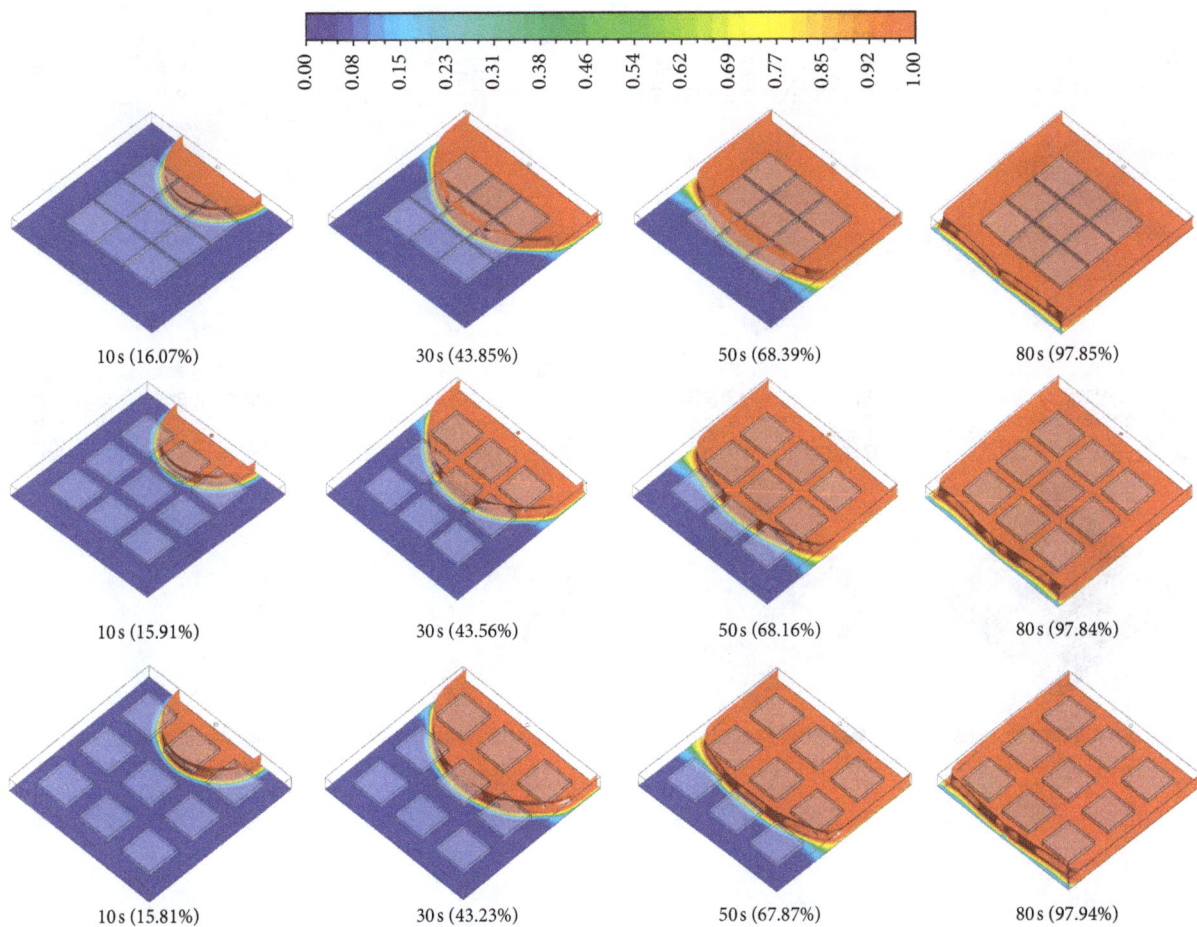

FIGURE 11: Distribution of the resin volumetric fraction inside the molds containing 9 inserts spaced apart with 2, 15, and 25 mm at different instants of process.

(e) Local gravitational acceleration: 9.81 m/s^2.

(f) Permeability of the porous medium in the x, y, and z directions: $3.89 \cdot 10^{-9}$ m^2. The porous medium structure consists of superimposed glass fibers layers within the mold cavity. Considering that the fibers are traced and that the distances between the layers are small, the permeability in the porous medium is treated as isotropic.

(g) Homogenous porosity: 0.82.

(h) Nonslip condition on the mold and insert walls and top, lateral, and bottom surfaces; this condition can be found as no casting occurs in the mold and the ceramic inserts are impermeable. The roughness of the surfaces was not considered because the flow inside the mold was laminar.

(i) Constant process temperature: 27°C.

In this research, it was considered that the mold is fully filled at room temperature before the curing process development. Under these conditions, the viscosity and temperature variations due to the resin hardening process are neglected. Thus, the fluid properties used on the simulation are described in Table 2.

3. Results and Discussion

In this research, fluid flow in porous media, with emphasis to polymer composite reinforced with fiber and ceramic inserts, is analyzed. Figure 5 illustrates the resin flow fronts at different times. These results show the contours of resin volumetric fraction on $y = 0$ mm plane and the resin-air interface traced throughout the mold volume during filling. From the analyses of this figure, we can see that the porous medium resists to the resin flow into the mold, but this resistance is overcome, due to the high vacuum pressure condition imposed, allowing the mold to be completely filled at 345 s of processing. From Figure 6, it can be seen that, within 80 s, the resin filling rate occurs in intensified way, causing the mold to be filled more than 90% of its capacity. After this period, the resin reaches the air outlets and is expelled from the mold, slowly loading the trapped air fractions, identified by the existence of resin-air interface within the mold, which is reduced in size between $t = 200$ s and $t = 345$ s of process.

The variation of the insert number in a fixed mold volume promotes two effects: (a) resistance to the flow due to the insert barriers; and (b) reduction in the useful cross-sectional area through which the fluid flows, which under small pressure variations promotes the increase in fluid velocity, and thus, the resin advances faster in to the mold.

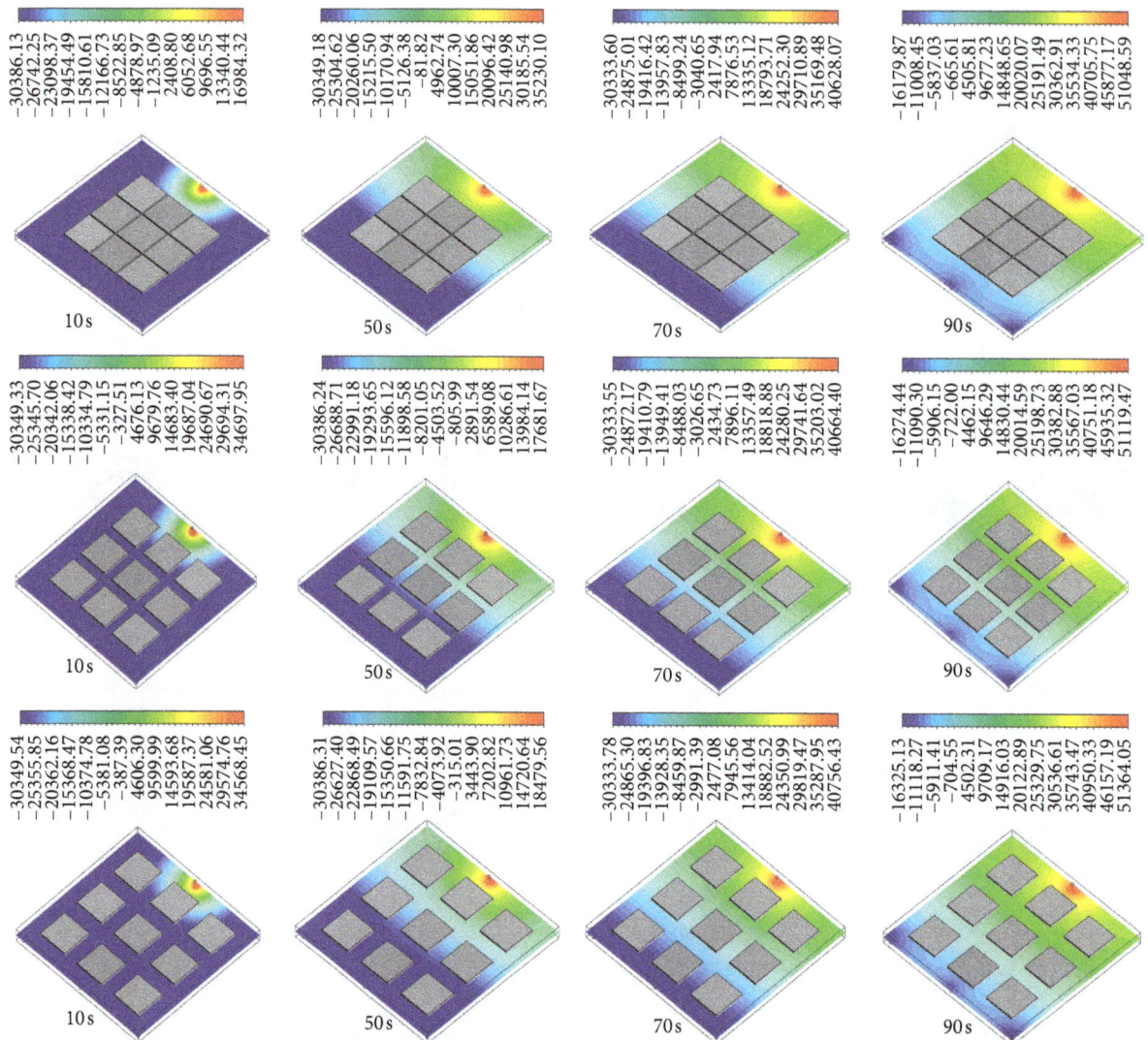

FIGURE 12: Pressure field in the $y = 6.5$ mm plane of the molds containing 9 inserts, spaced apart with 2, 15, and 25 mm at different instants of process.

Qualitatively in Figure 5 and quantitatively in Figure 6, it can be seen that inserts application strongly influences the resin advancement in the interval between 10 and 90 s, a period ahead resin flow takes to pass inserts, centralized in the mold. Thus, it can be observed in Figures 5 and 6, for the case with one insert, the resin velocity is reduced because of the flow resistance effect overriding the geometry effect. For five inserts, there is an increase in the geometric reduction effect compared with the resistance effect, making the 5-insert curve present a higher filling velocity than the case with 1 insert and closing the case with 0 insert. In this sense, when applying nine inserts in the mold cavity, the greatest filling velocity is observed. The number in parenthesis corresponds to percentage of resin into the mold at different process times.

The streamlines describe the intensity, orientation, and direction of a specific flow. Figure 7 shows the results obtained in this research. In order to distinguish the regions of the plane with different levels of flow intensity, the velocities' magnitude

was fixed in a range between 0 and 0.01 m/s, described in the single legend of the figure. In these streamlines, regions with velocities' magnitude higher than the 0.01 m/s are not distinguished. However, the maximum velocity at each time, which is above the described range, is presented next to each figure. Thus, the velocity differences between the mold regions, which lie within the given range, indicate important physical characteristics of the flow. In 1 s of process, a streamline structure intensifies towards the mold outputs. This flow comes from the air reaction to the resin injection at the inlet, which occurs from the set pressure conditions at the inlet and outlet. Then, follows the resin advancement from the mold inlet at considerably lower velocities due to its higher viscosity and density. In parallel to that, recirculation of air regions are performed around the mold outputs due to the narrow space that air is confined as the resin advances on the mold. When the mold volume is about 90% filled with the resin, the recirculation regions are extinguished, which occurs because the resin reaches the outlets in this period.

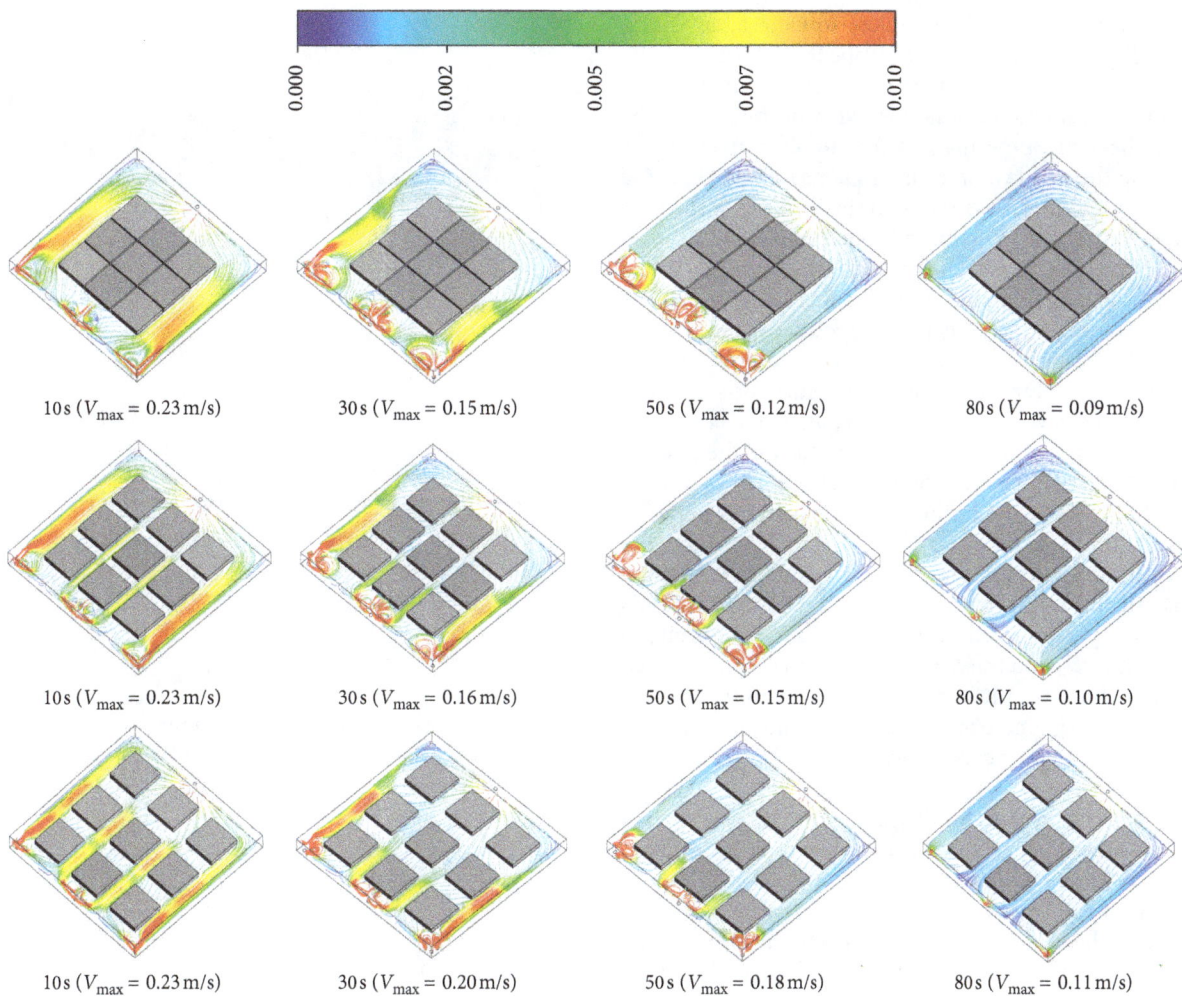

FIGURE 13: Streamlines with velocities (m/s) inside the molds containing 9 inserts and spaced apart at 2, 15, and 25 mm at different instants of process.

With the ceramic insertion structures in the composites, it is observed that, as the number of inserts is increased, the streamlines are affected by the reduction effects on the cross section area of the mold and the flow resistance. An increase in the recirculation intensity is observed as the inserts are added into the mold until 50 s of processing. Already at 80 s, the velocities of the resin flow are becoming smaller, due to the resistance effect coming from the inserts overlapping at that moment.

Figure 8 illustrates the pressure distribution inside the mold at different moments of molding process. In this figure, the pressure distribution along the central plane of the mold is observed as the mold is being filled by the resin. Initially, the effect of the vacuum pressure (−30386.73 Pa), established at the outputs and homogeneously distributed throughout mold, causes the entering of the resin. It is found that the reddish-toned contours advance in the mold over time, as far as the filling occurs. The advancement of the higher-pressure contours becomes linear ($t > 10$ s) with a small distance from the inlet, due to the effect of the flow resistance increasing, already mentioned. Upon reaching 90 s of processing, the variation of amount of resin into the mold, has been

relatively low and, thus, pressure contour profile remains constant until the full-filling the mold. Pressure conditions with small variations are observed when inserts are applied in the mold, which can be observed on the pressure contours legends. This pressure conditions associated with the reductions in the flow areas, due to de presence of inserts into the mold, promoting the different filling times in for each described case.

Resin injection into the mold with constant pressure promotes a reduction in resin flow rate at the mold inlet over time as shown in Figure 9. This is due to the fact that the resistance to scaling increases as the volume of resin, which needs to be moved inside the mold, increases. Because of the low viscosity and density, the air does not offer a high injection resistance, and therefore, the resin flow in the mold inlet at the initial times is relatively high. Subsequently, it decreases, due to the addition of resin, with high viscosity and density, in the porous cavity. This falling rate period occurs, until reaching the permanent regime, around 80 s, when the amount of resin present in the mold is almost constant, as can be observed comparing Figures 9 and 10. In the process, it possible to observe that the resin mass flow rates in the inlet and outlet are close to 0.006 kg/s. Quantitatively, Figures 9 and 10 also describe the influence of the ceramic structures application on the resin mass flow rates at the mold inlet and outlet along the filling process. It will be seen for the same time that, as the number of inserts is increased in the mold volume, the resin mass flow rate at the mold inlet is reduced. This is due to the increase in the flow resistance imposed by impermeable ceramic structures. Figure 10 illustrates that the resin achieves the mold outputs more rapidly as the number of inserts increases, due to the geometric reduction of the porous volume inside mold, under few variations on the pressure conditions, as the inserts number is varied. In the same figure, it is also observed that the levels at which the resin mass flow rates remain constant at the outputs are greater for the smaller number of inserts inside the mold.

Figure 11 shows the influence of the distance between the inserts in the mold filling time. It is noted that, as the distance between the inserts is reduced, there is a small increase in the resin volumetric fraction at the same instant of time as compared with previous case. This is due to the reducing effect on the cross section area through which multiphase flow occurs, under pressure conditions not sensitive to the geometric variations, as shown in Figure 12, thus increasing the fluid velocity in these narrower regions.

Figure 13 shows that, for the same time, as the distances between the inserts are increased, an increase in the maximum velocity, measured in the plane $y = 6.5$ mm, is observed. What can be seen in streamline structures are small increases on the flow intensity, especially in the initial times, 10 and 30 s, these variations do not influence the filling process because the mass flow rate at the inlet and outlets has no changes as the distances between the inserts are varied, as described in Figures 14 and 15.

4. Conclusions

In the work, fluid flow in porous media has been studied with particular reference to resin flow in a glass fiber preform. From the presented results, the following can be concluded:

The multiphase VOF model, used by Ansys FLUENT® software, is suitable for studying the resin transfer molding process.

The adaptation of the mesh, placing voids instead of the solid ceramic inserts, promoted a reduction in the computational time that does not interfere in the results, since the fluid dynamics process is affected only by the surfaces of the inserts.

Ceramic inserts influence the flow behaviour during the filling of the mold. An increase in the mold filling velocity was verified, as the number of inserts is increased.

The resin mass flow rate at the mold inlet is reduced as it is being filled by resin. In about 80 s of processing, the resin touches the mold outlets and the resin flow in this region is increased until reaching the value of the inlet resin mass flow rate. Both the resin fluxes at the mold inlet and at the mold

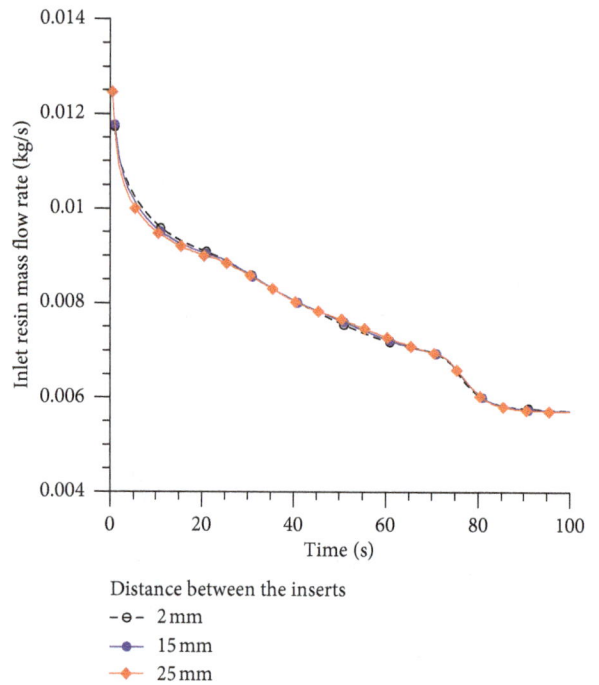

FIGURE 14: Mass flow rates at the mold inlet with the mold containing 9 inserts spaced apart with 2, 15, and 25 mm at different instants of process.

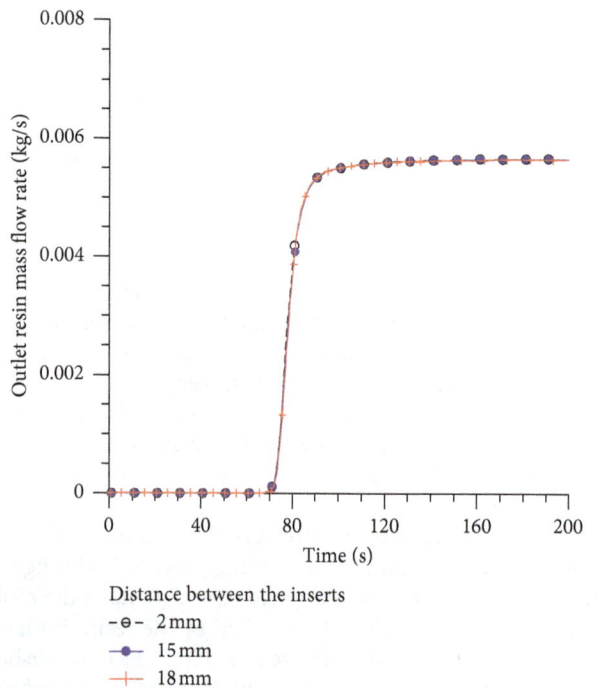

FIGURE 15: Mass flow rates at the mold outlets with the mold containing 9 inserts spaced apart with 2, 15, and 25 mm at different instants of process.

outlet are inversely proportional, affected as the insert number is increased and are not affected considerably when the distances between them are varied.

Conflicts of Interest

The authors declare that there are no conflicts of interest regarding the publication of this paper.

Acknowledgments

The authors are grateful to CNPq, CAPES, and FINEP (Brazilian Research Agencies) for financial support and to the authors cited in the text who helped in the improvement of this work.

References

[1] W. D. Callister Jr. and D. G. Rethwisch, *Materials Science and Engineering: An Introduction*, John Wiley & Sons, New Jersey, NY, USA, 8th edition, 2010.

[2] S. Tsanzalis, P. Karapappas, A. Vavouliotis et al., "Enhancement of the mechanical performance of an epoxy resin and fiber reinforced epoxy resin composites by the introduction of CNF and PZT particles at the microscale," *Composites Part A: Applied Science and Manufacturing*, vol. 38, no. 4, pp. 1076–1081, 2007.

[3] V. V. Vasilie and E. V. Morozov, *Mechanics and Analysis of Composite Materials*, Elsevier Science, Amsterdam, Netherlands, 1st edition, 2001.

[4] M. C. Rezende and E. C. Botelho, "The use of structural composites in the aerospace industry," *Polímeros*, vol. 10, no. 2, pp. e4–e10, 2000.

[5] Q. Wang, Z. Chen, and Z. Chen, "Design and characteristics of hybrid composite armor subjected to projectile impact," *Materials & Design*, vol. 46, pp. 634–639, 2013.

[6] U. K. Vaidyac, C. Ulven, and H. Ricks, "Acoustic impact evaluation of ballistic damage in VARTM composites," in *Proceedings of Ninth International Congress on Acoustics and Vibration (IIAV)*, pp. 1–8, Kampala, Uganda, July 2002.

[7] C. H. Lee, C. W. Kim, S. U. Yang, and B. M. Ku, "A development of integral composite structure for the ramp of infantry fighting vehicle," in *Proceedings of 23rd International Symposium on Ballistics*, pp. 1–6, Tarragona, Spain, April 2007.

[8] B. K. Fink, "Performance metrics composite integral armor," *Journal of Thermoplastic Composite Materials*, vol. 13, no. 1, pp. 417–431, 2000.

[9] F. A. L. Dullien, "Capillary and viscous effects in porous media," in *Handbook of Porous Media*, K. Vafai, Ed., pp. 53–112, Marcel Dekker, Inc., NewYork, NY, USA, 2000.

[10] C. K. Chen and S. W. Hsiao, "Transport phenomena in enclosed porous cavities," in *Transport Phenomena in Porous Media*, D. B. Inghan and I. Pop, Eds., pp. 40–65, Elsevier Science Ltd., Amsterdam, Netherlands, 1998.

[11] C. M. R. Franco and A. G. B. Lima, "Intermittent drying of porous media: a review," *Diffusion Foundations*, vol. 7, pp. 1–13, 2016.

[12] C. T. Hsu, "Dynamic modeling of convective heat transfer in porous media," in *Handbook of Porous Media*, K. Vafai, Ed., pp. 40–79, Taylor & Francis Group, Didcot, UK, 2005.

[13] M. A. Al-Nimr and M. K. Alkan, "Basic fluid flow problems in porous media," *Journal of Porous Media*, vol. 3, no. 1, pp. 45–59, 2000.

[14] S. Mall, D. W. Katwyk, R. L. Bolick, A. D. Kelkar, and D. C. Davis, "Tension–compression fatigue behaviour of a H-VARTM manufactured unnotched and notched carbon/epoxy composite," *Composite Structures*, vol. 90, no. 2, pp. 201–207, 2009.

[15] S. G. Advani and K. Hsiao, "Transport phenomena in liquid composites molding processes and their roles in process control and optimization," in *Porous Media*, K. Vafai, Ed., pp. 573–606, Taylor & Francis Group, Didcot, UK, 2005.

[16] A. Shojaei, S. R. Ghaffarian, and S. M. H. Karimian, "Modeling and simulation approaches in the resin transfer molding process: a review," *Polymer Composites*, vol. 24, no. 4, pp. 525–544, 2003.

[17] M. K. Yoon, J. Baidoo, J. W. Gillespie Jr., and D. Heider, "Vacuum assisted resin transfer molding (VARTM) process incorporating gravitational effects: a closed-form solution," *Journal of Composite Materials*, vol. 39, no. 24, pp. 2227–2241, 2005.

[18] A. C. Garay, V. Heck, A. J. Zattera, J. A. Souza, and S. C. Amico, "Influence of calcium carbonate on RTM and RTM light processing and properties of molded composites," *Journal of Reinforced Plastic and Composites*, vol. 30, no. 14, pp. 1213–1221, 2011.

[19] R. Chaudhary, M. Pick, O. Geiger, and D. Schimidt, "Compression RTM- a new process for manufacturing high volume continuous fiber reinforced composites," in *Proceedings of 5th International CFK-Valley State Convention*, Stade, Germany, June 2011.

[20] W. K. Anderson and D. L. Bonhaus, "An implicit upwind algorithm for computing turbulent flows on unstructured grids," *Computers & Fluids*, vol. 23, no. 1, pp. 1–21, 1994.

[21] B. P. Leonard and S. Mokhtary, "Ultra-Sharp nonoscillatory convection schemes for high-speed steady multidimensional flow," NASA Technical Memorandum 102568, ICOMP-90-12, NASA, Washington, DC, USA, 1990.

[22] T. J. Barth and D. C. Jespersen, "The design and application of upwind schemes on unstructured meshes," in *Proceedings of 27th Aerospace Sciences Meeting*, Reno, NV, USA, January 1989.

[23] Ansys, *Fluent-Theory Manual*, Ansys, Inc., Canonsburg, PA, USA, 2015.

[24] S. V. Patankar and D. B. Spalding, "Computer analysis of the three-dimensional flow and heat transfer in a steam generator," *Forschung im Ingenieurwesen*, vol. 44, no. 2, pp. 47–52, 1978.

14

Properties of Ceramic Substrate Materials for High-Temperature Pressure Sensors for Operation above 1000°C

YanJie Guo (iD),[1,2] **Fei Lu,**[1,2] **Lei Zhang,**[1,2] **HeLei Dong,**[1,2] **QiuLin Tan** (iD),[1,2] **and JiJun Xiong**[1,2]

[1]*Key Laboratory of Instrumentation Science and Dynamic Measurement, Ministry of Education, North University of China, Tai Yuan 030051, China*
[2]*Science and Technology on Electronic Test and Measurement Laboratory, North University of China, Tai Yuan 030051, China*

Correspondence should be addressed to QiuLin Tan; tanqiulin@nuc.edu.cn

Academic Editor: Marco Cannas

In order to identify suitable substrate materials for sue in high-temperature pressure sensors that can operate above 1000°C, the high-temperature properties of four high-performance ceramics (99% pure Al_2O_3 ($99Al_2O_3$), 97% pure Al_2O_3 ($97Al_2O_3$), sapphire, and ZrO_2) were investigated. Three-point bend testing was used to measure the flexural strengths and flexural moduli of these ceramics, and transient laser emission was used to measure their thermal conductivities. The samples were prepared by hot-press sintering: plates with the dimensions of $3.5 \times 5 \times 50 \, mm^3$ for the bend testing and rods of $\varphi 12.5 \times 1.5 \, mm^3$ for the thermal conductivity measurements. Curves showing the dependence of flexural strength, flexural modulus, and thermal conductivity on temperature were obtained. The results show that the flexural strength and thermal conductivity of sapphire are much greater than those of the other ceramics tested. Thus, we conclude that sapphire is the most appropriate of these materials for use in high-temperature pressure sensors for operation at up to 1000°C.

1. Introduction

Recently, high-temperature sensors have attracted much attention due to their excellent performance in harsh environments, such as high temperatures, high pressures, and corrosive atmospheres, and when subjected to strong mechanical shocks. For example, pressure monitoring in aircraft engines is of great importance for the correct functioning of aircraft. Currently, high-temperature pressure sensors that have been studied in-depth include silicon on insulator (SOI) high-temperature pressure sensors [1, 2], SiC high-temperature pressure sensors [3], silicon-sapphire pressure sensors [4], fiber optic pressure sensors [5], surface acoustic wave (SAW) pressure sensors [6], resonance circuit (LC circuit) pressure sensors [7–9], and microwave-scattering resonance pressure sensors [10]. Among these high-temperature pressure sensors, the LC circuit pressure sensors based on high-temperature cofired ceramic (HTCC) alumina substrates can operate at temperatures above 800°C. In the literature [11], ceramic-based temperature sensors

have shown to operate above 1000°C. By considering the various substrate materials of these high-temperature sensors, it can be concluded that the base materials of the high-temperature sensor must maintain stable performance in harsh environments. Four common high-temperature ceramics, $99Al_2O_3$, $97Al_2O_3$, sapphire and ZrO_2 have good oxidation and corrosion resistance, even in harsh environments. Since the materials processing technology for these ceramics is very mature, it is anticipated that it will be possible to use them to prepare high-temperature sensors that can operate above 1000°C. Thus, we have chosen these four refractory ceramics for this study.

The mechanical and thermal properties of refractory ceramics are of great importance for their use as the substrate materials for high-temperature pressure sensors. Any variation in the temperature can significantly affect the sensor performance, particularly under contact conditions, where stress levels are especially high. There are several methods which can be used to assess mechanical properties. Three-point bend testing has been used to measure the strength of composite ceramics [11–14]; while the Hertz indentation test has been used to measure the stress-strain properties of alumina and zirconia ceramics [15]. For the Hertz indentation test, three conditions are required: firstly, the deformation of the contact surface should be low; secondly, the contact surface must be oval; and, thirdly, the contact objects should behave as elastic half-spaces. Only if all three of these conditions are satisfied, the measurements can be regarded as Hertz contacts and the test can be valid. For the three-point bend method, the loading is relativity simple and the required sample dimensions are also small; however, due to the highly localized loading, the sample does not experience uniform force. This may result in defects in some parts of the sample escaping detecting and affect the accuracy of the measurement. Since the samples in this study are relatively uniform and the defect distribution is homogeneous, the authors chose to use three-point bend testing to investigate the mechanical properties of the ceramics.

For the measurement of thermal properties (such as thermal conductivity), commonly used tests include steady-state methods and unsteady-state methods. Steady-state methods include the heat flow meter method and the hot plate method; unsteady-state methods include the hot wire method and transient laser emission method. Steady-state methods are limited to measuring longitudinal thermal conductivity, and the effective temperature ranges are limited. Additionally, these methods are primarily suited to low thermal conductivity materials and thermal insulation materials. In a previous paper [16], three methods for thermal conductivity measurement were presented, and each method had its own advantages and disadvantages. In another study [17], a transient short hot wire technique was developed for the simultaneous measurement of thermal conductivity and thermal diffusivity; the effects of the thermophysical properties and the size of the hot wire, insulation coating, and samples were investigated. In the hot wire method, it is necessary to insert the hot wire into the sample before testing, and a larger sample size is required. Taking into account that the Mohs hardness of these four kinds of ceramics is relatively large, this method is not very convenient to measure these ceramic samples in this study. Laser flash methods, such as transient laser emission, have been employed to measure the thermal properties of ceramics and metals and may also be suitable for measuring the thermal properties of polymers [18–21]. The advantage of laser flash methods is that the sample can be small, the detection speed is high, and the effective temperature range is wide. During the test, only the relative temperature is measured, obviating the need to calibrate the instrument for absolute temperature measurements. Because of these advantages, the authors chose the transient laser emission method to measure the thermal conductivity in this study.

By measuring the mechanical and thermal properties of high-temperature ceramics over a temperature range from 25°C to 1500°C, we can assess their suitability for use in the preparation of sensor devices for ultra-high-temperature environments.

2. Determination of the Test Parameters

Owing to the operational mechanism of pressure sensors, the magnitude of the flexural modulus of the substrate directly affects the sensitivity of the pressure sensor. For resonant pressure sensors, when a given amount of pressure is applied to the surface of the sensor, a structure with a larger flexural modulus will produce a stronger signal. Further, the magnitude of the flexural strength determines the maximum measurement range of the sensor, and a pressure sensor with a lager flexural strength will be able to measure higher pressures.

Thermal conductivity is another important property of temperature-resistant materials; this is defined as the energy transferred per unit cross-sectional area per unit time when the vertical temperature gradient is 1°C/m. In high-temperature pressure sensors, the magnitude of the thermal conductivity directly affects the response time of the sensor and the accuracy of the measurement. The greater the thermal conductivity of the substrate, the smaller the temperature difference between the sensor and its environment, resulting in shorter response times and higher measurement accuracies.

3. Test Principles and Equipment

3.1. Flexural Test. In this study, three-point bend testing was used to measure the flexural strengths and moduli of the ceramics at different temperatures; this involves applying a load between two points of the sample until the sample is crushed. The measurements were carried out using WKM-2200, which is developed by Ukraine Strength Research Institute, shown in Figure 1(a). The schematic of this system is shown in Figure 2(a). Using a tungsten-rhenium thermocouple and the heating wire, the temperature around the test specimen can be measured and controlled. The force applied to the specimen is adjusted via the motor, which is controlled by the motor driver and computer. The load applied to the specimen is recorded by the pressure sensor, and the load-displacement curves were recorded, as shown in Figure 2(b). The samples, namely, $99Al_2O_3$, $97Al_2O_3$, sapphire, and ZrO_2, with the dimensions of $3.5 \times 5 \times 50 \, mm^3$, prepared by hot-press sintering, are shown in Figure 3(a).

During the measurement, the temperature is raised to the measurement temperature and held for 10 min to stabilize. Then, the load is applied gradually so that the specimen undergoes bending deformation. In the initial deformation stage, the test sample will bend elastically. From the intercept of this linear region of the curve, and (1) and (2), the flexural modulus E_b of the sample can be obtained:

$$E_b = \frac{L_s^3}{48I} \cdot \frac{\Delta F}{\Delta d},\tag{1}$$

$$I = \frac{1}{12} \cdot (bh^3).\tag{2}$$

In (1) and (2), L_s is the span of the three-point bending test, which refers to the distance between two support points below the specimen; and b and h are the width and height of the specimens, respectively. The maximum bending force, F_b

(a)   (b)

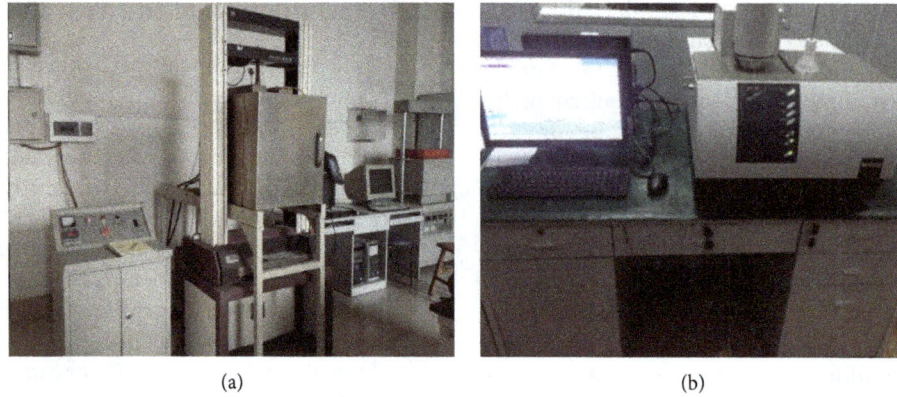

FIGURE 1: Photos of the test equipment: (a) WKM-2200 three-point bend testing device and (b) LFA-427 laser thermal conductivity meter.

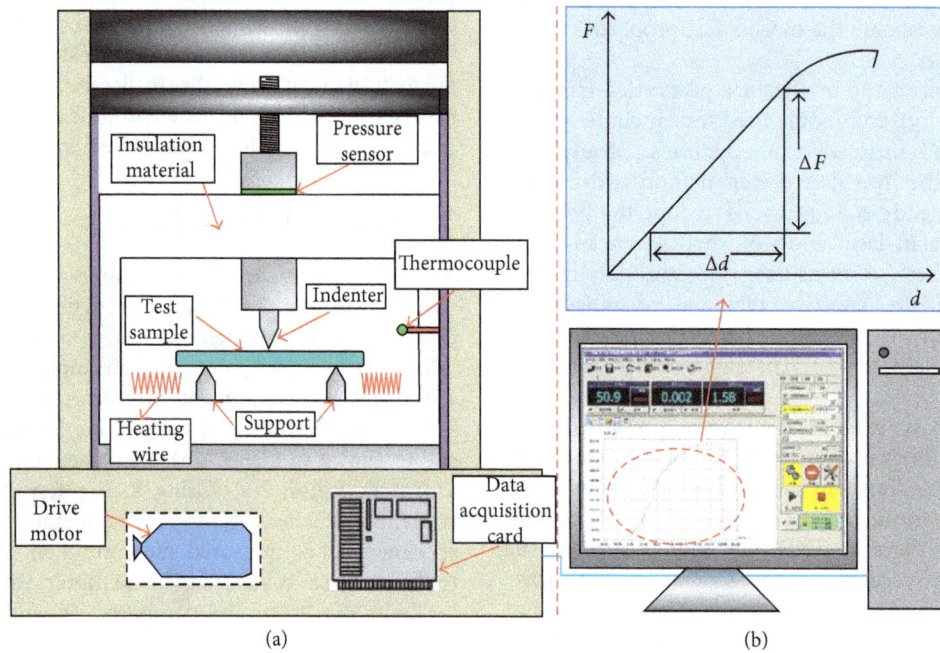

(a)   (b)

FIGURE 2: The schematic of the flexural test system: (a) three-point bend test structure and (b) software processing section.

is extracted from the measured load-deflection curve; the bending strength σ_b can be calculated using the following equations:

$$\sigma_b = \frac{F_b L_s}{4W},\qquad(3)$$

$$W = \frac{1}{6}\cdot bh^2.\qquad(4)$$

3.2. Thermal Conductivity Test.

For the thermal conductivity measurements, an LFA-427 (Netzsch) laser thermal conductivity meter were used, shown in Figure 1(b). The samples for the thermal conductivity measurements had the dimensions of $\varphi 12.5 \times 1.5 \text{ mm}^3$ and were prepared by hot-press sintering, as shown in Figure 3(b). LFA-427 comprises a laser, a vacuum heating furnace, and temperature measurement, and data acquisition and processing units.

According to the definition of thermal conductivity, the relationship between thermal conductivity k, thermal diffusivity α, specific heat capacity C, and material density ρ can be expressed as

$$k = \alpha \cdot C \cdot \rho.\qquad(5)$$

In order to obtain the thermal conductivity of the material, it is necessary to measure the thermal diffusivity α and specific heat capacity C of the material. At a specified temperature, the sample is uniformly irradiated by a laser pulse, resulting in an instantaneous increase in the surface temperature. This surface serves as the hot end of the temperature gradient, and energy transfer to the cold end (upper surface) occurs via one-dimensional heat transfer. An infrared detector is used to measure the temperature of the upper surface. The half-temperature rise time ($t_{1/2}$) when the temperature of the upper surface rises to half of the maximum value T_M is recorded. The thermal diffusivity α can

(a)

(b)

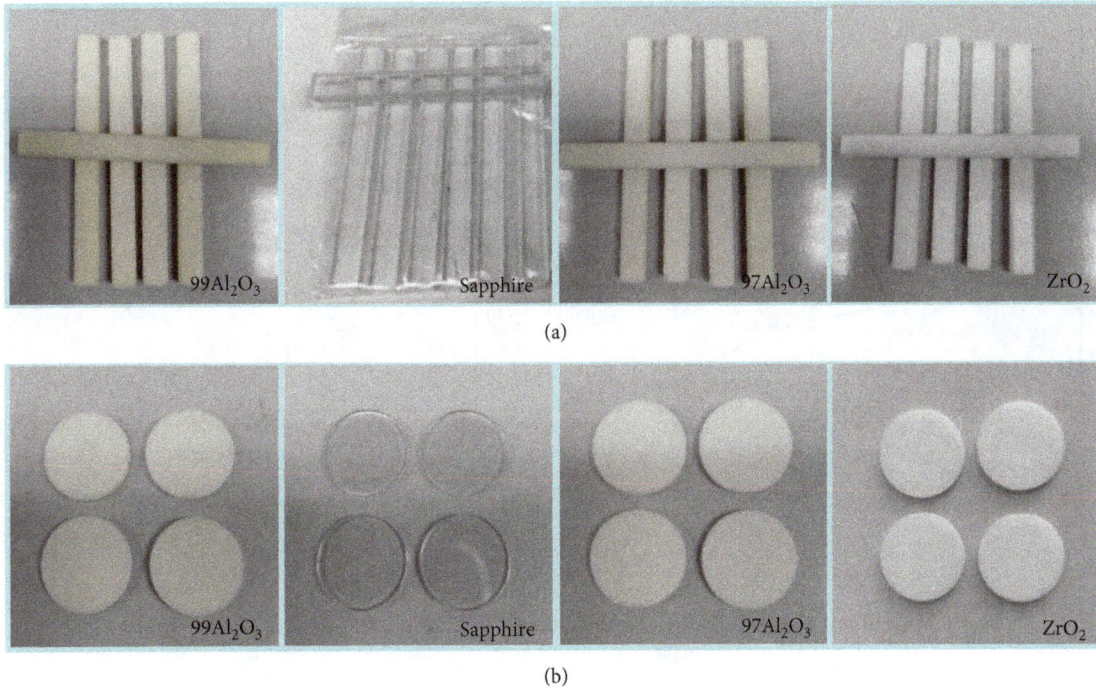

FIGURE 3: Test specimen: (a) size of $3.5 \times 5 \times 50 \, \text{mm}^3$ and (b) size of $\varphi 12.5 \times 1.5 \, \text{mm}^3$.

then be calculated using the following Fourier heat transfer equation:

$$\alpha = 1.38 L^2 \cdot \left(\pi^2 \cdot t_{1/2} \right)^{-1}. \tag{6}$$

To measure the specific heat capacity C of the materials, a standard reference sample, with similar cross-sectional shape, thickness, thermal properties, surface roughness, and a known specific heat capacity $C_{(\text{std})}$ is required. The reference sample and the test sample were subjected to surface coating at the same time. The definition of the specific heat capacity is shown in the following equation:

$$C = Q \cdot (\Delta T \cdot m)^{-1}, \tag{7}$$

where Q denotes the energy absorbed by the sample, ΔT denotes the temperature change after laser irradiation, and m is the mass of the material. From this, it is possible to obtain the following equation:

$$\frac{C_{\text{std}}}{C_{\text{sam}}} = \frac{Q_{\text{std}}}{\Delta T_{\text{std}} \cdot m_{\text{std}}} \cdot \left(\frac{Q_{\text{sam}}}{\Delta T_{\text{sam}} \cdot m_{\text{sam}}} \right)^{-1}, \tag{8}$$

when the energy absorbed by the reference and test sample is the same, that is, $Q_{\text{std}} = Q_{\text{sam}}$, (8) can be replaced by the following equation:

$$C_{\text{sam}} = C_{\text{std}} \cdot \Delta T_{\text{std}} \cdot m_{\text{std}} \cdot \left(\Delta T_{\text{sam}} \cdot m_{\text{sam}} \right)^{-1}. \tag{9}$$

4. Test Results and Analysis

The flexural properties of $99\text{Al}_2\text{O}_3$, $97\text{Al}_2\text{O}_3$, sapphire, and ZrO_2 at 25°C, 800°C, 1200°C, and 1500°C were measured using three-point bend testing, and load-displacement curves were obtained. Combining (1–4), the flexural strengths and flexural moduli of these ceramics at each temperature can be obtained. Plots of flexural strength and flexural modulus versus temperature are shown in Figures 4(a) and (b), respectively.

For the thermal conductivity measurements, graphite was selected as the standard reference sample. A temperature ramp of 20°C/min from 25°C to 1500°C was used. The thermal diffusivity and specific heat capacity were measured directly, and (3) was used to calculate thermal conductivity; these results are shown in Figure 5.

As shown in Figure 4(a), the flexural strengths of sapphire and ZrO_2 at 25°C are 740.8 MPa and 603.1 MPa, respectively, while the flexural strengths of $99\text{Al}_2\text{O}_3$ and $97\text{Al}_2\text{O}_3$ are 292.7 MPa and 272.8 MPa, respectively. Above 900°C, the flexural strengths of the ceramics decrease to almost 50% of their values at 25°C. At 1500°C, the reduction in the flexural strengths of sapphire and ZrO_2, which maintains strengths of around 150 MPa, is less than that for the alumina ceramics.

As shown in Figure 4(b), at 25°C, the flexural moduli of the alumina samples are greater than those of the sapphire and ZrO_2 sample. At 900°C, the flexural moduli of the alumina ceramics and sapphire decrease to approximately 46% of their value at 25°C, whereas the flexural modulus of ZrO_2 reduces to 25.414% of its value at 25°C. At 1200°C, the flexural modulus of $99\text{Al}_2\text{O}_3$ is 12.12 GPa, greater than that of $97\text{Al}_2\text{O}_3$ (9.21 GPa), sapphire (9.93 GPa), and ZrO_2 (6.19 GPa).

As shown in Figure 5(a), the thermal conductivities of these ceramics rapidly decrease as the temperature increases from 25°C to 800°C and remain stable at temperature above 1000°C. Although the thermal conductivity of ZrO_2 at

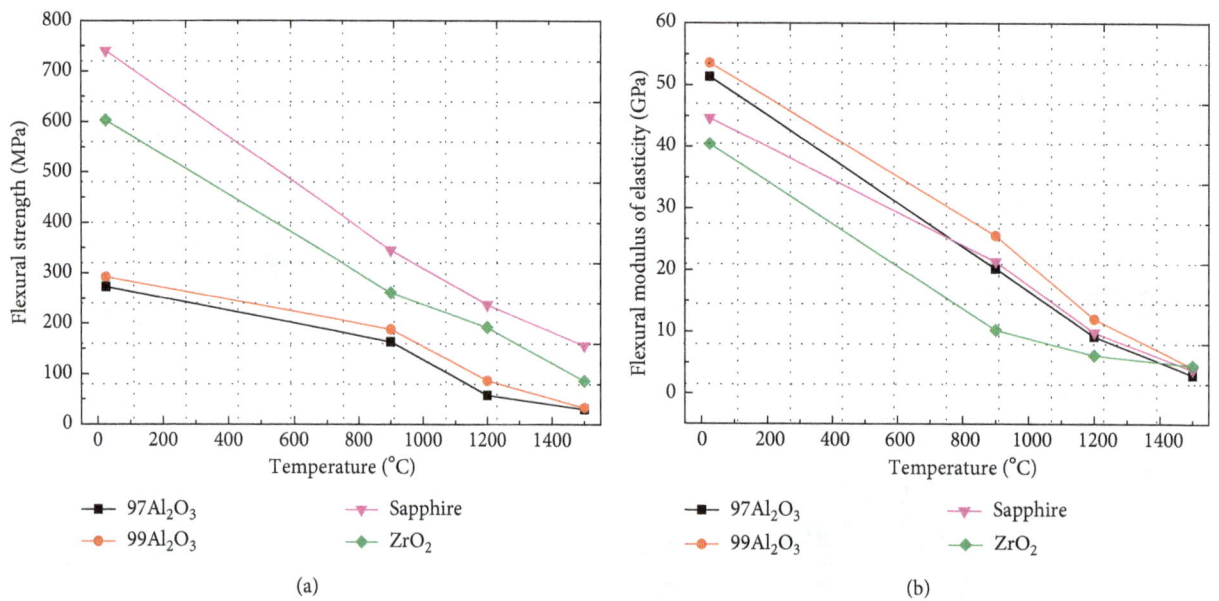

FIGURE 4: Flexural test result at different temperatures: (a) flexural strengths and (b) flexural moduli.

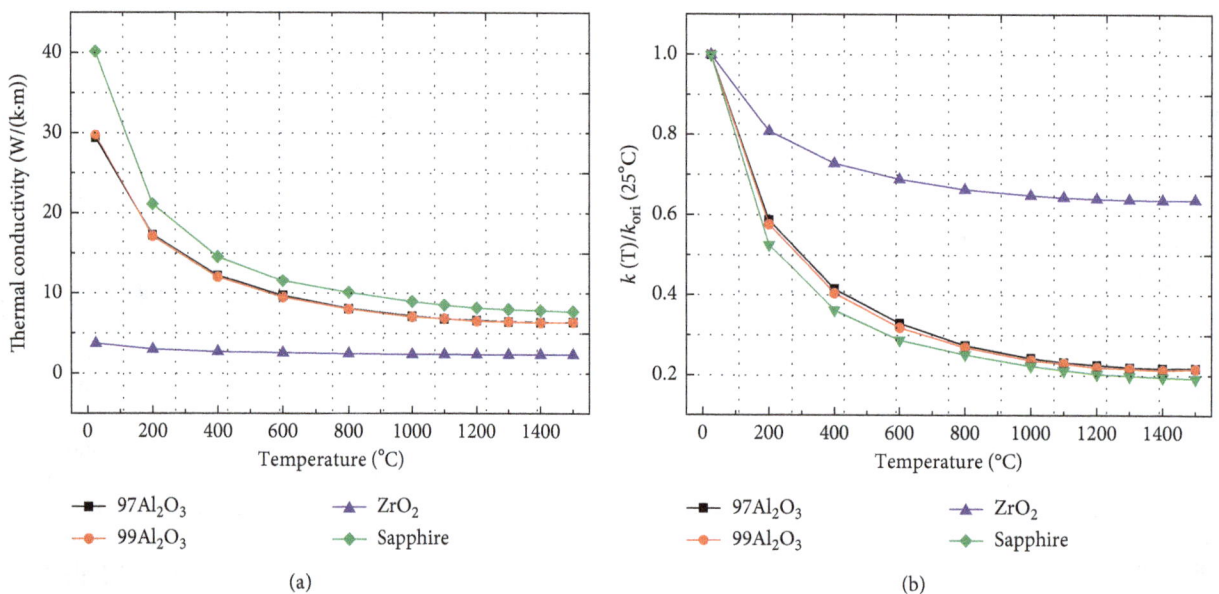

FIGURE 5: (a) Thermal conductivity change with the temperature. (b) The percentage of the thermal conductivity change with temperature.

1500°C is still over 60% of its value at 25°C, it is still significantly lower than the thermal conductivities of the alumina ceramics and sapphire. At all temperatures, sapphire has the highest thermal conductivity among the four ceramics tested.

5. Conclusion

In this work, three-point bend testing and transient laser emission were used to determine whether four ceramics are suitable for use as substrate materials for high-temperature pressure sensors; $99Al_2O_3$, $97Al_2O_3$, sapphire, and ZrO_2 were the ceramics investigated. The flexural moduli and the flexural strengths of the alumina ceramics and sapphire make them suitable for use as substrate materials in high-temperature pressure sensors for operation above 1000°C. Sapphire is suitable for use as a substrate in sensors operating in the environments above 1200°C. In terms of the thermal conductivity, high-temperature pressure sensors made from sapphire can achieve shorter response times and higher test accuracies than those made from other materials. The thermal conductivity of ZrO_2 is too low and will lead to large errors if used as a substrate material for high-temperature pressure sensors. However, this low thermal conductivity renders it suitable for use in thermal insulation.

Conflicts of Interest

The authors declare that they have no conflicts of interest.

Authors' Contributions

All works in relation to this paper have been accomplished by all authors' efforts. The idea and test scheme were proposed by YanJie Guo. The preparation of the test sample was completed by Fei Lu. The experiments of the flexural parameters were completed with the help of Lei Zhang and HeLei Dong. At last, every segment relate to this paper is accomplished under the guidance of QiuLin Tan and JiJun Xiong.

Acknowledgments

This work was supported by the National Natural Science Foundation of China (Grant nos. 61471324 and 51425505), the Outstanding Young Talents Support Plan of Shanxi province, and the Fund for Shanxi "1331 Project" Key Subject Construction.

References

[1] M. Narayanaswamy, R. J. Daniel, K. Sumangala, and C. Antony Jeyasehar, "Computer aided modelling and diaphragm design approach for high sensitivity silicon-on-insulator pressure sensors," *Measurement*, vol. 44, no. 10, pp. 1924–1936, 2011.

[2] S. Li, T. Liang, W. Wang, Y. Hong, T. Zheng, and J. Xiong, "A novel SOI pressure sensor for high temperature application," *Journal of Semiconductors*, vol. 36, no. 1, p. 014, 2015.

[3] T. H. Lee, S. Bhunia, and M. Mehregany, "Electromechanical computing at 500 C with silicon carbide," *Science*, vol. 329, no. 5997, pp. 1316–1318, 2010.

[4] D. A. Mills, D. Alexander, G. Subhash, and M. Sheplak, "Development of a sapphire optical pressure sensor for high-temperature applications," in *Proceedings of Sensors for Extreme Harsh Environments*, International Society for Optics and Photonics, vol. 9113, p. 91130H, Baltimore, MD, United States, 2014.

[5] J. Xu, G. Pickrell, X. Wang, W. Peng, K. Cooper, and A. Wang, "A novel temperature-insensitive optical fiber pressure sensor for harsh environments," *IEEE Photonics Technology Letters*, vol. 17, no. 4, pp. 870–872, 2005.

[6] A. Binder, G. Bruckner, N. Schobernig, and D. Schmitt, "Wireless surface acoustic wave pressure and temperature sensor with unique identification based on LiNbO$_3$," *IEEE Sensors Journal*, vol. 13, no. 5, pp. 1801–1805, 2013.

[7] Q. Tan, C. Li, J. Xiong et al., "A high temperature capacitive pressure sensor based on alumina ceramic for in situ measurement at 600°C," *Sensors*, vol. 14, no. 2, pp. 2417–2430, 2014.

[8] Q. Tan, T. Wei, X. Chen et al., "Antenna-resonator integrated wireless passive temperature sensor based on low-temperature co-fired ceramic for harsh environment," *Sensors and Actuators A: Physical*, vol. 236, pp. 299–308, 2015.

[9] H. Zhang, Y. Hong, T. Liang et al., "Phase interrogation used for a wireless passive pressure sensor in an 800°C high-temperature environment," *Sensors*, vol. 15, no. 2, pp. 2548–2564, 2015.

[10] S. Su, F. Lu, G. Wu et al., "Slot antenna integrated re-entrant resonator based wireless pressure sensor for high-temperature applications," *Sensors*, vol. 17, no. 9, p. 1963, 2017.

[11] C. Li, Q. Tan, W. Zhang, C. Xue, and J. Xiong, "An embedded passive resonant sensor using frequency diversity technology for high-temperature wireless measurement," *IEEE Sensors Journal*, vol. 15, no. 2, pp. 1055–1060, 2015.

[12] K. Wei, R. He, X. Cheng, R. Zhang, Y. Pei, and D. Fang, "Fabrication and mechanical properties of lightweight ZrO$_2$ ceramic corrugated core sandwich panels," *Materials & Design*, vol. 64, pp. 91–95, 2014.

[13] D. Li, W. Li, R. Wang, and H. Kou, "Influence of thermal shock damage on the flexure strength of alumina ceramic at different temperatures," *Materials Letters*, vol. 173, pp. 91–94, 2016.

[14] M. Marrelli, C. Maletta, F. Inchingolo, M. Alfano, and M. Tatullo, "Three-point bending tests of zirconia core/veneer ceramics for dental restorations," *International Journal of Dentistry*, vol. 2013, Article ID 831976, 5 pages, 2013.

[15] L. Yu, Y. Feng, J. Yang, T. Qiu, and L. Pan, "Mechanical and thermal physical properties, and thermal shock behavior of $(ZrB_2 + SiC)$ reinforced $Zr_3 [Al (Si)]_4C_6$ composite prepared by in situ hot-pressing," *Journal of Alloys and Compounds*, vol. 619, pp. 338–344, 2015.

[16] E. Sánchez-González, J. Meléndez-Martínez, A. Pajares, P. Miranda, F. Guiberteau, and B. R. Lawn, "Application of Hertzian tests to measure stress–strain characteristics of ceramics at elevated temperatures," *Journal of the American Ceramic Society*, vol. 90, no. 1, pp. 149–153, 2007.

[17] J. Cha, J. Seo, and S. Kim, "Building materials thermal conductivity measurement and correlation with heat flow meter, laser flash analysis and TCi," *Journal of Thermal Analysis and Calorimetry*, vol. 109, no. 1, pp. 295–300, 2012.

[18] H. Xie, H. Gu, M. Fujii, and X. Zhang, "Short hot wire technique for measuring thermal conductivity and thermal diffusivity of various materials," *Measurement Science and Technology*, vol. 17, no. 1, pp. 208–214, 2005.

[19] S. Min, J. Blumm, and A. Lindemann, "A new laser flash system for measurement of the thermophysical properties," *Thermochimica Acta*, vol. 455, no. 1-2, pp. 46–49, 2007.

[20] V. Casalegno, P. Vavassori, M. Valle, M. Ferraris, M. Salvo, and G. Pintsuk, "Measurement of thermal properties of a ceramic/metal joint by laser flash method," *Journal of Nuclear Materials*, vol. 407, no. 2, pp. 83–87, 2010.

[21] B. K. Jang and Y. Sakka, "Influence of microstructure on the thermophysical properties of sintered SiC ceramics," *Journal of Alloys and Compounds*, vol. 463, no. 1-2, pp. 493–497, 2008.

15

Investigations on the Wear Rate of Sintering Diamond Core Bit during the Hole Drilling Process of Al$_2$O$_3$ Bulletproof Ceramics

Lei Zheng [iD], Chen Zhang, Xianglong Dong, Shitian Zhao, Weidong Wu, Zhi Cui, and Yaoqing Ren

School of Mechanical Engineering, Yancheng Institute of Technology, No. 1 Middle Road, Xiwang Avenue, Tinghu District, Yancheng 224051, China

Correspondence should be addressed to Lei Zheng; alei611@163.com

Academic Editor: Zhonghua Yao

Bulletproof ceramics are usually hard and brittle with high elastic modulus, high compressive strength, and low tensile strength. While machining bulletproof ceramics, severe tool wear makes it difficult to obtain desired machining quality and efficiency, especially in hole drilling. In this work, an intensive experimental study on the overall wear rate of the sintering diamond thin-wall core bit during the hole drilling of Al$_2$O$_3$ bulletproof ceramics (99 wt.%) has been carried out. The quality loss of the bit after each hole drilled was selected for representing the overall wear rate of the bit. Based on experimental data, the influences of the main bit performance and machining process parameters on the overall wear rate of the bit have been analyzed. According to the results discussed, under the test conditions, finer diamond grit, higher diamond concentration, lower number of water gaps, thinner wall thickness, or lower bit load all can decrease the wear rate of the bit. However, within a certain range, the spindle speed has little influence on the overall wear resistance of the bit, but when the spindle speed increases, the machining efficiency can be significantly improved. The results obtained in this work can offer a valuable reference for the use of sintering diamond thin-wall core bits in the hole drilling of bulletproof ceramics.

1. Introduction

Engineering ceramics used in the field of armor protection (bulletproof ceramics) are usually hard and brittle with a very high elastic modulus and have high compressive strength and low tensile strength. During machining engineering ceramics, which are typical difficult-to-machine materials, severe tool wear makes it difficult to guarantee the machining quality and efficiency, especially in the hole machining. This greatly limits the widespread application and popularization in the field of armor protection.

At present, special processing technologies are usually adopted for the machining of hard and brittle materials such as engineering ceramics, including laser processing [1, 2], electrical discharge machining [3, 4], ultrasonic machining [5, 6], and so on. Bharatish et al. [1] performed CO$_2$ laser drilling tests of 2 mm-thick Al$_2$O$_3$ ceramic plates to examine the effects of laser parameters such as pulse frequency, laser power, scanning speed, and hole diameter on entrance circularity, exit circularity, heat affected zone, and taper. Rihakova and Chmelickova [2] made a review about the laser micromachining of glass, silicon, and ceramics. Interaction of these materials with laser radiation and the mechanisms of laser micromachining of materials were provided. In the study by Munz et al. [3], electrically conductive ceramic ZTA-TiC composites were machined by EDM (electrical discharge machining) drilling with variation of pulse shape, discharge current, discharge time, and flushing conditions. Yadav et al. [4] investigated the effects of varied voltage, electrolyte concentration, wire velocity, pulse on-time, and pulse off-time for the EDM of alumina epoxy nanocomposite. Guo et al. [5] designed a novel ultrasonic vibration apparatus and performed the experimental investigation of ultrasonic vibration-assisted grinding of SiC microstructures.

However, special processing technologies are mainly suitable for machining of microholes or microstructures, when it turns to larger holes of about ten mm or even tens of millimeters in diameter, the machining efficiency is very low. When machining a relatively large hole, it is very difficult to drill engineering ceramics with solid bits; therefore, diamond core bits are commonly used [7–11]. Using thin-wall impregnated diamond bits, Zheng et al. [7, 8] performed drilling trials in the ceramics/GFRP/aluminum alloy composite armor formed with 8 mm-thick Al_2O_3 engineering ceramics and only 1.5 mm-thick GFRP and aluminum alloy on each side [7] and the ceramics/KFRP double-plate composite armor formed with 6 mm thick KFRP backboard and 7 mm thick Al_2O_3 ceramic faceplate [8]. Bit parameters, feeding mode, compressive prestress process equipment, drilling efficiency, hole quality, and machining mechanism have been investigated experimentally. Gao and Yuan [9] carried out a comprehensive experimental study on hole drilling of Al_2O_3 armor ceramics by using impregnated diamond bits. The results showed that through selecting the reasonable drill parameters (including matrix composition, diamond type, grain size, diamond concentration, number of slots and wall thickness) and technological parameters (such as axial force and spindle speed), higher efficiency and surface quality can be obtained. Tan et al. [10] designed a new composite-impregnated diamond bit to solve the slipping problem and conducted laboratory drilling test in granite and field drilling application in crystal tuff. Zhang et al. [11] carried out a comparative study on core drilling of silicon carbide and alumina engineering ceramics with monolayer brazed diamond bit. The effect of coolant type and concentration on drilling torque and drilling efficiency and the morphologies of machined surface of ceramics were investigated.

The above analysis indicates that, for large holes machining of engineering ceramics, a higher machining efficiency and better machining quality can be achieved through the optimization of processing parameters and bit performance parameters. Thus, core drilling with diamond bit is an efficient method for machining of holes in engineering ceramics. However, few studies have been reported about the tool wear during the machining of engineering ceramics when employing diamond core bits. Noticeably, Zhang et al. [11] also examined the morphologies of the wear of diamond grains of monolayer brazed diamond bit and found that the wear of brazed diamond grains for drilling of silicon carbide is much severer than that of alumina.

According to the existing literature, investigations on the wear of diamond tools during the machining of engineering ceramics are mainly focused on the abrasive machining of diamond grinding wheels and diamond burs. Kizaki et al. [12] proposed the laser-assisted machining of zirconia ceramics using diamond burs. The results revealed that the grinding force and the tool damage could be reduced in the assisted processes. In the study by Shen et al. [13], the alumina ceramics were ground with and without the ultrasonic vibration-assisted grinding to investigate the wear characteristics of the diamond wheel. The changes of the diamond wheel surface topography were captured during

the grinding process. Zeng et al. [14], Liang et al. [15], and Ding et al. [16] also investigated the wear behaviors of diamond grinding wheels during the vibration-assisted grinding process of engineering ceramics in order to improve the machining quality and efficiency and found that the ultrasonic vibration caused greater wear but also improved the self-sharpening property of the diamond wheel. do Nascimento et al. [17] tested the viability of minimum quantity lubrication (with and without water) in grinding of advanced ceramics using a hybrid-bonded diamond wheel. The results showed that the minimum quantity lubrication with water (1:1) could reduce wheel wear greatly.

In this study, a typical high-purity alumina bulletproof ceramics (99 wt.% Al_2O_3) was selected as a machining object. Taking the drill quality variation measured as the indicator for reflecting the overall wear degree of the drill, a systematical experimental investigation on the wear rate of the sintering diamond thin-wall core drill has been conducted. The influences of the main bit performance and machining process parameters on the overall wear rate of the bit have been analyzed. The results obtained in the present paper can offer a valuable reference regarding the use of sintering diamond thin-wall core bits to machine bulletproof ceramics.

2. Experimental Procedure

2.1. Sintering Diamond Core Bit. According to the experimental research on the ceramic composite components [7] and double-plate composite components [8], and by combining with the machining features of the bulletproof ceramics, the sintering diamond core bit used in this study is illustrated as Figure 1. The diamond bit is mainly composed of a diamond layer (work layer), a transition layer, basal body, a narrow slot, several water gaps, and a drill shank (clamping handle). The basal body material was made from $45^{\#}$ seamless steel pipe, the diamond grade was SMD_{35}, and the copper-based matrix was fabricated through hot pressing sintering of the metallic binders with volume fractions of 48% Cu, 30% Co, 6% Ni, 5% WC, 5% Ti, 4% Sn, and 2% Cr. The sintering temperature was 900°C, the heat preservation time was 2 min, and the sintering pressure was 20 MPa [7, 8].

2.2. Experimental Set-Up. In order to reflect the overall wear loss variation of the diamond bit accurately, the bit was taken down after every hole was drilled on the bulletproof ceramics and then was washed cleanly and dried for weighing with a precise analytical balance. The overall wear rate of the bit was represented by the quality loss of the bit after each hole was drilled, which reflected the whole wear loss of the diamond grains and the matrix binding agent of the bit. When the drilling operation became significantly difficult or the bit sintering body broke up considerably, the normal drilling operation could not be continued and the bit should be scrapped.

The test was conducted on a ZXL-20 vertical drilling-milling machine (with a spindle power of 750 W) equipped with water cooling device. A constant pressure feed mode

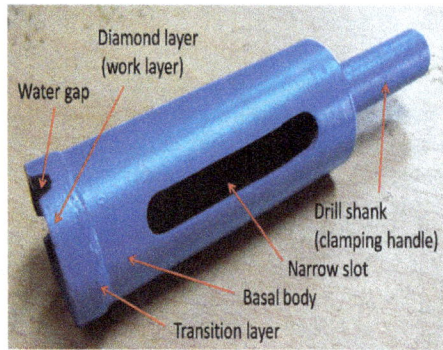

FIGURE 1: The sintering diamond core bit.

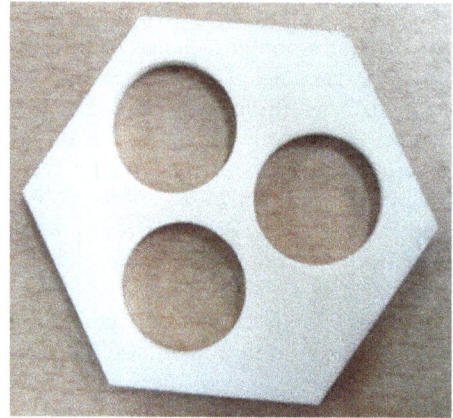

FIGURE 2: Al$_2$O$_3$ bulletproof ceramic brick drilled.

was adopted. The bulletproof ceramics used was a hexagonal high-purity alumina engineering ceramic brick (99 wt.% Al$_2$O$_3$) with a thickness of 10 mm. One Al$_2$O$_3$ bulletproof ceramic brick drilled is presented in Figure 2. The outer diameter of the bit was Φ 24 mm, and the work layer height was basically the same that of the water gaps, or even slightly smaller. The TG-328B precise analytical balance (as presented in Figure 3) was used for weighing, the reading precision of which was 0.001 g.

With the given matrix formula and diamond grade, the grit size, grit concentration, number of water gaps at the crown, and bit wall thickness are the main bit performance parameters which can affect the wear conditions of the tools. Under the constant pressure feed mode, the bit load and the spindle speed are two important process parameters that affect the wear conditions of the bit. According to the previous experimental studies on the Al$_2$O$_3$ ceramic composite armors [7, 8], in the wear test, the performance parameters of the bit and the machining technological parameters have been determined, as listed in Table 1, along with the number of holes that each bit have accomplished until scrapped.

3. Results and Discussion

3.1. Influence of Bit Performance Parameters on Wear Rate

3.1.1. Diamond Grit Size. Figure 4 shows the comparison results of the bit overall wear rate of the bit when the diamond grit size is 35/40 mesh (Test 1) and 50/60 mesh (Test 2), respectively. When the grit size is 35/40 mesh, the diamond layer is completely exhausted after machining the 23rd hole, and accordingly, the bit begins to skid. This means that the work layer has been fully utilized. When the grit size is 50/60 mesh, the bit begins to skid when drilling the 13th hole (not drilled through), and the work layer has not been fully taken use of.

As can be seen from Figure 4, when other conditions are kept constant, the decrease of the grit size can reduce the overall wear rate of the bit, thus enhancing the wear resistance of the drill. However, it is also found that the service life of the bit has not been extended accordingly. When the grit concentration is constant, the increase of the grit size can decrease the number of grinding grains on the diamond bit crown. Therefore, under the constant pressure feed mode, the cutting depth of a single diamond grit increases, making

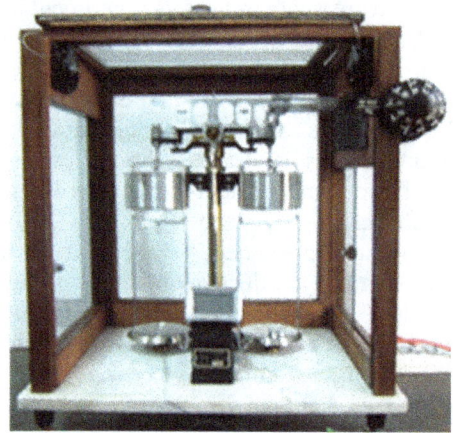

FIGURE 3: The TG-328B analytical balance.

the ceramic material easy to be fractured and cracked. Meanwhile, more large-sized ceramic abrasive debris is produced, which has a more severe wear effect on the bit.

In addition, the cutting load on a single diamond grit increases, which makes the diamond easy to break or fall off. Thus, new diamond grits can be constantly exposed to the working surface, and the performance of diamond bit can be fully taken use of. When the grit size becomes smaller, diamond grits exposed on the lip surface of the drill is increased, the cutting depth of a single diamond grit decreases (under a constant drilling load), the size of ceramic abrasive debris becomes smaller, and the wear of bit also becomes slighter; meanwhile, the cutting load on a single diamond grit decreases, and the diamond is easy to be worn into a planar shape, which results in the occurrence of the skidding phenomenon [7], such as the bit in Test 2. Therefore, a finer diamond grit does not always lead to a longer bit life, and it seems that an optimal diamond grit exists under certain machining conditions.

3.1.2. Diamond Concentration. Figure 5 shows the comparison results of the overall wear rate of the diamond bit when the diamond concentration is 75% (Test 1) and 125% (Test 3), respectively. When the diamond concentration is

TABLE 1: Test conditions.

| Test no. | Bit performance parameters | | Number of water gaps | Wall thickness (mm) | Technological parameters | | Number of machined holes |
	Grit size (mesh)	Concentration (%)			Bit load (N)	Spindle speed (rpm)	
1	35/40	75	3	2	352	3200	23
2	50/60	75	3	2	352	3200	13
3	35/40	125	3	2	352	3200	26
4	35/40	75	2	2.5	278	3200	33
5	35/40	75	4	2.5	278	3200	21
6	45/50	125	4	2	390	3200	15
7	45/50	125	4	2.5	390	3200	9
8	35/40	75	3	1.5	352	3200	9
9	45/50	125	4	2.5	278	3200	23
10	35/40	75	3	2	352	1750	21

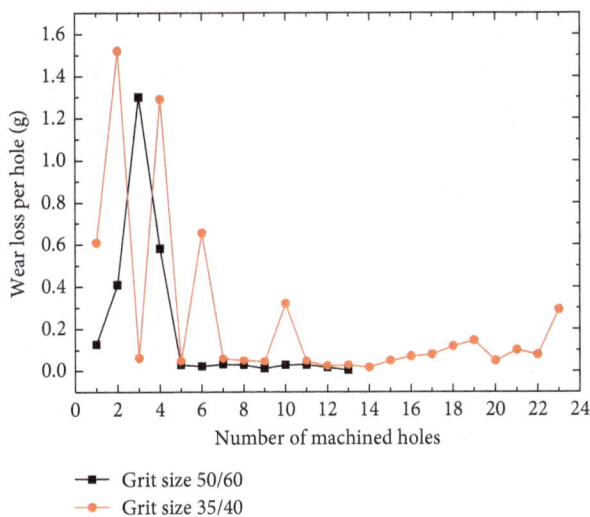

FIGURE 4: Wear rates of the bit under different diamond grits.

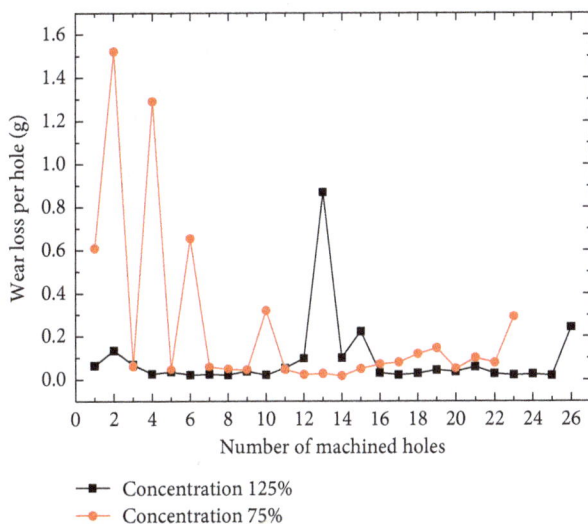

FIGURE 5: Wear rates of the bit under different diamond concentrations.

125%, massive breakup of the sintering body occurs after drilling through the 26th hole, and the work layer has not been fully taken use of.

Further, it can be seen from Figure 5 that when the concentration increases, the wear rate of the diamond bit can obviously decrease. When the concentration increases, the density of diamond grits on the bit crown surface becomes larger, and the bare area of the matrix reduces; thus, the wear resistance of the bit matrix increases. Under a constant pressure feed, the cutting depth of a single diamond reduces and the particle size of ceramic abrasive debris becomes smaller, the wear effect of the ceramic abrasive debris on the bit thus reduces and the wear resistance of the bit increases. At the same time, the cutting load of a single diamond grit is reduced, and more diamond grits are prone to be worn and exhausted. When the wear of the diamond grits proceeds to a certain level, a large number of the exposed diamond grits are ground and polished, i.e., the occurrence of skidding phenomenon. In addition, the increase of diamond concentration decreases the bond strength of the matrix and probably induces the breakup of the sintering body during the drilling process, thus leading to the bit being scrapped prematurely, such as the bit in Test 3.

When the concentration is relatively low, the cutting load acting on a single diamond grit becomes higher and the cutting and breaking of the ceramics becomes easier. Accordingly, the diamond grits on the bit crown are more prone to fragmentation wear, and the abrasion wear of diamond grits reduces. In this way, the cutting ability of the bit can be preserved. However, if the concentration is too low, the massive fragmentation of diamond grits will increase rapidly, which may cause severe wear and premature scrap of the bit. However, if the concentration is too high, the holding strength of the matrix binding agent to the diamond grits is not high enough and the diamonds are prone to fall off prematurely during the machining process, where the wear resistance of the bit is reduced.

3.1.3. Number of Water Gaps on the Crown. Figure 6 shows the comparison results of the overall wear rate of the diamond bit when the number of water gaps on the crown of bit matrix is 2 (Test 4) and 4 (Test 5), respectively. When the number of water gaps is 2, the bit begins to skid when drilling the 33rd hole (not drilled through), and the work layer is not completely used. When the number of water gaps

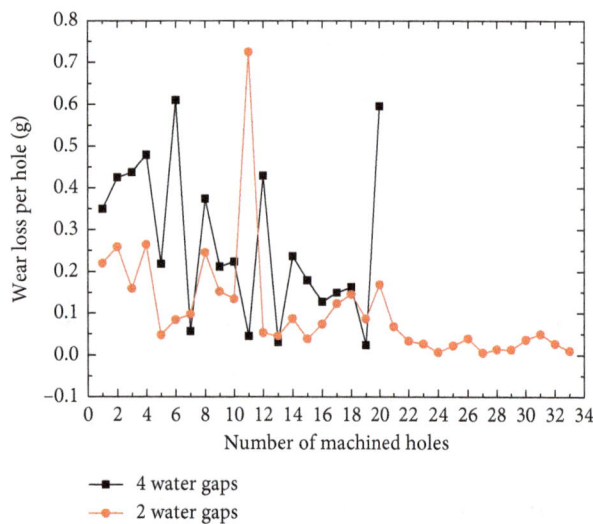

FIGURE 6: Wear rates of the bit under different number of water gaps.

is 4, at the beginning of drilling the 21st hole, the noise sharply increased, the bit jittered considerably, and the clamping shank twisted off quickly. It can be found that the water gaps have been ground to disappear, and that no diamond grits remain on the crown, namely, the diamond layer has been completely exhausted. At this moment, the bit work layer has been also fully used.

According to Figure 6, it can be found that, on the whole, the bit wear resistance is stronger when a smaller number of water gaps is employed. With the increase of the number of water gaps on the crown, the surface area of the work layer on the bit crown is reduced, and the number of exposed diamond grits is also reduced. Under the constant pressure feed mode, the cutting load of a single diamond grit increases, and the depth of the diamond cutting into the ceramic workpiece increases as well, the ceramic materials are thus more likely to be broken into large pieces. In addition, the particle size of ceramic abrasive debris being relatively large, it enhances the wear effect on the bit and leads to a poor wear resistance of the bit, as presented in Test 5. Meanwhile, a greater number of water gaps increases the frequencies of mechanical and thermal shocks. Frequent impacts decrease the binding strength of the diamond grits and the matrix, and accelerate the fall-off rate of the diamond grits; moreover, they are likely to cause massive fragmentation wear of diamond grits, thus reducing the wear resistance of the bit.

In addition, when the number of water gaps increases, the difference between the cutting loads on the diamonds at the inner and outer cutting circular lines increases, which further aggravates the uneven wear of the bit crown and degrades the wear resistance of the bit. In contrast, when there are fewer water gaps, the work layer surface area of the bit crown increases and the exposed diamond grits on the crown increases; the cutting load of a single diamond grit is thus reduced, and the proportion of diamond abrasion wear is increased. This probably leads to the skidding phenomenon, for example, the bit in Test 4. Therefore, the number of

water gaps should be determined according to specific processing conditions. According to the machining tests on bulletproof ceramics carried out in this investigation, the number of water gaps should be set to 2-3 (when the bit diameter is Φ 24 mm).

3.1.4. Bit Wall Thickness. Figure 7 shows the comparison results of the overall wear rate of the diamond bit when the wall thickness of the bit is 2 mm (Test 6) and 2.5 mm (Test 7), respectively. When the wall thickness of the bit is 2 mm, after machining the 15th hole, massive breakup of the diamond layer occurs, and the bit is scrapped. This means that the bit work layer has not been fully used. When the wall thickness is 2.5 mm, the clamping drill shank twists off when machining to the 9th hole (not drilled through). It can be found that no water gaps and no diamond grits remain on the crown, indicating that the diamond layer has been completely exhausted.

As can be seen in Figure 7, when the wall thickness increases, the wear rate of the bit increases obviously as well. This is because when the wall thickness increases, the surface area of the work layer of the bit crown also increases and accordingly the wear loss increases. However, it is also found that when the wall thickness increases, the service life of the bit is not extended accordingly; on the contrary, it is the bit with thinner wall thickness that shows a longer service life. It is observed that when the bit load is 390 N, bits with larger wall thickness show a very obvious inner trumpet-shaped wear shape on the crown. This uneven crown wear decreases the matrix strength and wear resistance of the bit. Further, the impact load per unit area on the crown matrix increases dramatically, and thus, the wear rate of the bit remarkably increases combined with screeching noise and bit bouncing, which decreases the service life of the bit tremendously. In contrast, when the wall thickness is thinner, although the volume of the diamond layer is reduced, the wear rate of the bit is relatively stable and the fluctuation is not very high, resulting in an improved service life of the bit.

On the whole, with the increase of wall thickness, the difference between the feed load and cutting load of the diamond at the inner and outer cutting circular lines is increased, so the uneven wear of the bit crown becomes more severe, which leads to a faster bit wear. In addition, when the wall thickness is relatively thin, the strength of the bit sintering body is reduced, so that cracking and even breakup of the diamond layer are easy to occur during machining, such as the bit in Test 6.

It is noticed that the bits in Test 3 (bit load is 352 N, and wall thickness is 2 mm) and Test 8 (bit load is 352 N, and wall thickness is 1.5 mm) are also scrapped prematurely because of the massive breakup of the sintering body. Therefore, the appropriate wall thickness and bit pressure has a crucial impact on the normal use of the bit. According to the machining tests of the bulletproof ceramics, the appropriate wall thickness should not be more than 2.5 mm and not less than 2 mm (when the bit diameter is Φ 24 mm). Too thick a wall thickness would reduce the service life of the bit due to the seriously uneven wear of the crown. Too thin a wall

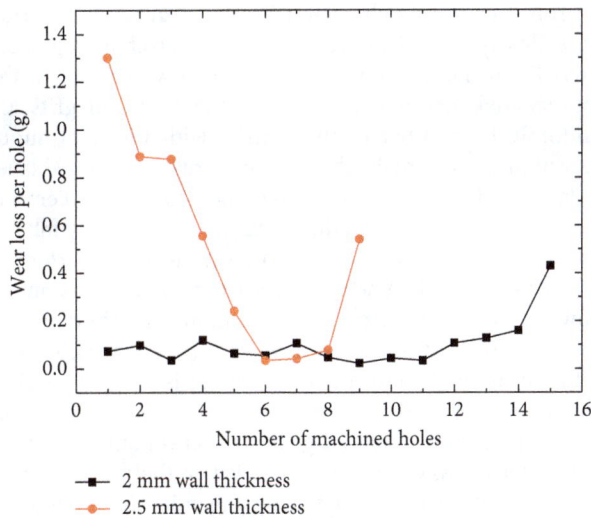

FIGURE 7: Wear rates of the bit under different wall thicknesses.

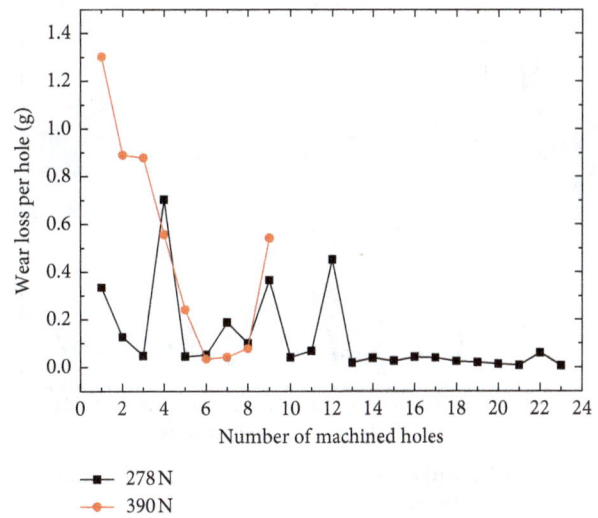

FIGURE 8: Wear rates of the bit under different bit loads.

thickness would lead to a poor strength and rigidity of the bit sintering body, which would result in the breakup of the work layer during machining process, finally causing the scrap of the bit prematurely. As a result, the wall thickness should match the bit load to obtain a longer bit service life.

3.2. Influences of Machining Technological Parameters on Wear Rate

3.2.1. Bit Load. Figure 8 shows the comparison results of the overall wear rate of the diamond bit when the bit load is 278 N (Test 9) and 390 N (Test 7), respectively. When the bit load is 278 N, after 23 holes are machined, the machining is not continued because this batch of ceramic bricks has been completely used out. At this moment, the depth of water gaps was 1.8 mm, and there were no cracks in the diamond layer. The bit preserved its good drilling performance and could still be used for further machining operations.

From Figure 8, it can be seen that reducing the bit load can greatly decrease the wear rate of the diamond bit. When the bit load is reduced, the cutting load applied on the diamond grits on the crown decreases and the impact effect to which the diamond grits are subjected is also reduced; thus, the massive fragmentation and fall-off of the diamond grits is reduced; at the same time, the depth of a single diamond grit cutting into the ceramics is decreased, and the particle size and the total amount of ceramic abrasive debris are both reduced, the wear effects of the abrasive debris on the matrix binder and the grits are therefore alleviated. Especially, the impact wear of the coarse ceramic hard debris particles on the diamond grits and the matrix are greatly reduced. The repeated impacts of these coarse ceramic hard debris particles can not only accelerate the fall-off of the diamond grits on the crown but also rapidly scratch and wear the matrix, which results in an extremely violent wear effect on the bit. Therefore, the overall wear rate of the bit can be greatly decreased by decreasing the bit load. Compared with the bit load of 390 N, under the bit load of 278 N, the

diamond bit can maintain stable wear for longer time, and consequently, the bit life can be extended considerably.

It can be observed that when the wall thickness is 2.5 mm (Tests 4, 5, 7, and 9), an inner trumpet-shaped wear of different degrees appears on the matrix crown during the machining process [7]. Especially when the bit load is relatively high (for example, the bit load is 390 N in Test 7), very obvious inner trumpet-shaped wear is found on the bit crown. This uneven wear of crown accelerates the overall bit wear rate greatly and considerably decreases the service life of the bit. In comparison, when the bit load is reduced to 278 N (Tests 4, 5, and 9), the uneven inner trumpet-shaped wear of the crown is greatly alleviated, and the service life of the bit is also extended significantly.

Therefore, the increase of wall thickness does not always enhance the wear resistance or extend the service life of the bit; the key is to match the wall thickness with the bit load. Only a reasonable bit load can make full use of the bit work layer and thereby achieve a longer service life.

3.2.2. Spindle Speed. Figure 9 shows the comparison results of the overall wear rate of the diamond bit when the spindle speed is 1750 rpm (Test 10) and 3200 rpm (Test 1), respectively. When the spindle speed is 1750 rpm, the diamond layer is completely exhausted after machining the 21st hole, and the bit work layer has been fully used.

It can be seen from Figure 9 that under test conditions, the change of the spindle speed has little effect on the overall wear resistance and service life of the bit, i.e., the average wear rate of the bit is basically kept constant. However, at the relatively high spindle speed, the wear rate of the bit exhibits a very large fluctuation, while at the relatively low spindle speed, the wear rate of the bit is relatively smooth. Compared with the spindle speed of 1750 rpm, when the spindle speed is increased to 3200 rpm, the average machining time per hole is about half of the time under the former speed. Hence, after increasing the spindle speed, the machining efficiency of the bit can be enhanced substantially. This is because

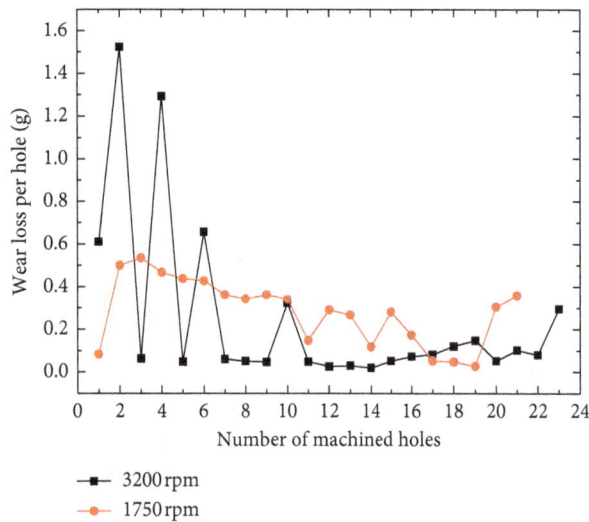

FIGURE 9: Wear rates of the bit under different spindle speeds.

when the speed is relatively low, the vibration of the bit during the machining process is moderate, which reduces the dramatic collision wear between the diamond grits and matrix and the ceramic materials. As a result, the massive fragmentation and fall-off phenomena of the diamond grits are alleviated, and the wear rate of the matrix is relatively stable. Hence, the wear rate of the bit is relatively smooth at low speeds. However, it should be noticed that the decrease of spindle speed will directly lead to the decrease of the linear velocity of the bit, which means that smaller volume of ceramic material can be cut by the single diamond grit per unit time. Therefore, the machining efficiency is decreased.

In addition, according to the study [18], the decrease of the spindle speed leads to a decrease in the axial cutting force acting on the diamond bit and accordingly the grinding force of a single diamond grit decreases. At this moment, the wear rate of the grits decreases; however, due to the decrease of the spindle speed, the machining time of per hole is significantly extended, which results in larger machining and wear times. Thus, the wear of grits during the entire hole machining process is not reduced. It can therefore be concluded that the increase of spindle speed within a certain range will not lower the overall wear resistance or accelerate the average wear rate of the bit significantly, but can increase the drilling rate substantially.

3.3. Relationship between Bit Wear Rate and Cumulative Drilling Depth. The cumulative drilling depth of the bit can be represented by the total number of machined holes. Hence, Figures 4–9 also reflect the relationship between the overall wear rate of the bit and the cumulative drilling depth. It can be found from Figures 4–9 that the wear rate of the bit shows an approximately periodic variation with the accumulated drilling depth, namely, the abrasion loss per hole varies from low (high) to high (low) and then from high (low) to low (high). Although the periodic trend is not very regular, the special wear characteristic of the bit can be qualitatively described.

This special wear characteristic of the sintering diamond bit is closely related to the periodical layer-changing characteristic of the diamond grits on the crown surface of the impregnated diamond tools. Since the diamond grits are randomly distributed in the matrix, with the continuous wear of the matrix body, the diamond grits are exposed layer by layer and act as microcutting edges to cut the ceramic material and are worn gradually, broken, and finally fall off, namely, continuously working in the form of periodical layer-changing [7]. When the exposed diamond grits on the crown of the matrix reach a certain amount and the diamond grits are mostly at the abrasion wear stage, the abrasion loss of the bit is small, and the wear process is relatively stable. When the diamond grits are at the massive fragmentation or fall-off stage, new diamond grits are continuously exposed, and the wear loss of the bit is high. When the diamond grits on the crown are dominated by the microfragmentation wear, the wear loss of the bit is between the above-mentioned two conditions. Hence, in Figures 4–9, the relationship between the overall wear rate of the bit and the cumulative drilling depth is reflected as the continuous variation of the wear rate with the number of cumulative machined holes. This working process of the diamond bit goes round and round until the diamond bit is scrapped by the skidding of the bit, the breakup of the sintering body, or the depletion of the diamond layer.

In general, if the wear loss of the bit after machining a certain hole is large, the bit should be in the layer-changing process of the diamond grits on the crown surface when the hole is being machined, whereas when the wear loss of the bit after machining a certain hole is small, the bit should be at the abrasion wear stage or microfragmentation wear stage. Thus, during the hole machining process, the bit is sometimes at the abrasion wear stage, sometimes at the microfragmentation stage, sometimes at the massive fragmentation or fall-off stage, and sometimes in the layer-changing process of the diamond grits on the crown surface. Hence, in the machining process of each hole, the state of the diamond grits on the crown surface is different, which leads to differences in the wear loss after machining different holes. Since the wear loss after machining each hole is used to represent the overall wear rate of the bit, the wear rate curve of the bit sometimes does not cover a complete cycle, such as in Test 7 (Figure 8). It should be noticed that this phenomenon does not mean that there is no periodical variation for the bit wear rate in the hole machining process.

4. Conclusions

In this work, an intensive experimental study on the wear rate of the sintering diamond thin-wall core bit during drilling Al_2O_3 bulletproof ceramics (99 wt.%) has been conducted. The influences of the main bit performance and machining process parameters on the overall wear rate of the bit have been analyzed. According to the experimental results, the main conclusions can be summarized as the following:

(1) Under the test conditions, finer diamond grit, higher diamond concentration, lower number of water gaps, thinner wall thickness, or lower bit load all can decrease the wear rate of the bit.

(2) When the diamond grits turns finer, the diamond concentration turns higher, or the number of water gaps is lowered, the bit can skid easily, whereas thinner wall thickness and higher diamond concentration can make the diamond layer to crack easily or even break up. The increase in wall thickness does not always increase the wear resistance or extend the service life of the bit, and the key is to ensure the matching of bit load and wall thickness.

(3) From the perspective of wear resistance, the number of water gaps should be 2-3, and the wall thickness should be kept between 2 and 2.5 mm (when the bit diameter is Φ 24 mm).

(4) Within a certain range, the spindle speed has little influence on the overall wear resistance of the bit, but when the spindle speed increases, the machining efficiency can be significantly improved.

(5) The wear rate of the bit shows an approximately periodical variation with the cumulative drilling depth. This is closely related to the periodical layer-changing characteristic of the diamond grains on the working surface of the impregnated diamond tools.

Conflicts of Interest

The authors declare that there are no conflicts of interest regarding the publication of this paper.

Acknowledgments

This work was supported by National Natural Science Foundation of China (51575470), Jiangsu Provincial Six-Big-Talent-Peak High Level Personnel Project of China (JXQC-029), Qing Lan Project of Jiangsu Higher Education of China (Document 15th of Jiangsu Education Department in 2016), Postgraduate Research & Practice Innovation Program of Jiangsu Province of China (SJCX17-0444 and SJCX17-YG01), and Shandong Provincial Natural Science Foundation of China (ZR2012EEL04). Thanks to all study participants for their contributions.

References

[1] A. Bharatish, H. N. N. Murthy, B. Anand, C. D. Madhusoodana, G. S. Praveena, and M. Krishna, "Characterization of hole circularity and heat affected zone in pulsed CO_2 laser drilling of alumina ceramics," *Optics & Laser Technology*, vol. 53, pp. 22–32, 2013.

[2] L. Rihakova and H. Chmelickova, "Laser micromachining of glass, silicon, and ceramics," *Advances in Materials Science & Engineering*, vol. 2015, Article ID 584952, 6 pages, 2015.

[3] M. Munz, M. Risto, R. Haas, R. Landfried, F. Kern, and R. Gadow, "Machinability of ZTA-TiC ceramics by electrical discharge drilling," *Procedia CIRP*, vol. 6, pp. 77–82, 2013.

[4] P. Yadav, V. Yadava, and A. Narayan, "Experimental investigation of kerf characteristics through wire electro-chemical spark cutting of alumina epoxy nanocomposite," *Journal of Mechanical Science and Technology*, vol. 32, no. 1, pp. 345–350, 2018.

[5] B. Guo, Q. L. Zhao, and M. J. Jackson, "Ultrasonic vibration-assisted grinding of micro-structured surfaces on silicon carbide ceramic materials," *Proceedings of the Institution of Mechanical Engineers, Part B: Journal of Engineering Manufacture*, vol. 226, no. 3, pp. 553–559, 2012.

[6] C. Nath, G. C. Lim, and H. Y. Zheng, "Influence of the material removal mechanisms on hole integrity in ultrasonic machining of structural ceramics," *Ultrasonics*, vol. 52, no. 5, pp. 605–613, 2012.

[7] L. Zheng, B. Wang, B. Feng, C. Gao, and J. Yuan, "Experimental study on trepanning drilling of the ceramic composite armor using a sintering diamond tool," *International Journal of Advanced Manufacturing Technology*, vol. 56, no. 5–8, pp. 421–427, 2011.

[8] L. Zheng, H. Zhou, C. Gao, and J. Yuan, "Hole drilling in ceramics/Kevlar fiber reinforced plastics double-plate composite armor using diamond core drill," *Materials & Design*, vol. 40, pp. 461–466, 2012.

[9] C. Gao and J. Yuan, "Efficient drilling of holes in Al_2O_3 armor ceramic using impregnated diamond bits," *Journal of Materials Processing Technology*, vol. 211, no. 11, pp. 1719–1728, 2011.

[10] S. Tan, X. Fang, K. Yang, and L. Duan, "A new composite impregnated diamond bit for extra-hard, compact, and nonabrasive rock formation," *International Journal of Refractory Metals and Hard Materials*, vol. 43, pp. 186–192, 2014.

[11] F. L. Zhang, P. Liu, L. P. Nie et al., "A comparison on core drilling of silicon carbide and alumina engineering ceramics with mono-layer brazed diamond tool using surfactant as coolant," *Ceramics International*, vol. 41, no. 7, pp. 8861–8867, 2015.

[12] T. Kizaki, Y. Ito, S. Tanabe, Y. Kim, N. Sugita, and M. Mitsuishi, "Laser-assisted machining of Zirconia ceramics using a diamond bur," *Procedia CIRP*, vol. 42, pp. 497–502, 2016.

[13] J. Y. Shen, J. Q. Wang, B. Jiang, and X. P. Xu, "Study on wear of diamond wheel in ultrasonic vibration-assisted grinding ceramic," *Wear*, vol. 332-333, pp. 788–793, 2015.

[14] W. M. Zeng, Z. C. Li, Z. J. Pei, and C. Treadwell, "Experimental observation of tool wear in rotary ultrasonic machining of advanced ceramics," *International Journal of Machine Tools & Manufacture*, vol. 45, no. 12-13, pp. 1468–1473, 2005.

[15] Z. Liang, X. Wang, Y. Wu, L. Xie, Z. Liu, and W. Zhao, "An investigation on wear mechanism of resin-bonded diamond wheel in elliptical ultrasonic assisted grinding (EUAG) of monocrystal sapphire," *Journal of Materials Processing Technology*, vol. 212, no. 4, pp. 868–876, 2012.

[16] K. Ding, Y. Fu, H. Su, X. Gong, and K. Wu, "Wear of diamond grinding wheel in ultrasonic vibration-assisted grinding of silicon carbide," *International Journal of Advanced Manufacturing Technology*, vol. 71, no. 9–12, pp. 1929–1938, 2014.

[17] W. R. do Nascimento, A. A. Yamamoto, H. J. de Mello,

R. C. Canarim, P. R. de Aguiar, and E. C. Bianchi, "A study on the viability of minimum quantity lubrication with water in grinding of ceramics using a hybrid-bonded diamond wheel," *Proceedings of the Institution of Mechanical Engineers, Part B: Journal of Engineering Manufacture*, vol. 230, no. 9, pp. 1630–1638, 2016.

[18] L. Zheng, B. Wang, B. Feng, J. Yuan, and Z. Wang, "Experimental study on drilling force of ceramic composite component," *Manufacturing Technology & Machine Tool*, vol. 68, no. 5, pp. 57–60, 2010, in Chinese.

16

Microstructure and Density of Sintered ZnO Ceramics Prepared by Magnetic Pulsed Compaction

Ji-Woon Lee [ID],[1] Changhyun Jin,[1] Soon-Jig Hong,[2] and Soong-Keun Hyun [ID][1]

[1]Department of Materials Science and Engineering, Inha University, Incheon 22212, Republic of Korea
[2]Division of Advanced Materials Engineering, Kongju National University, Kongju 32588, Republic of Korea

Correspondence should be addressed to Soong-Keun Hyun; skhyun@inha.ac.kr

Academic Editor: Mikhael Bechelany

Three different sintered ZnO ceramics were prepared by magnetic pulsed compaction (MPC) and other conventional methods. The microstructures of the sintered ZnO ceramics prepared by MPC at sintering temperatures ranging from 900 to 1300°C showed a homogeneous grain growth compared to those of the samples prepared using other methods under the same sintering conditions. This implies that interpowders and/or intergrains can induce the minimization of wall friction effect because of the application of a high compaction pressure for a short process time in the MPC method. In addition, the microstructure of the sample obtained using the cold isostatic pressing method showed the presence of heterogeneous regions, indicating its low quality even though the densities of the three different samples were almost similar in the range of 97–99% at sintering temperatures of 900, 1100, and 1300°C. Therefore, different methods used for the compaction of ZnO ceramics may result in different microstructural and physical properties of the product.

1. Introduction

Recently, ZnO has attracted increasing attention in various research areas, such as solar cells, transparent electrodes, varistors, gas sensors, and piezoelectric devices [1–4]. This is because ZnO exhibits high transmittance in the visible region, and its electrical resistivity can be easily controlled by doping with elements such as Al [5–8] and Ga [9–11]. ZnO is also advantageous from the economic and the environmental viewpoints owing to its low cost and nontoxic properties [12–14]. However, most of the recent studies are focused on low-dimensional ZnO nanostructures because the 0- and 1-dimensional nanoscale is relatively free from electric and magnetic barriers, and consequently, they exhibit diverse and complicated functions when compared to the bulk materials. In other words, studies on the comparison of sintered ZnO bulk ceramics prepared using different synthetic methods are very scarce because of the time-consuming multimanipulations. Most importantly, the sintered ZnO ceramics should exhibit high relative density and microstructural homogeneity to ensure the reliability of

the product. Sintered ZnO ceramics are generally prepared via a series of optimization processes consisting of compacting, sintering, and thin-film deposition. Among these steps, the compacting process is the most critical because the physical properties of the compact mainly determine the final optical and electrical properties of the prepared ZnO samples. For instance, at the first stage, if there are defects in a compact or the microstructure is not uniform, they critically affect the subsequent processes. Therefore, detailed analyses are required to synthesize high-quality compacts and to understand the effects of compacting conditions on the physical properties of the sintered ZnO ceramics [15–19]. In this study, three different compacting methods, namely, (1) conventional uniaxial pressing, (2) cold isostatic pressing (CIP), and (3) magnetic pulsed compaction (MPC), were used for fabricating high-quality sintered ZnO ceramics, and the properties of the three different samples were compared. In particular, the MPC technique is more advantageous than other methods because an extremely high pressure (~5 GPa) can be applied to the powders within a short period of time (~500 μs) [20]. The physical properties of the ZnO ceramics, including green and relative densities corresponding to the functional and microstructural changes in the samples, were evaluated. The comparative results were used to establish the effect of compacting conditions on the properties of the sintered ZnO ceramics.

2. Materials and Methods

The ZnO raw powders (99.9% purity) used in the three different synthetic methods were purchased from Kojundo Chemical Laboratory, Japan. The average particle size of the ZnO raw powder was approximately 3.285 μm. Polyvinyl alcohol (OCI, Korea) used as a binder for the ZnO powder was mixed with distilled water. Then, the mixture of the slurry binder and the ZnO powder was completely dried to obtain bulk ZnO.

2.1. Compacting Methods. The three different compacting methods used for preparing the sintered ZnO ceramics are described as follows:

(1) In uniaxial pressing, the ZnO powders were pressed into a steel mold after lubricating the inside of the steel mold with a BN spray for easy detachment of the ZnO ceramics from the steel mold. The mold diameter was 20 mm, and the maximum pressure used for compacting ZnO was 165 MPa, which was applied for 1 min.

(2) In CIP, the ZnO powders were compacted under a pressure of 180 MPa for 10 min after uniaxial pressing at 119 MPa for 1 min.

(3) In MPC, the ZnO powders were compacted by MPC uniaxial pressing at 800 MPa for approximately 500 μs. The notable feature of the MPC method is that the applied pressure and time are much higher and shorter, respectively, than those of the uniaxial pressing and the CIP methods. Table 1 presents the details of the compacting conditions, such as mass, lubrication condition, diameter, pressure, and time used for the different methods.

2.2. Sintering. All the specimens were preheated at 400°C for 1 h in air to burn off the binder from the ZnO samples. Then, continuous sintering was performed at 900, 1100, and 1300°C for 1, 2, and 4 h without the use of supporting gases, respectively. The heating rate for all the samples was set at 5°C·min^{-1}. After the sintering process, the samples were cooled in the furnace.

2.3. Characterization. The green densities and the relative densities of the cylindrical ZnO samples were calculated from the weight to volume ratio. Furthermore, while calculating the relative density, Archimedes' principle was applied using distilled water for factual accuracy. For studying the microstructural properties, all the sintered samples were gradually polished with SiC paper and 1 μm diamond paste. Subsequently, the smooth surfaces of the

TABLE 1: Compacting conditions used for uniaxial pressing (Uni), cold isostatic pressing (CIP), and magnetic pulsed compaction (MPC) processes.

Compact mass		11.22 g
Lubrication condition		BN sprayed
Mold diameter	Uni	20 mm
	MPC	
Compacting pressure	Uni	165 MPa
	CIP (Uni)	180 (119) MPa
	MPC	800 MPa
Compacting time	Uni	1 min
	CIP	10 min
	MPC	~500 μs

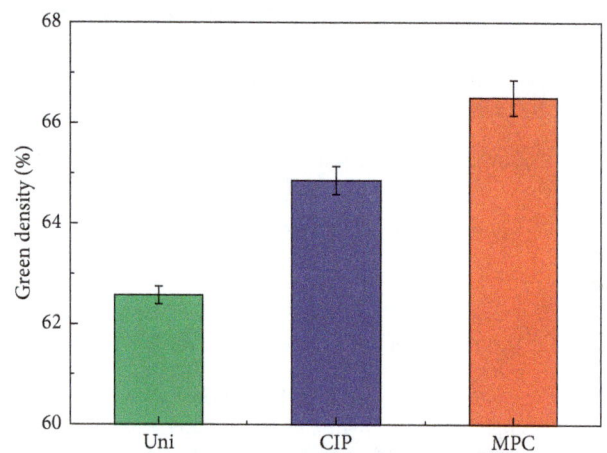

FIGURE 1: Variations in the green densities of the ZnO ceramics prepared using the three compacting methods (uniaxial pressing, CIP, and MPC) with respect to the sintering temperature and duration.

specimens were etched with hydrochloric acid for 30 s. The microstructural observations of parallel and perpendicular cross sections were performed using a Keyence VHX-900 optical microscope and a Hitachi S-4300SE scanning electron microscope. The mean grain sizes were measured using the image analysis software.

3. Results and Discussion

3.1. Compacting Properties. Figure 1 shows the comparison of the green densities of the ZnO compacts prepared by the three compacting methods. The green densities of the samples prepared using the different methods are approximately in the range of 60 to 68%. It appears that there was no significant difference in the value of green density even though the compacting pressure used in MPC was 4.8 times larger than the compacting pressure used in uniaxial pressing (Table 1). The green density of the uniaxially pressed compacts was 62.6%, and that of the CIP compacts with uniaxial pressing was 64.9%. The maximum green density of 66.5% was obtained using the MPC method. Based on this result, we surmise that the relationship between the compacting pressure and the green density is not linear. In addition, it also seems to be converging to a certain maximum value. Nevertheless, it is important to determine the best method among the three because the green density can generally affect the final relative density at the same sintering conditions [21].

3.2. Sintering Properties. The relative densities of the three different ZnO ceramics sintered at different temperatures and durations are shown in Figure 2. The densities of all the sintered ZnO ceramics increased with increase in the sintering temperature. This is correlated with the number, and the size of pores decreases as the sintering temperature increases [22]. For example, the samples compacted by the three different methods exhibited densities greater than 97, 98, and 99%, at 900, 1100, and 1300°C, respectively. However, the relative density values of the samples compacted using the MPC method at 1300°C were slightly lower than those of the samples compacted using the uniaxial pressing and CIP methods. While the effect of sintering temperature on the relative density of the compacted sample is consistent, there is no direct relationship between the sintering time and the density except for samples compacted using MPC, as can be seen from Figure 2. This is in contrast to a previous report, in which the relative density of ZnO ceramics was found to be dependent on sintering time in the isothermal sintering test [23]. Therefore, the inconsistent results related to the sintering time observed in the present study can be associated with the nonisothermal sintering test arising from the densification and the consequent grain growth phenomena during the sintering processes.

3.3. Microstructural Properties. Figure 3 shows the variations in the grain sizes of the different sintered ZnO ceramics. At 900°C, no grain growth was observed regardless of the compacting method and the sintering time. Only the densification phenomena occurred because the supplied thermal energy at 900°C was not sufficient for atomic diffusion from one site to the other. However, at 1100 and 1300°C, notable grain growth occurred, and the grain size increased with increase in the temperature. The accelerating tendency of the grain growth was more evident in the samples compacted using the MPC method. In other words, the grain size of the sintered ZnO ceramics prepared using the MPC method was larger than that of the samples prepared using the other compacting methods at the same sintering conditions. However, Ahn et al. [24] studied the effects of green density on grain growth and reported that a high green density can suppress the grain growth during sintering. Therefore, the unusual trend in the grain growth observed in the samples prepared using the MPC method is believed to be caused by the aggregation of minute amounts of grains with sizes smaller than the critical grain size.

Figures 4 and 5 show the results of microstructural observations at the central (marked as (a)) and the outer (marked as (b)) regions of the parallel and the perpendicular cross sections of the samples prepared by CIP (Figure 4) and

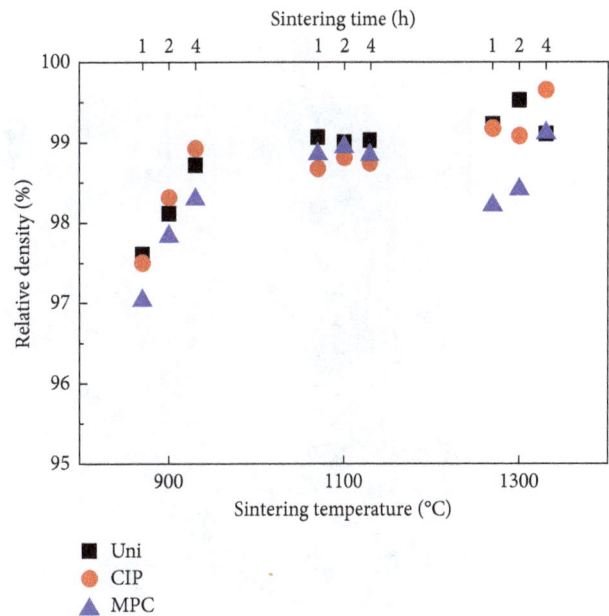

FIGURE 2: Variations in the relative densities of the sintered ZnO ceramics prepared using the three compacting methods (uniaxial pressing, CIP, and MPC) with respect to the sintering temperature and duration.

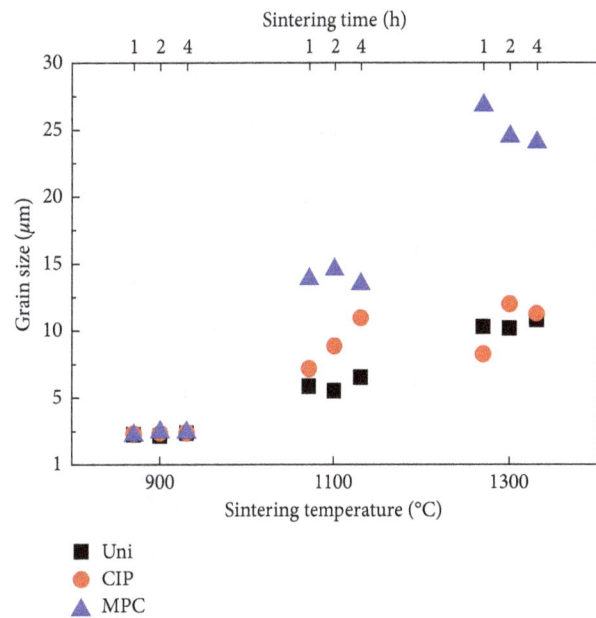

FIGURE 3: Variations in the grain sizes of the sintered ZnO ceramics prepared using the three compacting methods (uniaxial pressing, CIP, and MPC) with respect to the sintering temperature and duration.

MPC (Figure 5), respectively. The sintered ZnO ceramics synthesized using the CIP method showed a significant difference in the grain size at the center and the outer regions irrespective of the direction of the cross section. In other words, the grain size and the density at the central region of the ZnO ceramics compacted using the CIP method are

	Sintered body	(a) Center	(b) Outside
Parallel cross section			
Perpendicular cross section			

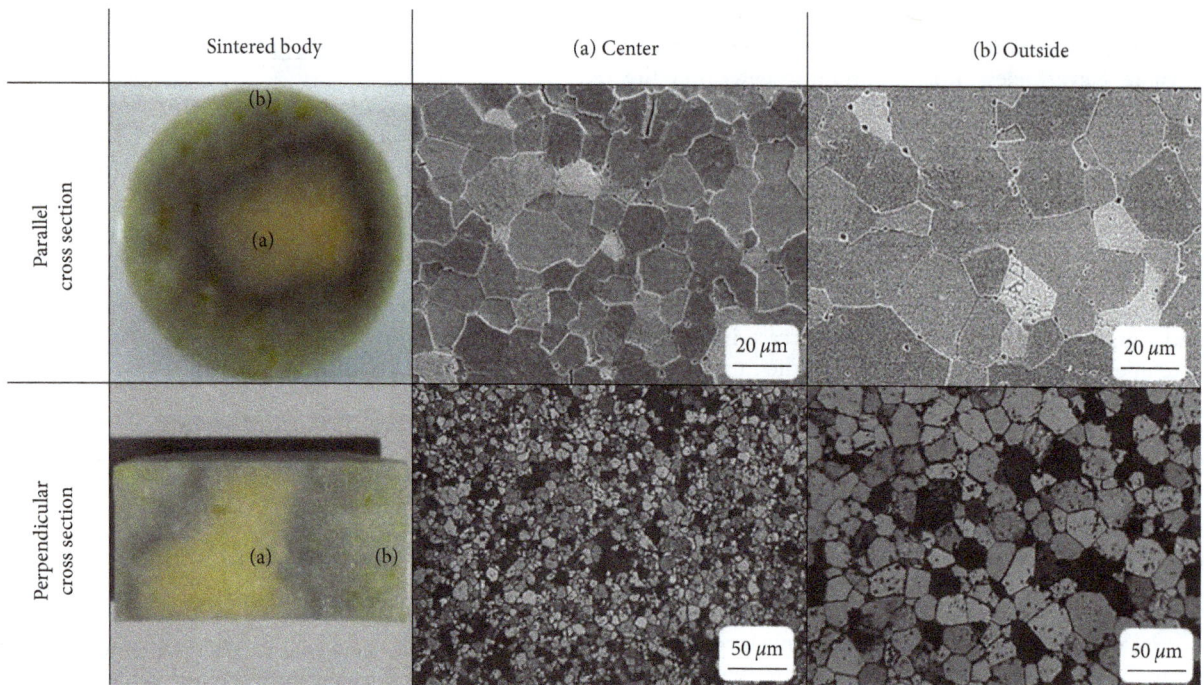

FIGURE 4: Perpendicular and parallel cross-sectional microstructures of the ZnO ceramics compacted by uniaxial pressing and CIP and sintered at 1300°C for 2 h recorded from (a) central region and (b) outer region.

	Sintered body	(a) Center	(b) Outside
Parallel cross section			
Perpendicular cross section			

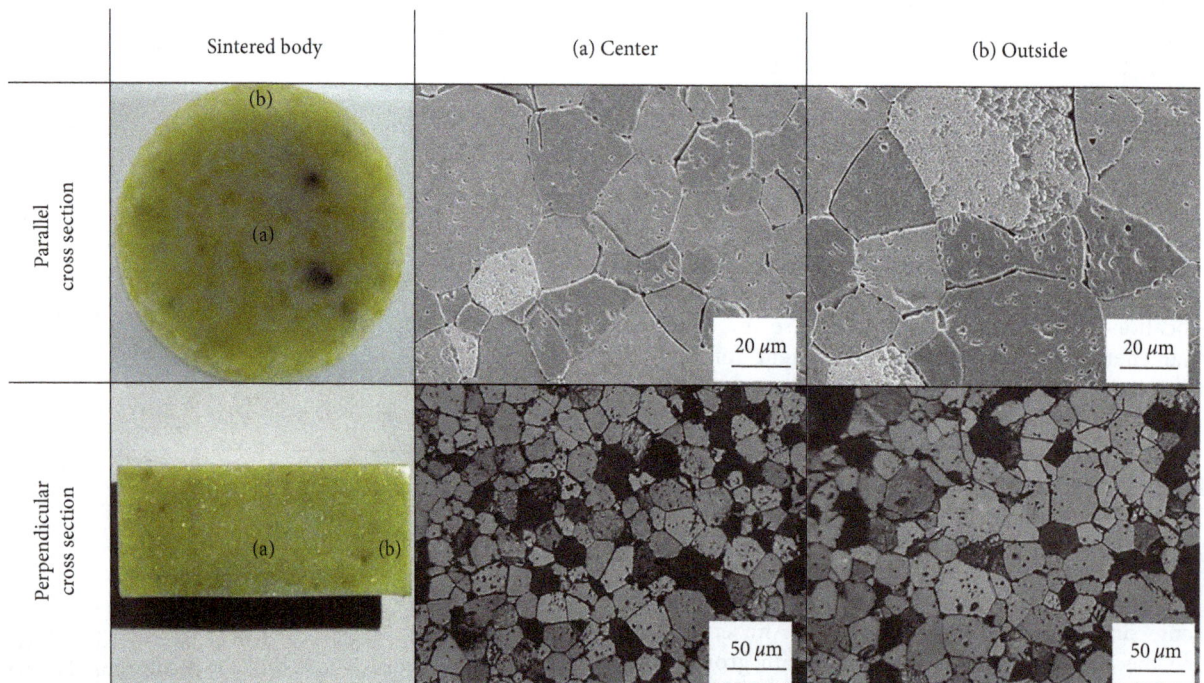

FIGURE 5: Perpendicular and parallel cross-sectional microstructures of the ZnO ceramics compacted by MPC and sintered at 1300°C for 2 h recorded from (a) central region and (b) outer region.

slightly smaller and higher, respectively, than those at the outer region (Figure 4). However, the sintered ZnO ceramics compacted using the MPC method showed a similar grain size and density at the central and the outer regions. This indicates increased grain growth and more homogeneous grain size distribution in the sintered ZnO ceramics compacted using the MPC method than the samples prepared using the other compacting methods at the same sintering conditions. The grain size distributions of all the samples are given in Table 2.

TABLE 2: Comparison of the grain sizes of the central and outer regions in the sintered specimens prepared using the three compacting methods (uniaxial pressing, CIP, and MPC).

Sintering temperature (°C)	Compacting method	Grain size (μm)					
		1 h		2 h		4 h	
		Center	Side	Center	Side	Center	Side
900	Uni	3.285 (no grain growth)					
	CIP						
	MPC						
1100	Uni	6.66	17.66	6.34	18.47	7.31	17.39
	CIP	7.97	16.42	9.59	17.97	11.64	19.41
	MPC	14.44	16.76	15.13	19.03	14.06	23.50
1300	Uni	10.98	18.21	10.85	20.27	11.46	25.89
	CIP	9.02	18.85	12.65	23.00	11.94	24.71
	MPC	26.94	29.29	24.71	37.37	24.29	34.29

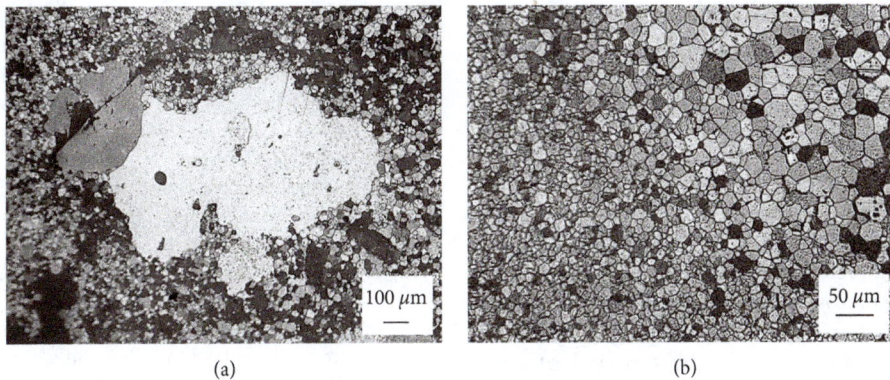

(a)   (b)

FIGURE 6: Perpendicular and parallel cross-sectional microstructures of the ZnO ceramics compacted by MPC and sintered at 1300°C for 2 h recorded from (a) central region and (b) outer region.

The microstructures of the sintered ZnO sample prepared by uniaxial pressing and CIP at 1300°C are presented in Figure 6. Compared with normal microstructures, two specialized regions existed: one corresponding to abnormal grain growth (Figure 6(a)) and the other corresponding to grain size transition (Figure 6(b)). The abnormal grain growth was not observed in the sintered ZnO ceramics prepared using the MPC method, whereas it was observed in the samples prepared by uniaxial pressing and CIP as shown in Figure 6(a). Furthermore, the grain size transition region was observed in the samples prepared by uniaxial pressing and CIP at 1300°C, while it did not exist in the sample prepared using the MPC method (Figure 6(b)). The main cause for the abnormal grain growth and the grain size transition is probably the relative density inhomogeneity arising from the friction between the mold wall and the ZnO powders during the compaction process as it has been reported previously [25, 26].

Figure 7 shows the effect of relative density inhomogeneity on the microstructure of ZnO samples prepared by uniaxial pressing and CIP. The grain growth tendency at the outer region is gradually increased because of its higher relative density. On the other hand, the onset of grain growth at the center is delayed. Owing to the instability arising from the differences in the grain growth at different regions, the region of abnormal grain growth was observed only at the

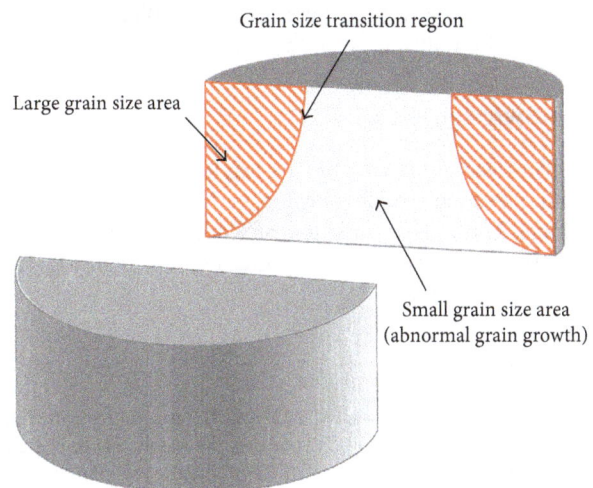

FIGURE 7: Schematic diagram of the sintered ZnO ceramic compacted by uniaxial pressing and CIP.

center. However, the grain size homogeneity observed in the MPC method is attributed to the high compacting pressure (800 MPa) applied for an extremely short period of time (~500 μs). In short, the MPC method minimizes the effects of wall friction that causes nonuniformity in the grain size.

4. Conclusions

In this study, pure ZnO ceramics were compacted by MPC, and their physical properties such as green and relative densities and the microstructure were compared with those of the ZnO samples compacted by uniaxial pressing and CIP. The following conclusions were drawn from the experimental results:

(1) The green density of the MPC compacts was 66.5%, which is the highest value when compared to the densities of the samples prepared by uniaxial pressing (65.6%) and CIP (64.9%). The relative density was greater than 97, 98, and 99%, respectively, at 900, 1100, and 1300°C regardless of the compacting methods.

(2) The grain growth tendency of the sintered ZnO ceramics increased with increase in the temperature from 1100 to 1300°C. This is attributed to the aggregation of minute amounts of grains with sizes smaller than the critical grain size.

(3) The sintered ZnO ceramics compacted using the MPC method showed a homogeneous grain size at the inner and outer regions of the samples without the formation of abnormal grain growth and grain size transition regions. The MPC method can be used to achieve microstructural homogeneity because of the application of a high compacting pressure (800 MPa) for an extremely short period of time ($\sim$500 μs).

Conflicts of Interest

The authors declare that there are no conflicts of interest regarding the publication of this paper.

Acknowledgments

This work was supported by The Leading Human Resource Training Program of Regional Neo Industry through the National Research Foundation of Korea (NRF) funded by the Ministry of Science, ICT and Future Planning (no. 2016H1D5A1910612).

References

[1] V. Tiron, L. Sirghi, and G. Popa, "Control of aluminium doping of ZnO:Al thin films obtained by high-power impulse magnetron sputtering," *Thin Solid Films*, vol. 520, no. 13, pp. 4305–4309, 2012.

[2] H. Kim, J. S. Horwitz, W. H. Kim, A. J. Makinen, Z. H. Kafafi, and D. B. Chrisey, "Doped ZnO thin films as anode materials for organic light-emitting diodes," *Thin Solid Films*, vol. 420-421, no. 2, pp. 539–543, 2002.

[3] L. Cai, G. Jiang, C. Zhu, and D. Wang, "High quality Al-doped ZnO thin films deposited using targets prepared by chemical coprecipitation," *Physica Status Solidi A*, vol. 206, no. 7, pp. 1461–1464, 2009.

[4] N. Neves, R. Barros, E. Antunes et al., "Sintering behavior of nano- and micro-sized ZnO powder targets for rf magnetron sputtering applications," *Journal of American Ceramic Society*, vol. 95, no. 1, pp. 204–210, 2012.

[5] D. Xu, Z. Deng, Y. Xu et al., "An anode with aluminum doped on zinc oxide thin films for organic light emitting devices," *Physics Letters A*, vol. 346, no. 1–3, pp. 148–152, 2005.

[6] Y. H. Chou, J. L. H. Chau, W. L. Wang, C. S. Chen, S. H. Wang, and C. C. Yang, "Preparation and characterization of solid-state sintered aluminum-doped zinc oxide with different alumina contents," *Bulletin of Materials Science*, vol. 34, no. 3, pp. 477–482, 2011.

[7] C. P. Liu and G. R. Jeng, "Properties of aluminum doped zinc oxide materials and sputtering thin films," *Journal of Alloys and Compounds*, vol. 468, no. 1-2, pp. 343–349, 2009.

[8] J. Zhang, W. Zhang, E. Zhao, and H. J. Jacques, "Study of high-density AZO ceramic target," *Materials Science in Semiconductor Processing*, vol. 14, no. 3-4, pp. 189–192, 2011.

[9] H. Makino, Y. Sato, N. Yamamoto, and T. Yamamoto, "Changes in electrical and optical properties of polycrystalline Ga-doped ZnO thin films due to thermal desorption of zinc," *Thin Solid Films*, vol. 520, no. 5, pp. 1407–1410, 2011.

[10] M. Miyazaki, K. Sato, A. Mitsui, and H. Nishimura, "Properties of Ga-doped ZnO films," *Journal of Non-Crystalline Solids*, vol. 218, pp. 323–328, 1997.

[11] Y. Sato, H. Makino, N. Yamamoto, and T. Yamamoto, "Structural, electrical and moisture resistance properties of Ga-doped ZnO films," *Thin Solid Films*, vol. 520, no. 5, pp. 1395–1399, 2011.

[12] J. I. Nomoto, M. Konagai, K. Okada, T. Ito, T. Miyata, and T. Minami, "Modeling of phosphorus diffusion in Ge accounting for a cubic dependence of the diffusivity with the electron concentration," *Thin Solid Films*, vol. 518, no. 11, pp. 2937–2940, 2010.

[13] C. S. Hsi, B. Houng, B. Y. Hou, G. J. Chen, and S. L. Fu, "Effect of Ru addition on the properties of Al-doped ZnO thin films prepared by radio frequency magnetron sputtering on polyethylene terephthalate substrate," *Journal of Alloys and Compounds*, vol. 464, no. 1-2, pp. 89–94, 2008.

[14] H. S. Huang, H. C. Tung, C. H. Chiu et al., "Highly conductive alumina-added ZnO ceramic target prepared by reduction sintering and its effects on the properties of deposited thin films by direct current magnetron sputtering," *Thin Solid Films*, vol. 518, no. 21, pp. 6071–6075, 2010.

[15] N. Neves, R. Barros, E. Antunes et al., "Aluminum doped zinc oxide sputtering targets obtained from nanostructured powders: processing and application," *Journal of European Ceramic Society*, vol. 32, no. 16, pp. 4381–4391, 2012.

[16] Y. H. Sun, W. H. Xiong, C. H. Li, and L. Yuan, "Effect of dispersant concentration on preparation of an ultrahigh density ZnO–Al$_2$O$_3$ target by slip casting," *Journal of the American Ceramic Society*, vol. 92, no. 9, pp. 2168–2171, 2009.

[17] T. Senda and R. C. Bradt, "Grain growth in sintered ZnO and ZnO-Bi$_2$O$_3$ ceramics," *Journal of American Ceramic Society*, vol. 73, no. 1, pp. 106–114, 1990.

[18] M. Mazaheri, S. A. Hassanzadeh-Tabrizi, and S. K. Sadrnezhaad, "Hot pressing of nanocrystalline zinc oxide compacts: densification and grain growth during sintering," *Ceramics International*, vol. 35, no. 3, pp. 991–995, 2009.

[19] X. J. Qin, G. J. Shao, R. P. Liu, and W. K. Wang, "Sintering characteristics of nanocrystalline ZnO," *Journal of Maters Science*, vol. 40, no. 18, pp. 4943–4946, 2005.

[20] S. J. Hong, R. Rumman, and C. K. Rhee, "Effect of magnetic pulsed compaction (MPC) on sintering behavior of materials," in *Sintering of Ceramics–New Emerging Techniques*, Chapter 8, InTech, Rijeka, Croatia, 2012.

[21] M. N. Rahaman, *Sintering of Ceramics*, CRC Press, New York, NY, USA, 2007.

[22] J. W. Lee, J. S. Lee, M. G. Kim, and S. K. Hyun, "Fabrication of porous titanium with directional pores for biomedical applications," *Materials Transactions*, vol. 54, no. 2, pp. 137–142, 2013.

[23] T. K. Gupta and R. L. Coble, "Sintering of ZnO: I, densification and grain growth," *Journal of American Ceramic Society*, vol. 51, no. 9, pp. 521–525, 1968.

[24] J. P. Ahn, J. K. Park, and M. Y. Huh, "Effect of green density on the subsequent densification and grain growth of ultrafine SnO_2 powder during isochronal sintering," *Journal of American Ceramic Society*, vol. 80, no. 8, pp. 2165–2167, 1997.

[25] H. M. Macleod and K. Marshall, "The determination of density distribution in ceramic compacts using autoradiography," *Powder Technology*, vol. 16, no. 1, pp. 107–122, 1997.

[26] B. J. Briscoe and S. L. Rough, "The effects of wall friction in powder compaction," *Colloids and Surfaces A: Physicochemical and Engineering Aspects*, vol. 137, no. 1–3, pp. 103–116, 1998.

17

Novel Low-Permittivity $(Mg_{1-x}Cu_x)_2SiO_4$ Microwave Dielectric Ceramics

Chengxi Hu ⑩,[1] Yuan Liu,[2] Wujun Wang,[1] and Bo Yang[3]

[1]*Faculty of Science, Xi'an Aeronautical University, No. 259, West 2nd Ring, Xi'an 710077, China*
[2]*School of Physics and Information Technology, Shaanxi Normal University, Xi'an 710062, China*
[3]*Surface and Interface Science Laboratory, RIKEN, 2-1 Hirosawa, Wako-shi, Saitama 351-0198, Japan*

Correspondence should be addressed to Chengxi Hu; huchengxi@163.com

Academic Editor: Francesco Ruffino

The effects of B_2O_3–LiF addition on the phase composition, microstructures, and microwave dielectric properties of $(Mg_{0.95}Cu_{0.05})_2SiO_4$ ceramics fabricated by a wet chemical method were studied in detail. The B_2O_3–LiF was selected as liquid-phase sintering aids to reduce the densification sintering temperature of $(Mg_{0.95}Cu_{0.05})_2SiO_4$ ceramics. The B_2O_3 6%–Li_2O 6%-modified $(Mg_{0.95}Cu_{0.05})_2SiO_4$ ceramics sintered at 1200°C possess good performance of $\varepsilon_r \sim 4.37$, $Q \times f \sim 36,700$ GHz and $\tau_f \sim 42$ ppm/°C.

1. Introduction

The rapid development in modern communications and Internet technology has created an urgent demand for the development of microwave ceramics [1–5]. Such microwave equipments require light weight, low loss, small size, and good temperature stability. Therefore, the materials used for the microwave substrates and dielectric resonators should have low dielectric constant (ε_r), high quality factor ($Q \times f$), and near-zero temperature coefficient of resonator frequency (τ_f). Mg_2SiO_4 is an essential material with low ε_r and high $Q \times f$. So it is considered to be a promising candidate as a low-dielectric-constant microwave material for applications in microwave substrates [2–9]. To our knowledge, the relation between the microwave dielectric properties and the structure for the Cu^{2+}-doped Mg_2SiO_4 system has not yet been investigated. Moreover, the sintering temperature of $(Mg_{1-x}Cu_x)_2SiO_4$ is too high in the recently reported microwave ceramics. Thus, 12 wt.% B_2O_3–LiF was used as sintering aids to reduce the densification sintering temperature down to a lower temperature range. In this work, we have reported the $(Mg_{1-x}Cu_x)_2SiO_4$ ceramics by a wet chemical process. The effects of additive B_2O_3–LiF on the phase compositions, microstructures, and microwave dielectric behavior were studied.

2. Experimental Procedure

SiO_2 nanospheres were prepared by the sol-gel method, and then the powders were used as raw material to synthesize the SiO_2-based ceramics. The microwave dielectric ceramics with high performance were successfully prepared in our previous work. The $(Mg_{1-x}Cu_x)_2SiO_4$ ceramics belong to SiO_2-based ceramics. Following the method of Hu et al. [1], it was quite expected to achieve success. All experiments were conducted in air. All reagents were analytical pure and used as received without further purification. $(Mg_{1-x}Cu_x)_2SiO_4$ powders were prepared using the sol-gel method. For this purpose, $Mg(NO_3)_2 \cdot 6H_2O$ and $CuSO_4 \cdot 5H_2O$ were dissolved individually in 100 ml ethanol and 10 ml distilled water and then added into the mixture under magnetic stirring, marked as A. Tetraethyl orthosilicate (TEOS) and $CuSO_4$ were used as received. The starting sol was prepared by hydrolysis of TEOS under magnetic stirring in the presence of 100 ml alcohol solution in 40°C water for 30 min, marked as B. Subsequently, the solution B was added to A by continuous vigorous magnetic stirring for 3 h at 40°C in water and then kept at 60°C overnight to allow gel formation. Finally, the obtained products were dried and calcined at 900°C for 4 h in an alumina crucible. The calcined powders were milled with ZrO_2 balls in ethanol for 12 h and dried to obtain

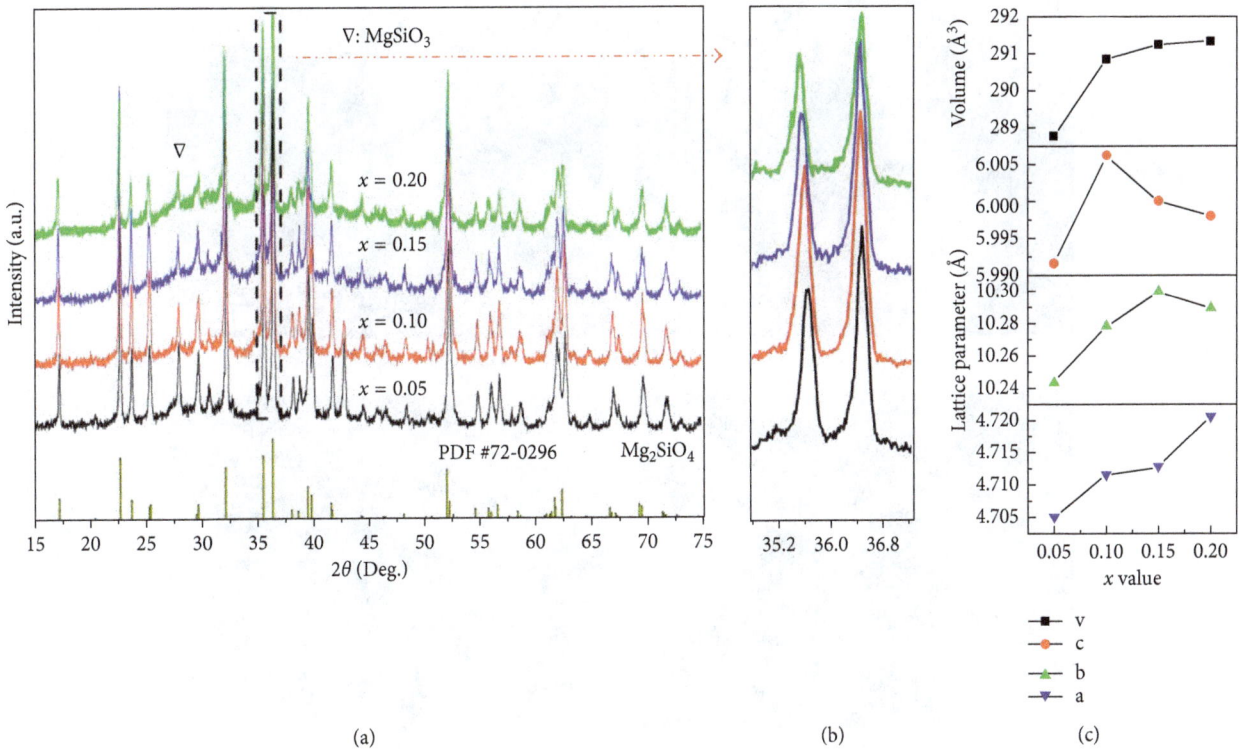

FIGURE 1: (a) XRD patterns of $(Mg_{1-x}Cu_x)_2SiO_4$ ($x = 0.05$–0.20) ceramics sintered at 1300°C for 4 h; (b) locally magnified peak profiles indicated in (a); (c) the lattice parameters as a function of the x value.

$(Mg_{1-x}Cu_x)_2SiO_4$ powders. The prepared powders were divided into two groups. The first group was not doped, while the second group was doped with B_2O_3–LiF. Both of them with polyvinyl alcohol water solution were pressed individually into cylindrical specimen with a diameter of 11.5 mm under a uniaxial pressure of 120 MPa. The undoped specimens were placed in an alumina crucible and heated to the sintering temperatures varying from room temperature to 1350°C with a rate of 5°C/min. After sintering for 4 h in air atmosphere, the specimens were freely cooled down to room temperature inside the furnace. The B_2O_3–LiF co-doped $(Mg_{1-x}Cu_x)_2SiO_4$ ceramics were sintered at 1000–1350°C for 4 h.

The crystalline-phase structure was identified by D/max-2550V/PC X-ray diffractometer (Rigaku, Tokyo, Japan) with Cu-$\kappa\alpha$ radiation (at 40 kv and 20 mA) at a scan rate $2\theta = 0.02\,s^{-1}$. The morphology of the samples was characterized by a scanning electron microscope (SEM, FEI-quanta 200, USA) and a high-resolution transmission electron microscope (JEM-2100, Japan). The X-ray photoelectron spectroscopy (XPS, Kratos Analytical Ltd., Japan) with a monochromatic Al was conducted to examine the microstructure for fracture surface of ceramics.

The bulk densities (ρ) of the samples were measured by the Archimedes method using distilled water as medium. The microstructures were observed on the as-sintered ceramic surfaces of samples by a scanning electron microscope (SEM, FEI-quanta 200, USA). The phase evolution of the $(Mg_{1-x}Cu_x)_2SiO_4$ ceramics was performed by XRD. The as-sintered cylindrical specimens with the thickness of about 5 mm and diameter of about 10 mm were used for evaluating the microwave dielectric

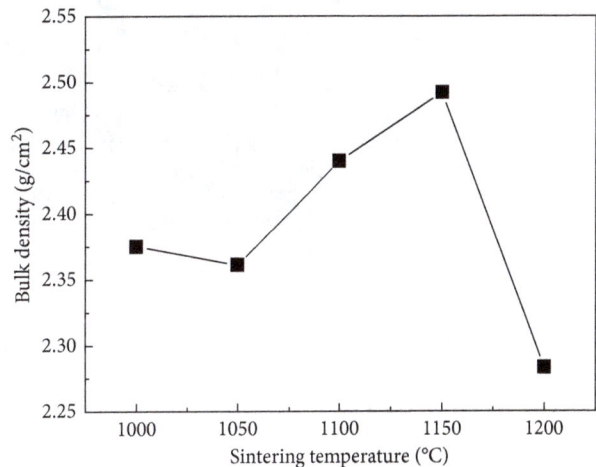

FIGURE 2: The variation of bulk density of $(Mg_{0.95}Cu_{0.05})_2SiO_4$ with 12% B_2O_3–LiF sintered at (a) 1000°C, (b) 1050°C, (c) 1100°C, (d) 1150°C, and (e) 1200°C.

properties. The dielectric constant and quality values were determined by the $TE_{01\delta}$ shielded cavity method using a vector network analyzer (ZVB20, Rohde & Schwarz, Munich, Germany). The temperature coefficient of the resonant frequency (τ_f) was calculated with the following formula:

$$\tau_f = \frac{f_{80} - f_{25}}{f_{25} \times (80 - 25)}, \tag{1}$$

where f_{25} and f_{80} represent the $TE_{01\delta}$ resonant frequency measured at 25°C and 80°C, respectively.

FIGURE 3: SEM images of $(Mg_{0.95}Cu_{0.05})_2SiO_4$ with 12% B_2O_3–LiF sintered at (a) 1000°C, (b) 1050°C, (c) 1100°C, (d) 1150°C, (e) 1200°C, and (f) EDS pattern of microdomain marked in Figure 3(d).

3. Results and Discussion

Figure 1(a) depicts the XRD patterns of $(Mg_{1-x}Cu_x)_2SiO_4$ ($x = 0.05$–0.20) ceramics sintered at 1350°C for 4 h. It could be seen that all the main diffraction peaks can be well indexed to the standard patterns of Mg_2SiO_4 (PDF#72-0296), indicating that the forsteritic-olivine solid solution with a single phase was formed. A very little protoenstatite Mg_2SiO_3 secondary phase appeared along with the main phase Mg_2SiO_4 in all compositions, which is similar to the results of Li et al. [10]. This result might be attributed to the fact that the amount of miscellaneous phase MgO is enough so that it could not react with $MgSiO_3$ to form Mg_2SiO_4. Moreover, as shown in Figure 1(b), with increasing the Cu content (x), the main peaks of Mg_2SiO_4 shift slightly toward lower angles, indicating an increase of unit-cell volume. Furthermore, the refined lattice parameters were also plotted as a function of Cu content (x) as

shown in Figure 1(c), a near linear dependence between the lattice parameters (except c) and x value can be found, which is in agreement with the Vegard's law and also confirms the formation of a solid solution.

Figure 2 showed the bulk density of the $(Mg_{0.95}Cu_{0.5})_2SiO_4$ with 12% B_2O_3–LiF ceramics sintered at 1000°C was low, about 2.38 g/cm^2, but increased with increasing sintering temperature to a maximum value of 2.48 g/cm^2 for the specimen sintered at 1150°C and then reduced with a further increase in the temperature. The 12 wt.% H_3BO_3 added ceramics shows relative densities of 89%. The low density of magnesium silicate ceramics can be caused by two factors: the first is its crystallographic anisotropy, and the other is the hollow formed by the sintering agent at high temperature.

The SEM micrographs of the $(Mg_{0.95}Cu_{0.5})_2SiO_4$ with 12% B_2O_3–LiF ceramics sintered at 1000–1200°C for 4 h in air are shown in Figures 3(a)–3(e). Figure 3 shows that the

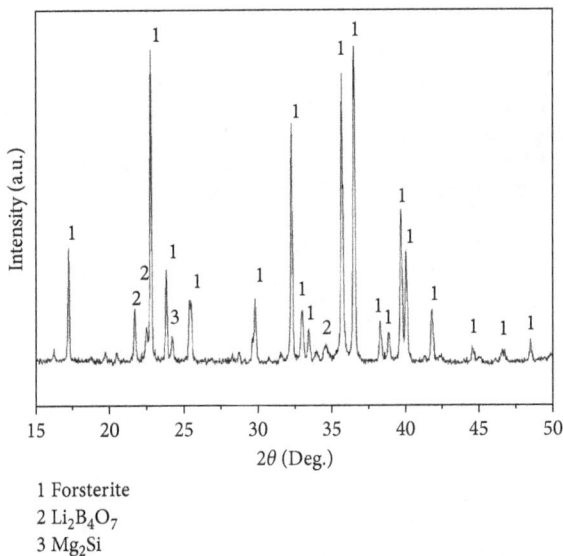

FIGURE 4: The XRD pattern of $(Mg_{0.95}Cu_{0.05})_2SiO_4$ with 12% B_2O_3–LiF sintered at 1150°C.

1 Forsterite
2 $Li_2B_4O_7$
3 Mg_2Si

grains became lager and denser with increasing sintering temperature. Three types of grains were observed in the specimens of B_2O_3–LiF-doped $(Mg_{0.95}Cu_{0.5})_2SiO_4$ ceramics. In order to understand the distribution of the elements in the sample, the EDS of $(Mg_{0.95}Cu_{0.5})_2SiO_4$ with 12% B_2O_3–LiF ceramics sintered at 1150°C for 4 h is shown in Figure 3(f). The results presented the ratio of Mg : Si : O of rod-shaped grain (sport A and B) which is approximately 2 : 1 : 4, which agreed with the composition of forsterite. Li and coworkers also observed Mg_2SiO_4 with similar shape [10]. It seems impossible for the EDS detector to detect boron and lithium ions, which led to the restricted detection of B_2O_3–LiF. The small grains are to be Mg_2Si and $Li_2B_4O_7$, which is in accordance with the XRD analysis as shown in Figure 4.

Figure 5 showed the dielectric constant, quality factor, and τ_f of the $(Mg_{0.95}Cu_{0.5})_2SiO_4$ with 12% B_2O_3–LiF ceramics sintered at various temperatures for 4 h, respectively. The ε_r is mainly dominated by the structural characteristics, dielectric polarizability, and relative density of the ceramics [11–13]. It can be seen from Figure 5(a) that the variation trend of ε_r with sintering temperatures was not in agreement with that of relative density. It is noted that the sample had lower ε_r (4.37) when the sintering temperature was 1150°C. The $Q \times f$ values of the $(Mg_{0.95}Cu_{0.5})_2SiO_4$ with 12% B_2O_3–LiF ceramics sintered at 1000°C were low (15,000 GHz at 12.36 GHz) due to the low density, small grain size, and porous microstructure (as shown in Figure 3(a)). But, it increased with increasing sintering temperature to a value of 36,700 GHz for the specimen sintered at 1150 then seldom decreased to 32,600 GHz. The optimum sintering temperature was reduced by 200°C compared to the results of Li et al. Based on the classical dielectric theory, the value should increase as the grain size increases, because a reduction in the number of grain boundaries per unit volume would result in materials with a lower loss [12–15]. In fact, the ceramics had the highest $Q \times f$ values when the mean grain size of the ceramics was about 8~17 μm as shown in Figure 3.

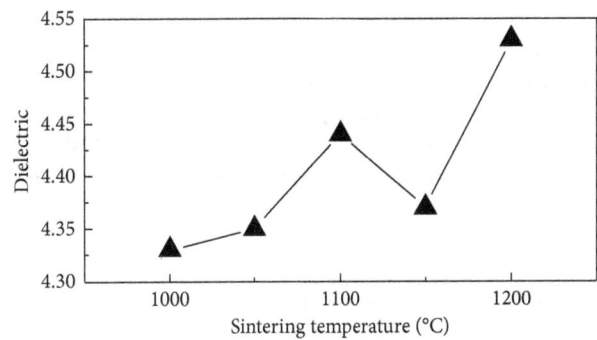

(a)

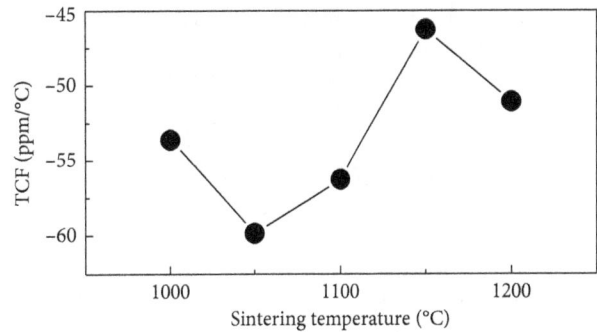

(b)

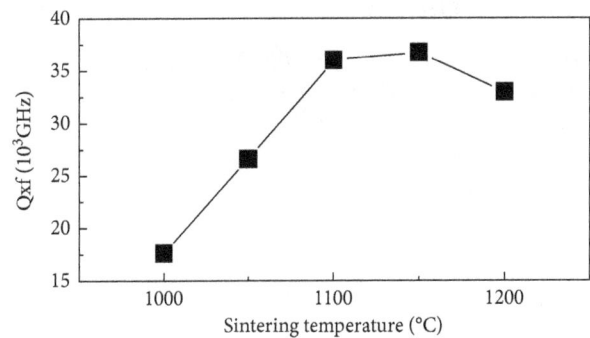

(c)

FIGURE 5: The dielectric constant ε_r, quality factor $Q \times f$, and temperature coefficient of resonant frequency τ_f of $(Mg_{0.95}Cu_{0.05})_2SiO_4$ with 12% B_2O_3–LiF as a function of sintering temperature.

Li et al. also observed similar result in our previous work [10]. This could be due to the interaction of various factors, such as distribution of element, porosity, grain size, presence of liquid-phase, and so on, which made it too difficult to reveal definitive remarks on grain size loss relationships [15–18]. The τ_f values of all ceramics, on the other hand, appeared to be a slight change (from −60~−40 ppm/°C) to sintering temperature, as shown in Figure 5. Generally, glass materials of lower melting temperature were mixed with the ceramic materials to degrade the sintering temperature. However, network formers contained in the glass ceramics may absorb the microwave power deeply at microwave frequency band, deteriorating the microwave performance for the material. Chang et al. reported that Li_3BO_3 ceramic possesses the microwave dielectric properties of $\varepsilon_r \sim 5$,

$Q \times f \sim 37{,}200\,\mathrm{GHz}$, and $\tau_\mathrm{f} \sim 3.1\,\mathrm{ppm/^\circ C}$ [19]. In this work, the addition of 12 wt.% of B_2O_3–LiF effectively reduced the sintering temperature but worsened the microwave performance of the ceramic. It is an important task to explore the low sintering temperature, high density, and high-performance Mg_2SiO_4 ceramics in future work.

4. Conclusions

The microwave performance of $(Mg_{1-x}Cu_x)_2SiO_4$ ($x = 0.05$–0.20) ceramics were studied in terms of their microstructure and structural characteristics, as well as the sintering behavior. The results show that the Cu substitution not only obviously enhances the sintering activity but also improves the microwave performance of the ceramics. The $(Mg_{0.95}Cu_{0.5})_2SiO_4$ with 12% B_2O_3–LiF ceramics sintered at 1150°C for 4 h achieved excellent microwave dielectric properties of $\varepsilon_\mathrm{r} = 4.37$, $Q \times f = 36{,}700\,\mathrm{GHz}$, and $\tau_\mathrm{f} = -42.6\,\mathrm{ppm/^\circ C}$. It can be used as a promising microwave substrate and LTCC material [20].

Conflicts of Interest

The authors declare that there are no conflicts of interests regarding the publication of this paper.

Acknowledgments

This work was supported by the National Natural Science Foundation of China (NSFC) (Project nos. 51272150, 51072110, and 51402235).

References

[1] C. Hu, Y. Liu, P. Liu, and B. Yang, "Novel low loss, low permittivity $(1 - x)SiO_2 - xTiO_2 + ywt\%\ H_3BO_3$, microwave dielectric ceramics for LTCC applications," *Journal of Alloys and Compounds*, vol. 712, pp. 804–810, 2017.

[2] H. Ohsato, T. Tsunooka, M. Ando, Y. Ohishi, Y. Miyauchi, and K. Kakimoto, "Millimeter-wave dielectric ceramics of alumina and forsterite with high quality factor and low dielectric constant," *Journal of the Korean Ceramic Society*, vol. 40, no. 4, pp. 350–353, 2003.

[3] T. Tsunooka, M. Androu, Y. Higashida, H. Sugiura, and H. Ohsato, "Effects of TiO_2, on sinterability and dielectric properties of high-Q, forsterite ceramics," *Journal of the European Ceramic Society*, vol. 23, no. 14, pp. 2573–2578, 2003.

[4] T. Tsunooka, H. Sugiyama, K. Kakimoto, H. Ohsato, and H. Ogawa, "Zero temperature coefficient τ_f and sinterability of forsterite ceramics by rutile addition," *Journal Ceramic Society of Japan*, vol. 112, pp. S1637–S1640, 2004.

[5] M. Ando, K. Himura, T. Tsunooka, I. Kagomiya, and H. Ohsato, "Synthesis of high-quality forsterite," *Japanese Journal of Applied Physics*, vol. 46, no. 10, pp. 7112–7116, 2007.

[6] M. Ando, H. Ohsato, I. Kagomiya, and T. Tsunooka, "Quality factor of forsterite for ultrahigh frequency dielectrics depending on synthesis process," *Japanese Journal of Applied Physics*, vol. 47, no. 9, pp. 7729–7731, 2008.

[7] H. Ohsato, "Millimeter-wave materials," in *Microwave Materials and Applications*, Chapter 5, M. T. Sebastian, R. Ubic, and H. Jantunen, Eds., Wiley, Hoboken, NJ, USA, 2017.

[8] T. S. Sasikala, C. Pavithran, and M. T. Sebastian, "Effect of lithium magnesium zinc borosilicate glass addition on densification temperature and dielectric properties of Mg_2SiO_4, ceramics," *Journal of Materials Science Materials in Electronics*, vol. 21, no. 2, pp. 141–144, 2010.

[9] C. Zhang, R. Zuo, J. Zhang, and Y. Wang, "Structure dependent microwave dielectric properties and middle temperature sintering of forsterite $(Mg_{1-x}Ni_x)_2SiO_4$ ceramics," *Journal of the American Ceramic Society*, vol. 98, no. 3, pp. 702–710, 2015.

[10] J. Li, P. Liu, Z. F. Fu, and Q. Q. Feng, "Microwave dielectric properties of low-fired $Mg_{1.9}Cu_{0.1}SiO_4$-$(La_{0.5}Na_{0.5})TiO_3$ composite ceramics," *Journal of Alloys and Compounds*, vol. 660, pp. 93–98, 2015.

[11] C. L. Pan, P. C. Chen, T. C. Tan, W. C. Lin, C. H. Shen, and S. H. Lin, "Low-temperature sintering and microwave dielectric properties of $CaWO_4$-Mg_2SiO_4 ceramics," *Advanced Materials Research*, vol. 933, pp. 12–16, 2014.

[12] K. X. Song and X. M. Chen, "Phase evolution and microwave dielectric characteristics of Ti-substituted Mg_2SiO_4, forsterite ceramics," *Materials Letters*, vol. 62, no. 3, pp. 520–522, 2008.

[13] N. W. J. Wan, H. Abdullah, and M. S. Zulfakar, "Effect of Zn site for Ca Substitution on optical and microwave dielectric properties of $ZnAl_2O_4$ thin films by sol gel method," *Advances in Materials Science and Engineering*, vol. 2014, Article ID 619024, 8 pages, 2014.

[14] Y. Wang, D. Zhou, Y. Zhang, and C. Chang, "Using multilayered substrate integrated waveguide to design microwave gain equalizer," *Advances in Materials Science and Engineering*, vol. 2014, Article ID 109247, 6 pages, 2015.

[15] C. Stergiou, "Microstructure and electromagnetic properties of Ni-Zn-Co ferrite up to 20 GHz," *Advances in Materials Science and Engineering*, vol. 2016, Article ID 1934783, 7 pages, 2016.

[16] L.-X. Pang and D. Zhou, "Microwave dielectric properties of low-firing Li_2MO_3 (M=Ti, Zr, Sn) ceramics with B_2O_3–CuO addition," *Journal of the American Ceramic Society*, vol. 93, no. 11, pp. 3614–3617, 2010.

[17] G. G. Yao, C. J. Pei, P. Liu, H. Y. Xing, L. X. Fu, and B. C. Liang, "Novel temperature stable $Ba_{1-x}Sr_xV_2O_6$, microwave dielectric ceramics with ultra-low sintering temperature," *Journal of Materials Science Materials in Electronics*, vol. 28, no. 18, pp. 13283–13288, 2017.

[18] J. Yang, X. Y. Deng, J. B. Li et al., "Broad and dielectric spectroscopy analysis of dielectric properties of barium titanate ceramics," *Advanced Materials Research*, vol. 744, pp. 323–328, 2013.

[19] S. Y. Chang, H. F. Pai, C. F. Tseng, and C. K. Tsai, "Microwave dielectric properties of ultra-low temperature fired Li_3BO_3, ceramics," *Journal of Alloys and Compounds*, vol. 698, pp. 814–818, 2017.

[20] M. T. Sebastian and H. Jantunen, "Low loss dielectric materials for LTCC applications: a review," *International Materials Reviews*, vol. 53, no. 2, pp. 57–90, 2008.

18

Microstructural and Electrical Properties of Sn-Modified BaTiO$_3$ Lead-Free Ceramics by Two-Step Sintering Method

Wichita Kayaphan[1,2] and Pornsuda Bomlai ⓘ[1,2]

[1]Department of Materials Science and Technology, Faculty of Science, Prince of Songkla University, Hat Yai, Songkhla 90112, Thailand
[2]Center of Excellence in Nanotechnology for Energy (CENE), Prince of Songkla University, Hat Yai, Songkhla 90112, Thailand

Correspondence should be addressed to Pornsuda Bomlai; pornsuda.b@psu.ac.th

Academic Editor: Joon-Hyung Lee

Ba(Ti$_{0.92}$Sn$_{0.08}$)O$_3$ lead-free ceramics were prepared using a two-step sintering (TSS) technique. Varying the first sintering temperature T_1 (1400 and 1500°C) and the dwell time t_1 (0, 15, and 30 min), we obtained dense ceramics which were then soaked at a constant temperature of 1000°C (T_2) for 6 h (t_2). The structural and electrical properties were investigated. XRD results indicated that all the ceramics showed a pure perovskite phase with tetragonal symmetry. Density and grain size increased with higher T_1 temperatures and increased t_1 dwell times. Enhanced electrical properties were achieved by sintering at the optimized T_1 sintering temperature and t_1 dwelling time. At the lower T_1 sintering temperature of 1400°C, the dielectric and piezoelectric properties and the Curie temperature of the ceramics were improved significantly by increasing t_1 dwell time. Further, increasing the sintering temperature T_1 to 1500°C, excellent properties were obtained at $t_1 = 15$ min which then deteriorated when t_1 was increased to 30 min. The electrical properties of the sample sintered under the $T_1/t_1/T_2/t_2$ condition of "1500/15/1000/6" showed the best values. For this sample the piezoelectric coefficient (d_{33}), dielectric permittivity (ε_r), loss factor (tanδ), and Curie temperature (T_C) were 490 pC/N, 4385, 0.0272, and 48°C, respectively.

1. Introduction

Because of the high toxicity of lead oxide, the use of Pb (Zr$_x$Ti$_{1-x}$)O$_3$-based materials has caused serious environmental problems. Hence, the development of piezoelectric materials has focused considerable attention in recent years on lead-free materials which exhibit highly piezoelectric properties [1–4]. Barium titanate (BaTiO$_3$) is a lead-free ferroelectric material currently used to replace Pb (Zr$_x$Ti$_{1-x}$)O$_3$ piezoelectric ceramics in electronic devices ($d_{33} = 191$ pC/N) [3]. It is a typical ferroelectric material with a tetragonal symmetry of perovskite structure at room temperature [5]. It is known that an appropriate amount of doping into the BaTiO$_3$ can produce piezoelectricity comparable with Pb(Zr$_x$Ti$_{1-x}$)O$_3$-based ceramics [5–9]. Among modified BaTiO$_3$ composites, barium stannate titanate (BaTi$_{1-x}$Sn$_x$O$_3$) can modify both the microstructure and the

electrical properties of BaTiO$_3$ [9]. Enhanced piezoelectric properties have been reported for Ba$_{0.90}$Ca$_{0.10}$Ti$_{1-x}$Sn$_x$O$_3$ ceramics with a super-high d_{33} (521 pC/N) at $x = 0.10$ [10]. However, the densification of Ba$_{0.90}$Ca$_{0.10}$Ti$_{1-x}$Sn$_x$O$_3$ ceramics requires sintering for 2 h at temperatures above 1400°C, which is too high for low-temperature cofired ceramics (LTCC) processing.

The two-step sintering process is heating rate-controlled and effectively densifies the ceramics at lower temperatures. For this method, samples are heated to a higher temperature (T_1) and held at T_1 for a short time t_1, then the temperature is immediately lowered to a soaking temperature, T_2. The process can eliminate pores and reduce the volatilization of the low melting point substances [11]. In the present work, Ba(Ti$_{1-x}$Sn$_x$)O$_3$ ceramics at $x = 0.08$ were fabricated using the two-step sintering method to control grain growth, and the effects of sintering temperature and dwell time on the

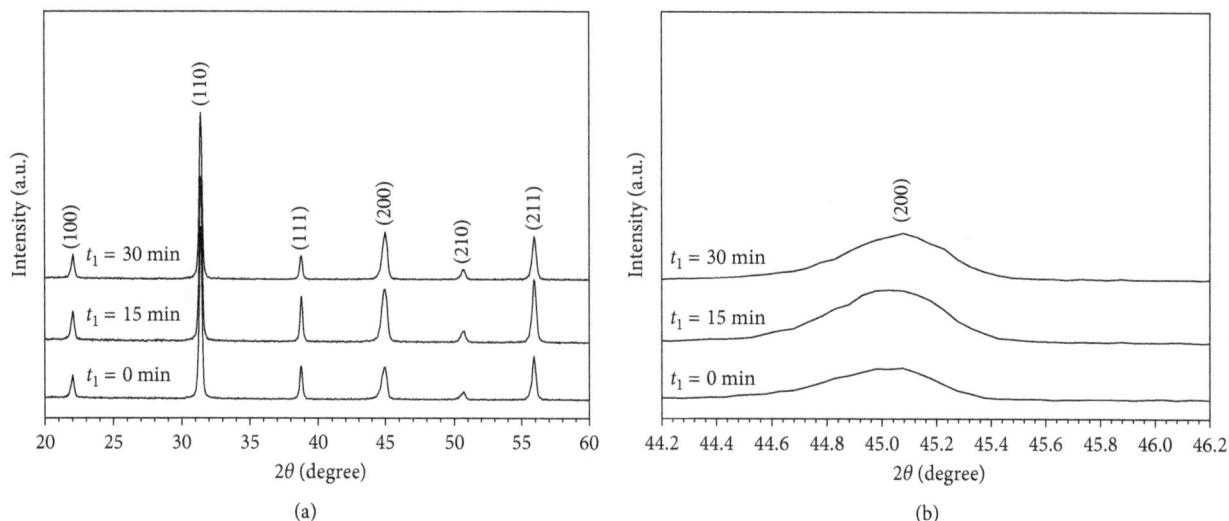

FIGURE 1: (a) XRD patterns of BTS samples sintered at different dwelling times, t_1, and (b) the expanded XRD patterns in the 2θ range of 44.2–46.2 of BTS ceramics.

phase structure, microstructure, and electrical properties of the fabricated ceramics were systematically investigated.

2. Materials and Method

$Ba(Ti_{1-x}Sn_x)O_3$ (abbreviated as BTS: $x = 0.08$) lead-free ceramics were synthesized by a solid-state method using $BaCO_3$ (99.9%), TiO_2 (99.9%), and SnO_2 (99.9%) as raw materials. All starting powders were ball milled in alcohol using zirconia balls for 24 h. After calcination at 1200°C for 2 h, the obtained powders were uniaxially pressed into disk-shaped compacts and then sintered using a two-step sintering method with the following profile. The green samples were first heated from room temperature to 900°C at a slow heating rate of 5°C/min and then to the sintering temperature T_1 of either 1400 or 1500°C at a heating rate of 10°C/min and held for a short dwell time (t_1) of 0, 15, or 30 min. To realize the desired final density and to control grain growth, the soaking temperature of the second stage, T_2, was set lower than T_1 at 1000°C with a cooling rate of 20°C/min, and the soaking time (t_2) at T_2 was 6 h. The symbols "$T_1/t_1/T_2/t_2$" are used to indicate the two-step sintering conditions.

The crystalline phases of the sintered samples were characterized with a powder X-ray diffractometer (XRD, X'Pert MPD, Philips) with CuKα radiation generated at 40 kV and 30 mA. The density of the sintered samples was measured using a water immersion method in accordance with Archimedes' principle. The microstructure of the ceramic surfaces was examined by scanning electron microscopy (SEM, Quanta400, FEI). The intercept method was used to determine the average grain size. For electrical characterizations, the ceramic samples were carefully polished and painted with silver paste on both sides. The dielectric properties were measured using a high precision LCR meter (LCR 821, GW INSTEK) from 25°C to 200°C at a heating rate of 3°C/min. To measure their piezoelectric

properties, the ceramic samples were poled at room temperature (25°C) for 20 min under an electric field of 1.0–4.0 kV/mm in a silicone oil bath. The piezoelectric coefficient d_{33} was measured by a piezo-d_{33} meter (YE2730A d_{33} meter, APC International, Ltd.).

3. Results and Discussion

Figure 1 shows the XRD patterns of BTS samples sintered under "1500/t_1/1000/6" conditions with different t_1 dwell times of 0, 15, and 30 min at room temperature. It can be observed that all sintered samples showed a single perovskite structure with a tetragonal phase indicating that Sn^{4+} diffused into $BaTiO_3$ lattices during sintering. The diffraction peaks showed slightly sharper at t_1 of 15 min indicating an enhancement of crystallinity in the samples. Moreover, the intensity of the (111) diffraction peak of the sample sintered at $t_1 = 15$ min was stronger than for other samples. The expanded XRD patterns of BTS ceramics in the 2θ range of 44.2–46.2° showed that the peak position shifted slightly to a higher angle with increasing dwell time t_1, Fig. 1(b). In part this difference may relate to strain effects. The lattice parameters for samples with dwell time $t_1 = 0$ and 15 min were $a = 4.0232$ and $c = 4.0402$ Å, whereas for samples with dwell time $t_1 = 30$ min, they were $a = 4.0190$ and $c = 4.0446$ Å. It can also be seen that the intensity of the XRD peaks slightly decreased in response to decreased crystalline phase content, as well as increased lattice strain when t_1 was increased to 30 min. Average crystallite size was calculated from the Debye–Scherrer formula [12]:

$$D = \frac{0.9\lambda}{\beta\cos\theta}, \qquad (1)$$

where D is crystallite size, λ is X-ray wavelength (0.15406 nm), β is full width at half maximum (FWHM) of peak in radians, and θ is Bragg's angle. The average crystallite size increased with increasing dwell time up to 15 min and

(a)

(b)

(c)

(d)

(e)

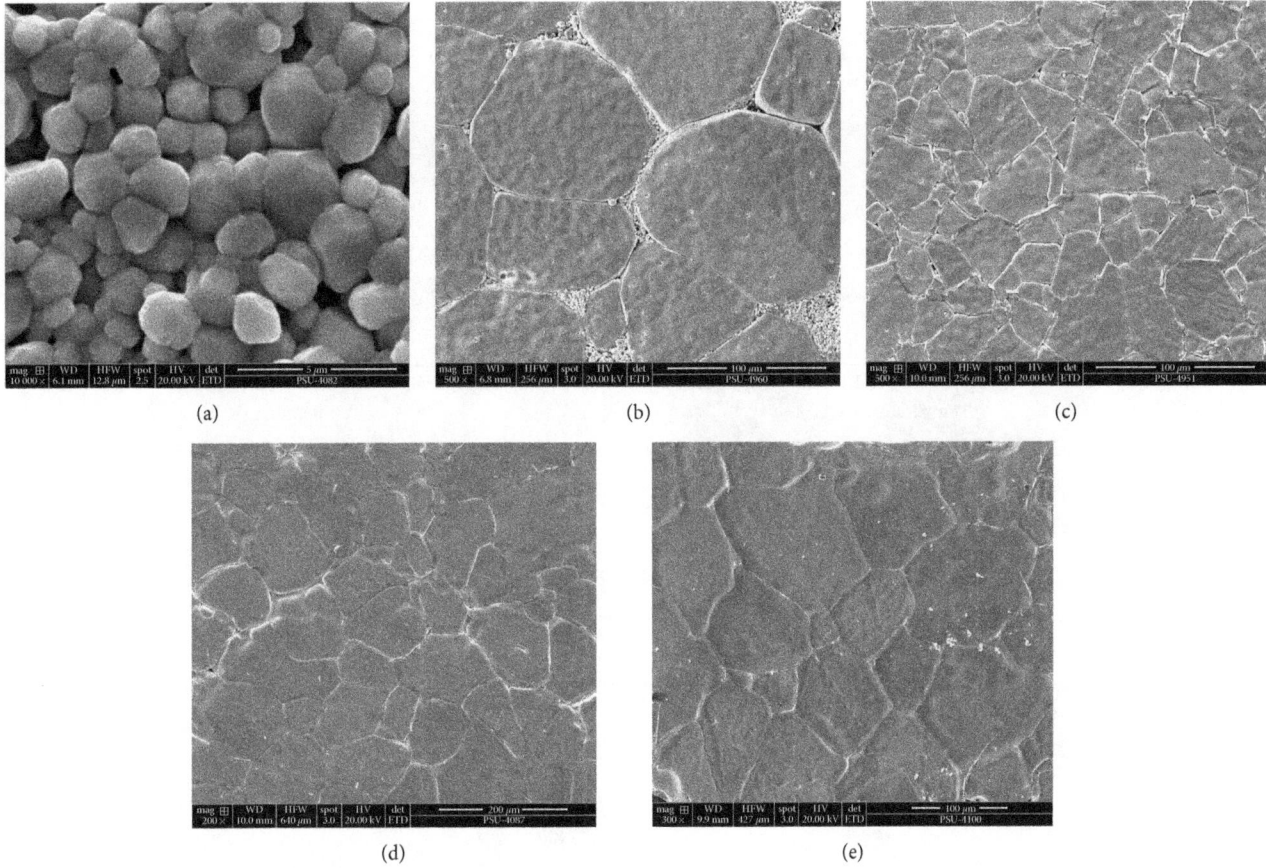

FIGURE 2: SEM micrographs of the BTS ceramics sintered under the various conditions: (a) "1400/15/1000/6"; (b) "1400/30/1000/6"; (c) "1500/0/1000/6"; (d) "1500/15/1000/6"; (e) "1500/30/1000/6".

TABLE 1: The bulk density, ε_r, and $\tan\delta$ at room temperature (RT), $\varepsilon_{r,\max}$ and $\tan_{\delta,\max}$ at Curie temperature (T_C), and T_C at 1 kHz, grain size, and piezoelectric coefficient (d_{33}) of the BTS samples.

Sintering conditions $T_1/t_1/1000/6$		Bulk density (g/cm^3)	$\varepsilon_{r,\mathrm{RT}}$	$\tan_{\delta,\mathrm{RT}}$	$\varepsilon_{r,\max}$	$\tan_{\delta,\max}$	T_C (°C)	Grain size (μm)	d_{33} (pC/N)
T_1 (°C)	t_1 (min)								
1400	0	4.86 ± 0.02	*	*	*	*	*	*	*
	15	5.60 ± 0.05	4085	0.0291	9729	0.0126	41	1.4 ± 0.4	35
	30	5.82 ± 0.05	4301	0.0776	16730	0.0248	44	43.2 ± 4.6	47
1500	0	5.88 ± 0.03	4091	0.0766	19832	0.0208	42	20.1 ± 3.4	260
	15	5.90 ± 0.01	4385	0.0272	18601	0.0203	48	30.1 ± 6.0	490
	30	5.85 ± 0.02	3564	0.0738	17950	0.0127	47	47.8 ± 7.9	301

*Ceramic sample could not be achieved by sintering under conditions of 1400/0/1000/6.

then remained constant as the value of t_1 further increased to 30 min. It was found to be 34.2, 42.8, and 42.8 nm for dwell time $t_1 = 0$, 15, and 30 min, respectively.

Figure 2 shows the microstructure of the BTS samples sintered under conditions of "$T_1/t_1/1000/6$" ($T_1 = 1400$ and 1500°C; $t_1 = 0$, 15, and 30 min). The SEM micrographs indicate that the microstructure is dependent on the sintering profile. At the lower sintering temperature ($T_1 = 1400$°C) without dwell time ($t_1 = 0$ min), well-sintered samples could

not be obtained (SEM micrograph is not shown). At a dwell time of 30 min, the grain size increased rapidly and exhibited an abnormal grain growth. A relatively uniform grain size of $20.1 \pm 3.4\,\mu$m could be achieved for samples sintered under the "1500/0/1000/6" condition. It is evident that the increase in the sintering temperature, T_1, and dwelling time, t_1, helped to promote grain growth. This is because the higher sintering temperature and dwell time can enlarge the diffusion coefficient and make grain boundary migration easier [8].

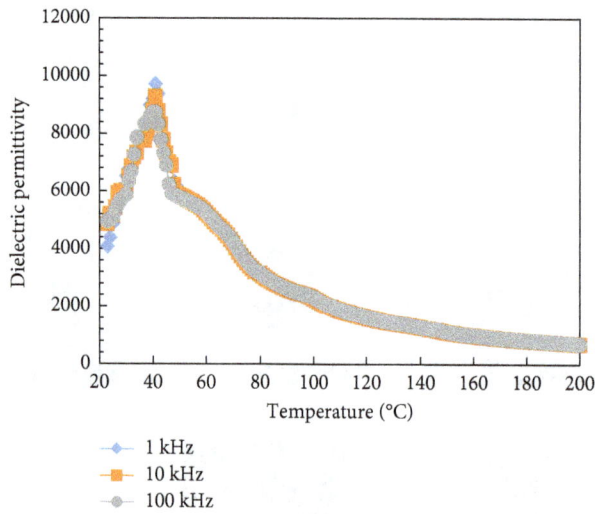

(a)

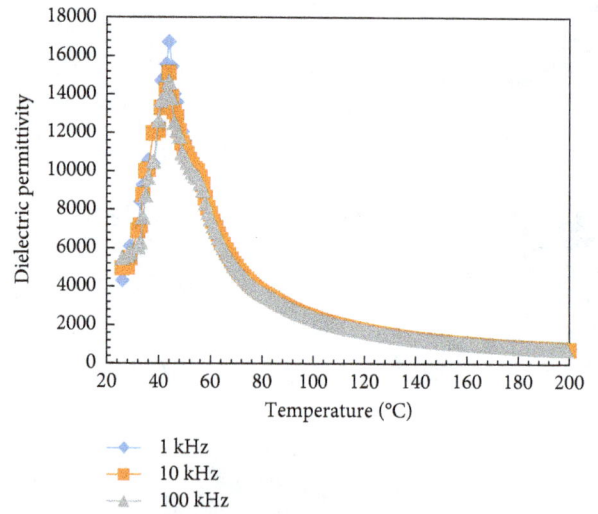

(b)

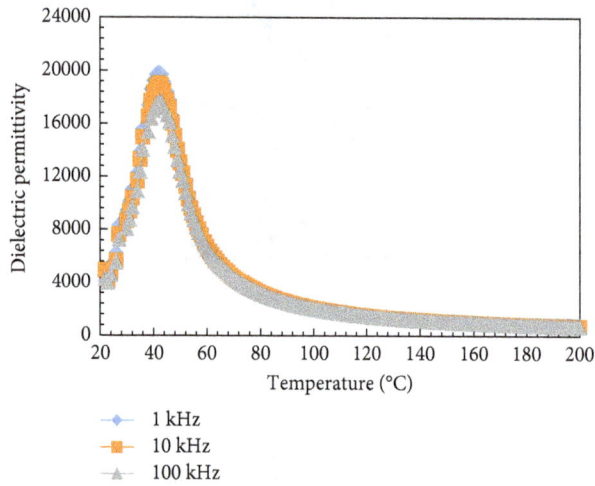

(c)

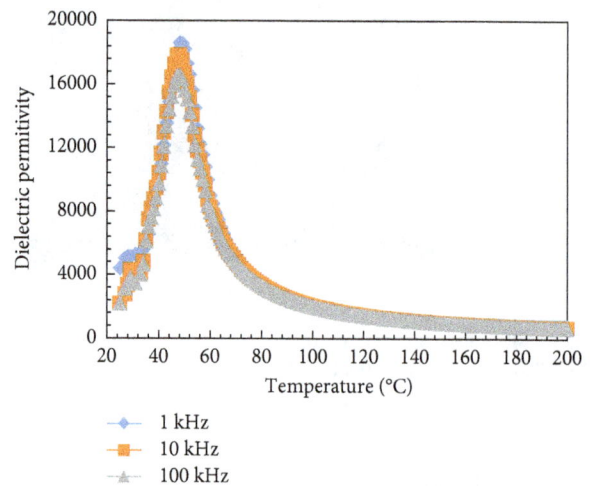

(d)

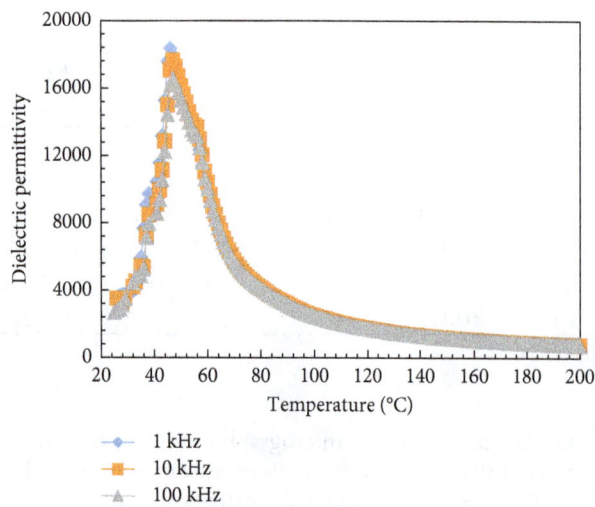

(e)

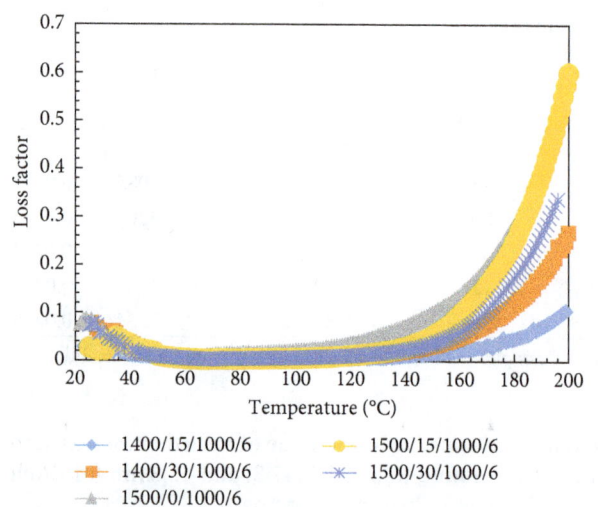

(f)

FIGURE 3: The temperature dependence of dielectric permittivity of BTS ceramics sintered at different conditions at frequencies of 1, 10, and 100 kHz: (a) "1400/15/1000/6"; (b) "1400/30/1000/6"; (c) "1500/0/1000/6"; (d) "1500/15/1000/6"; (e) "1500/30/1000/6"; (f) loss factor (tanδ) of the BTS ceramics at 1 kHz.

In the present case, increasing the sintering temperature and dwell time improved densification as well as the grain growth. It can be seen in Table 1 that when the sintered samples underwent the same dwell time, t_1, the bulk densities were increased by increasing the sintering temperature T_1, whereas when the samples were sintered at the same sintering temperature T_1, improvements in the relative density were recorded with correspondingly longer dwell times, t_1. The optimal value of t_1 dwell time depended significantly on the sintering temperature, T_1. The highest density of 5.90 ± 0.01 g/cm^3 was obtained when $t_1 = 15$ min, and the density was slightly lower when $t_1 = 30$ min. The densification of the BTS samples improved significantly at the higher sintering temperature T_1 with the longer dwell time t_1. In general, the two-step sintering technique not only effectively densified the ceramics, but also enabled sintering at a reduced temperature. For this work, the soaking temperature T_2 was set at a low temperature of 1000°C. At higher temperatures and longer dwell times, grain boundary migration causes grain growth, while grain boundary diffusion can still be activated causing enhanced density [13]. However, the density of the samples decreases if the dwell time exceeds the dwell time of the optimum sintering condition ("1500/30/1000/6"). Longer dwell time greatly influenced the porosity of the surface oxide layer with the surface getting more porous.

Figure 3 shows the temperature dependence of dielectric properties of the BTS ceramics as a function of sintering temperature, T_1, and dwell time, t_1. Increasing the sintering temperature and dwell time clearly increased the values of $\varepsilon_{r,RT}$ up to the maximum value of 4385, which was attained when T_1 was 1500°C and the dwell time 15 min. The value of $\varepsilon_{r,RT}$ then decreased when t_1 was increased to 30 min. This is due to an increase in grain size. The low dielectric loss of $\leq 2\%$ indicated that these BTS ceramics sintered under two-step conditions are promising for practical applications. The temperature dependence of the dielectric permittivity (ε_r) of the BTS ceramics was also determined in order to characterize their phase transition behavior, as shown in Figure 3. The sintering temperature and dwell time also affected the Curie temperature. It can be seen that the tetragonal-cubic (or Curie temperature T_C) phase transition temperatures all increased slightly at the higher sintering temperatures, T_1, and longer dwell times, t_1. The T_C is shown in Table 1 as T_{max} which showed the maximum value of dielectric permittivity. Since the sintering temperature and the dwell time (T_1, t_1) change the microstructure of ceramics, they have an important influence on the tetragonal-cubic phase transition temperature. It can be seen that the samples with bigger grain sizes exhibited higher T_C values. This is attributed to the increased proportion of tetragonal phase and increased ferroelectricity of the ceramics. This result is consistent with other perovskite structure ceramics [14, 15]. The piezoelectric coefficient d_{33} as a function of sintering temperature, T_1, and dwell time, t_1, for the BTS ceramics, is also shown in Table 1. The d_{33} value increased with increasing sintering temperatures and dwell times, reached a maximum value of 490 pC/N at the "1500/15/1000/6" condition and then decreased at a dwell time of 30 min. At the longer dwell time, deflection in the stoichiometric composition and deterioration in the microstructure degraded the piezoelectric properties of the sample.

4. Conclusions

Two-step sintering was used to fabricate dense Ba(Ti, Sn)O$_3$ ceramics at low sintering temperatures. The microstructural and electrical properties of the Ba(Ti$_{0.92}$Sn$_{0.08}$)O$_3$ ceramics are obviously dependent on the sintering temperature (T_1) and the dwelling time (t_1). Under the optimal condition of "1500/15/1000/6," the dense samples had an average grain size of approximately 30 μm and showed good dielectric and piezoelectric properties. Values for the piezoelectric constant (d_{33}), relative permittivity (ε_r), loss factor (tanδ), and Curie temperature (T_C) were 490 pC/N, 4385, 0.0272, and 48°C, respectively.

Conflicts of Interest

The authors declare that there are no conflicts of interest regarding the publication of this paper.

Acknowledgments

This research was financially supported by Prince of Songkla University under Contract no. SCI580905S.

References

[1] J. Rödel, W. Jo, K. T. P. Seifert, E. M. Anton, T. Granzow, and D. Damjanovic, "Perspective on the development of lead-free piezoceramics," *Journal of the American Ceramic Society*, vol. 92, no. 6, pp. 1153–1157, 2009.

[2] Y. Saito, H. Takao, T. Tani et al., "Lead-free piezoceramics," *Nature*, vol. 432, no. 7013, pp. 84–87, 2004.

[3] W. F. Liu and X. B. Ren, "Large piezoelectric effect in Pb-free ceramics," *Physical Review Letters*, vol. 103, no. 25, p. 257602, 2009.

[4] K. Datta and P. A. Thomas, "Structural investigation of a novel perovskite-based lead-free ceramics: xBiScO$_3$–(1–x)BaTiO$_3$," *Journal of Applied Physics*, vol. 107, p. 043516, 2010.

[5] Y. C. Lee and C. S. Chiang, "High dielectric constant of (Ba$_{0.96}$Ca$_{0.04}$)(Ti$_{0.85}$Zr$_{0.15}$)O$_3$ multilayer ceramic capacitors with Cu doped Ni electrodes," *Journal of Alloys and Compounds*, vol. 509, no. 24, pp. 6973–6979, 2011.

[6] H. Ogihara, C. A. Randall, and S. Trolier-McKinstry, "Weakly coupled relaxor behavior of BaTiO$_3$–BiScO$_3$ ceramics," *Journal of the American Ceramic Society*, vol. 92, no. 1, pp. 110–118, 2009.

[7] C. Ma and X. Tan, "Morphotropic phase boundary and electrical properties of lead-free (1–x)BaTiO$_3$–xBi(Li$_{1/3}$Ti$_{2/3}$)O$_3$ ceramics," *Journal of Applied Physics*, vol. 107, no. 12, p. 124108, 2010.

[8] D. J. Kim, M. H. Lee, J. S. Park, M.-H. Kim, and T. K. Song, "Effects of sintering temperature on the electric properties of Mn-modified BiFeO$_3$-BaTiO$_3$ bulk ceramics," *Journal of the Korean Physical Society*, vol. 66, no. 7, pp. 1115–1119, 2015.

[9] V. V. Shvartsman, J. Dec, Z. K. Xu, J. Banys, P. Keburis, and W. K. Leemann, "Crossover from ferroelectric to relaxor behavior in BaTi$_{1-x}$Sn$_x$O$_3$ solid solutions," *Phase Transitions*, vol. 81, no. 11-12, pp. 1013–1021, 2008.

[10] D. Lin, K. W. Kwok, and H. L. W. Chan, "Structure, dielectric and piezoelectric properties of $Ba_{0.90}Ca_{0.10}Ti_{1-x}Sn_xO_3$ lead-free ceramics," *Ceramics International*, vol. 40, no. 5, pp. 6841–6846, 2014.

[11] D. L. Wang, K. J. Zhu, H. L. Ji, and J. H. Qiu, "Two-step sintering of the pure $K_{0.5}Na_{0.5}NbO_3$ lead-free piezoceramics and its piezoelectric properties," *Ferroelectrics*, vol. 392, no. 1, pp. 120–126, 2009.

[12] J. Langford and A. Wilson, "Scherrer after sixty years: a survey and some new results in the determination of crystallite size," *Journal of Applied Crystallography*, vol. 11, no. 2, pp. 102-103, 1978.

[13] H. L. Cheng, W. C. Zhou, H. L. Du, F. Luo, and D. M. Zhu, "Effects of dwell time during sintering on electrical properties of $0.98(K_{0.5}Na_{0.5})NbO_3$–$0.02LaFeO_3$ ceramics," *Transactions of Nonferrous Metals Society of China*, vol. 23, no. 10, pp. 2984–2988, 2013.

[14] T. A. Kamel and G. de With, "Grain size effect on the poling of soft $Pb(Zr, Ti)O_3$ ferroelectric ceramics," *Journal of European Ceramic Society*, vol. 28, no. 4, pp. 851–861, 2008.

[15] C. A. Randall, N. Kim, J. P. Kucera, W. W. Cao, and T. R. Shrout, "Intrinsic and extrinsic size effects in fine-grained morphotropic-phase boundary lead zirconate titanate ceramics," *Journal of the American Ceramic Society*, vol. 81, no. 3, pp. 677–688, 1998.

19

Template-Free Synthesis of Star-Like ZrO$_2$ Nanostructures and their Application in Photocatalysis

Xue-Yu Tao [ID],[1,2,3] Jie Ma,[1] Rui-Lin Hou,[1] Xiang-Zhu Song,[1] Lin Guo,[1] Shi-Xiang Zhou,[1] Li-Tong Guo [ID],[1] Zhang-Sheng Liu,[1] He-Liang Fan,[1] and Ya-Bo Zhu[1]

[1]School of Materials Science and Engineering, China University of Mining and Technology, Xuzhou, Jiangsu 221116, China
[2]Jiangsu Province Engineering Laboratory of High Efficient Energy Storage Technology and Equipments, China University of Mining and Technology, Xuzhou, Jiangsu 221116, China
[3]Xuzhou City Key Laboratory of High Efficient Energy Storage Technology and Equipments, China University of Mining and Technology, Xuzhou, Jiangsu 221116, China

Correspondence should be addressed to Xue-Yu Tao; taoxueyu@cumt.edu.cn and Li-Tong Guo; guolitong810104@163.com

Academic Editor: Marián Palcut

Star-like nano-ZrO$_2$ has been synthesized using Zr(NO$_3$)$_4$·5H$_2$O as zirconium source by a hydrothermal process without any template and surfactant. The structure of the as-prepared ZrO$_2$ powder was investigated by multiple advanced analytical methods. The results showed that CH$_3$COO$^-$ and NO$_3$$^-$ had great effects on the formation of star-like ZrO$_2$ nanostructures. The as-prepared ZrO$_2$ had a superior catalytic activity, and the reason for it was analyzed by UV-Vis diffuse reflectance spectroscopy. The effect of raw material ratios on the photocatalytic property of ZrO$_2$ was studied. The synthesized ZrO$_2$ showed a narrow bandgap (3.50–3.85 eV) and an excellent photocatalytic activity, and the degradation of RhB was up to nearly 100% in 30 min with this photocatalyst.

1. Introduction

Zirconia (ZrO$_2$) is one of the most important ceramic materials with three different phases: monoclinic stable below 1175°C, tetragonal stable at 1175–2370°C, and cubic stable at 2370–2680°C, respectively [1, 2]. Nanosized zirconia has specific optical and electrical properties which suits it for prospective applications in transparent optical devices, electrochemical capacitor electrodes, fuel cells, catalyst, and advanced ceramics [3–8]. Numerous synthetic strategies have been developed to obtain zirconia nanostructures including solution combustion synthesis [9], microwave-hydrothermal [10], sol-gel [11], spray pyrolysis [12], chemical vapour synthesis [13], and precipitation approach [14]. Among them, the hydrothermal method has attracted much attention because of its simple operation, mild experimental conditions, and high product purity. Catalytic properties of inorganic nanomaterials not only are related to their phase structure and chemical composition but also depend on their morphology [15]. The morphology of ZrO$_2$ has a significant effect on its properties [16–23] because it can control a variety of physical and chemical properties at the same time. For instance, with special morphology, flower-like zirconia nanomaterials [24] showed an excellent photocatalytic activity on the degradation of rhodamine B. In addition, the spinous ZrO$_2$ core-shell morphology [25] exhibited a superior hydrogen storage performance, reaching a hydrogen uptake of 1.521 wt.% at 298 K under 5 MPa.

Wastewater pollution has become a serious problem in many countries [26]. The removal of dyes from wastewater through heterogeneous photocatalysis has drawn an increasing attention over the last few decades. For degradation of organic pollutants, many studies on the heterogeneous photocatalysis were performed with oxide semiconductors such as TiO$_2$ [27], ZnO [28], Fe$_2$O$_3$ [29], ZrO$_2$ [30], and CuO [31] being applied. These nanomaterials showed an excellent photocatalytic activity on the degradation of organic dyes. In recent years, ZrO$_2$-based materials have gained a considerable scientific and technological attention in heterogeneous catalysis. They have been used in the photodegradation of dye compound due to their high photocatalytic activity in the ultraviolet range, high thermal stability, chemical stability, low cost, nontoxicity, and environmentally friendly nature [32].

In recent years, many different approaches [17, 33–35] have been utilized to prepare zirconia nanomaterials with different morphologies using suitable templates and surfactants. For example, ZrO$_2$ nanowires and nanobelts were prepared by an alumina template and pyrolysis of Zr(OH)$_4$: RE particles, respectively. ZrO$_2$ mesopore microfibers have been prepared with a pluronic P-123 template [17]. A ZrO$_2$ hollow fiber membrane was successfully synthesized employing a polypropylene hollow fiber as the template, and the prepared zirconia hollow fiber was demonstrated to be a highly selective adsorbent for the phosphonic acid-containing compounds with high sensitivity [35]. Flake-like ZrO$_2$ nanocrystallites were prepared using cetyltrimethyl ammonium bromide (CTAB) as the surfactant, and the use of surfactant led to the formation of stabilized tetragonal ZrO$_2$ nanoparticles (15 nm) [21]. These studies demonstrated successful synthesis of ZrO$_2$ nanostructures with special morphology; however, most of these approaches required either suitable templates or surfactants to prepare ZrO$_2$ nanostructures and also included the removal of the template in the process. Therefore, searching for a simple but effective method to get a special morphology of ZrO$_2$ in the absence of template remains a challenge.

In this paper, a facile route was employed to synthesize the ZrO$_2$ nanostructures with Zr(NO$_3$)$_4$·5H$_2$O and CH$_3$COONa as starting materials without any template or surfactant. Starting materials Zr(NO$_3$)$_4$·5H$_2$O and CH$_3$COONa are cheap and affordable. Furthermore, Zr(NO$_3$)$_4$·5H$_2$O is water soluble and suitable for hydrothermal synthesis. Also NO$_3^-$ is easy to remove from the system, which prevents the contamination of the products. CH$_3$COO$^-$ anions tend to adsorb on the surface of ZrO$_2$ and play an important role in the formation of ZrO$_2$ particles with special morphology. In this work, the degradation of rhodamine B (RhB) with nanosized ZrO$_2$ was studied in aqueous solution. The novel nanostructure of ZrO$_2$ may lead to superior performance in the photodegradation of RhB.

2. Experimental

2.1. Materials.
The chemicals used were of analytical grade and purchased from Aladdin Chemistry Co. Ltd. The chemicals were used as received without further purification.

2.2. Synthesis.
ZrO$_2$ nanostructures were synthesized using the hydrothermal method. Figure 1 shows the synthesis route of nanosized ZrO$_2$. Typically, 0.123 g (0.0015 mol) of CH$_3$COONa was added into the solution of Zr(NO$_3$)$_4$·5H$_2$O under magnetic stirring. Then, the solution was transferred into a 25 mL beaker in a Teflon-lined stainless steel autoclave

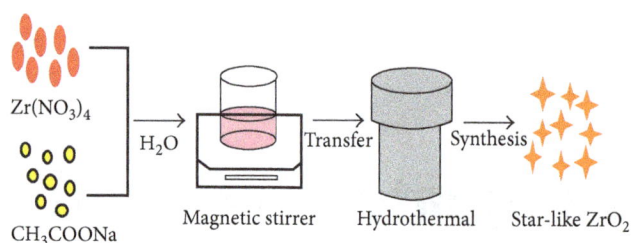

FIGURE 1: Flowchart of the preparation route of star-like ZrO$_2$.

and heat treated at 180°C for 6 h. After reaction, the autoclave was left to cool down to room temperature. The products were centrifuged and collected. Then, the products were washed for several times with deionized water and ethanol. In the end, the ZrO$_2$ products were obtained by drying at 90°C for 8 h.

2.3. Characterization.
The morphology was investigated using the high-resolution transmission electron microscope (HR-TEM, Tecnai G2 F20) working with an accelerating voltage of 200 kV. The phase constitution of the products were analyzed by an X-ray diffractometer (Rigaku D/M4X 2500, Rigaku Co., Japan) with Cu Kα radiation ($\lambda = 0.15418$ Å). The infrared (FTIR) spectra were measured by a Nicolet is 35 using the KBr pellet technique in the range of 4000–400 cm^{-1}. Thermogravimetric (TG) analysis was recorded on a Netzsch STA 449 F3 at a heating rate of 10°C·min^{-1} in a flowing air. The UV-visible spectrum of ZrO$_2$ was measured using the Cary 300 UV-visible spectrophotometer. The X-ray photoelectron spectrum (XPS) was recorded on an ESCALAB 250Xi spectrometer with an energy analyzer working in the pass energy mode at 20.0 eV, and the Al Kα line was used as the excitation source. The binding energy reference was taken at 284.8 eV for the C1s peak arising from surface hydrocarbons. Brunauer–Emmett–Teller (BET) surface area was obtained with N$_2$ adsorption by using a Micromeritics ASAP 2020 nitrogen adsorption apparatus via determination of nitrogen adsorption isotherm at 77 K.

2.4. Photocatalytic Activity Test.
The photocatalytic experiments were carried out by following the RhB degradation under UV irradiation in a separate chamber. Prior to irradiation, the suspensions were magnetically stirred in a complete darkness for 20 min to attain adsorption equilibrium. During the RhB photodecomposition, the samples were withdrawn at regular intervals and centrifuged to separate solid particles for analysis. The concentration of RhB was determined by a UV-Vis spectroscopy at its maximum absorption wavelength (about 554 nm).

3. Results and Discussion

3.1. Characterizations of ZrO$_2$ Nanostructures.
The XRD results of the products, prepared at 180°C with different reaction times, are shown in Figure 2. It can be seen that the

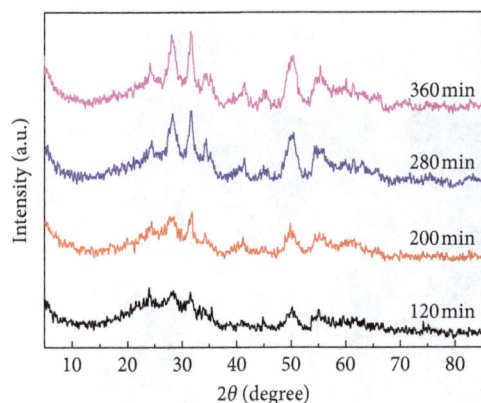

FIGURE 2: XRD patterns of the products prepared at different reaction times.

TABLE 1: Estimated crystallite sizes of ZrO_2 synthesized at different reaction times.

Sample	Crystal face	FWHM (rad)	Crystallite size (nm)
ZrO_2 (120 min)	[−1 1 1]	0.0215	6.97
ZrO_2 (200 min)	[−1 1 1]	0.0205	7.31
ZrO_2 (280 min)	[−1 1 1]	0.0182	8.23
ZrO_2 (360 min)	[−1 1 1]	0.0166	9.02

product heat treated for 120 min was crystalline, but the peaks were not sharp. The onset of monoclinic ZrO_2 was observed in the product heat treated for 200 min. The crystallinity of the products increased with the reaction time. When the hydrothermal time was prolonged to 360 min, the products changed into monoclinic ZrO_2 completely (JCPDS number 37–1484), and no other phase was observed.

The mean crystallite size of the products was calculated by the following Scherrer equation:

$$\tau = \frac{k\lambda}{\beta \cos \theta}, \tag{1}$$

where τ is the mean size, k is the shape factor, β is the full width at half the maximum (FWHM) intensity, θ is the Braggs angle, and λ is the wavelength of X-ray source applied in XRD. The calculated average crystallite size of ZrO_2 at 120 min reaction time was 6.97 nm. When the reaction time was 360 min, the crystallite size of ZrO_2 was found to be approximately 9.02 nm (Table 1). These results suggest that the crystallite size of ZrO_2 increases with increasing reaction time. The small crystallite size of ZrO_2 may be due to the groups adsorbed on the surface as well as on the grain boundary. These adsorbed groups may prevent the continuous growth of zirconium oxide nanocrystals, which can be achieved by reducing their surface energy and surface activity.

The morphologies of ZrO_2 were investigated by FE-SEM and TEM. From Figures 3(a) and 3(b), it can be seen that ZrO_2 exhibited a beautiful star-like shape, and stars were in the range of 30–80 nm. The HR-TEM image in Figure 3(c) shows that ZrO_2 stars were composed of short nanorods of ca. 15–20 nm in length and 3–5 nm in diameter. The corresponding selected area electron diffraction (SAED) pattern of a single structure is a ring pattern (Figure 3(d)), suggesting that ZrO_2 has a short-range crystalline structure on the nanoscale. The HR-TEM image shown in Figure 3(f) is the magnification at the area denoted by the black arrow in Figure 3(e). The lattice fringe with an interplanar spacing of 0.28 nm is consistent with the value of the (111) lattice planes of ZrO_2 (JCPDS number 37-1484).

Figure 4(a) presents the FTIR spectrum of the star-like ZrO_2. The absorption bands at 751 cm^{-1}, 679 cm^{-1}, and

504 cm^{-1} are assigned to the vibration of Zr-O [36]. The absorption bands between 1000 and 1400 cm^{-1} are due to NO_3^- anions. However, nitrate groups were no longer coordinated in a chemical bond fashion, and nitrate anions only remained on the ZrO_2 nanostructure surface. The absorption at 1531 cm^{-1} is due to the symmetric vibration absorption of COO^- [15]. The bands at 3234 cm^{-1} and 1638 cm^{-1} are attributed to the surface hydroxyl groups or adsorbed water strongly bound to the ZrO_2 surfaces [37]. It is reported that surface hydroxyl groups play an important role in heterogeneous photocatalysis and act by capturing light-induced holes thereby, producing reactive hydroxyl radicals with high oxidation capacity [38, 39]. This result suggests that the surface of ZrO_2 was probably covered by acetate groups, hydroxyl, and adsorbed water.

XPS measurements were performed on the product (Figure 4(b)), and the signals of Zr, O, C, and N were detected in the survey XPS of ZrO_2. The signals at 181.6 and 183.9 eV correspond to Zr3d5/2 and Zr3d3/2 of ZrO_2 (Figure 4(c)), which are found to be those related to the presence of zirconium in the composite, that is, Zr^{4+} of ZrO_2 as reported in [40]. The O1s peaks at 529.4, 531.1, and 532.4 eV can be ascribed to the lattice oxygen, the adsorbed oxygen (−OH, H_2O), and acetate groups, respectively [41] (Figure 4(d)). The high-resolution C1s XPS of ZrO_2 shows a strong peak at 284.2 eV, which can be attributed to C-C, and another relatively weak C1s peak is also observed, which can be ascribed to −COOH adsorbed on the surface of ZrO_2 in the form of acetate [42] (Figure 4(e)). In addition, Figure 4(f) shows a weak N1s peak at 406.6 eV, which may result from a small amount of residual nitrate as confirmed by checking the binding energy table [42]. According to literature results [43], XPS peaks of N in N-doped ZrO_2 are at 396.8 eV(Zr-N) and 400.0eV (N-O), respectively. Therefore, there is no nitrogen doping in the as-prepared ZrO_2 (406.6 eV) (Figure 4(f)). The XPS results indicate that the sample possesses a surface-adsorbed water, hydroxyl, and acetate groups, which is consistent with the FTIR spectra.

The TG curve of the ZrO_2 product is shown in Figure 5. The TG curve shows three-stage weigh loss events at 25–165°C, 165–500°C, and 500–1200°C, respectively. There is ca. 1.94% weight loss upon heating from room temperature to 165°C, which is due to the removal of adsorbed water on the surface of the product. The weight loss from 165°C to 500°C may be ascribed to the decomposition of the CH_3COO^- and NO_3^- anions, as the acetate and nitrate anions generally combust at about 350°C [44] and 192°C [45], respectively. The weight loss between 500°C and 1200°C is a result of the elimination of hydroxyls adsorbed

FIGURE 3: SEM image (a), TEM image (b, c), SAED patterns (d), TEM image (e) of the synthesized ZrO_2, and (f) HR-TEM image of the area denoted by the black arrow in (e).

on the surface. From the analysis of the TG curve and IR spectrum, the final product is covered by adsorbed water, CH_3COO^-, NO_3^-, and hydroxyl, respectively. However, it was difficult to analyze the TG curve quantitatively due to the overlap of the weight loss.

3.2. Photocatalytic Activities of ZrO_2. The degradation of RhB under UV irradiation was carried out to evaluate the

photocatalytic activity of as-prepared ZrO_2 nanostructures. For comparison, a blank experiment without catalyst was also conducted under identical conditions. Figure 6(a) indicates that the RhB concentration decreases with increasing irradiation time. When the solution was irradiated for 30 min without catalyst, a small amount of RhB was degraded (<20%). This result is similar to the results reported by other authors [24]. When the as-prepared star-like ZrO_2 was added into the solution, the degradation of RhB increased

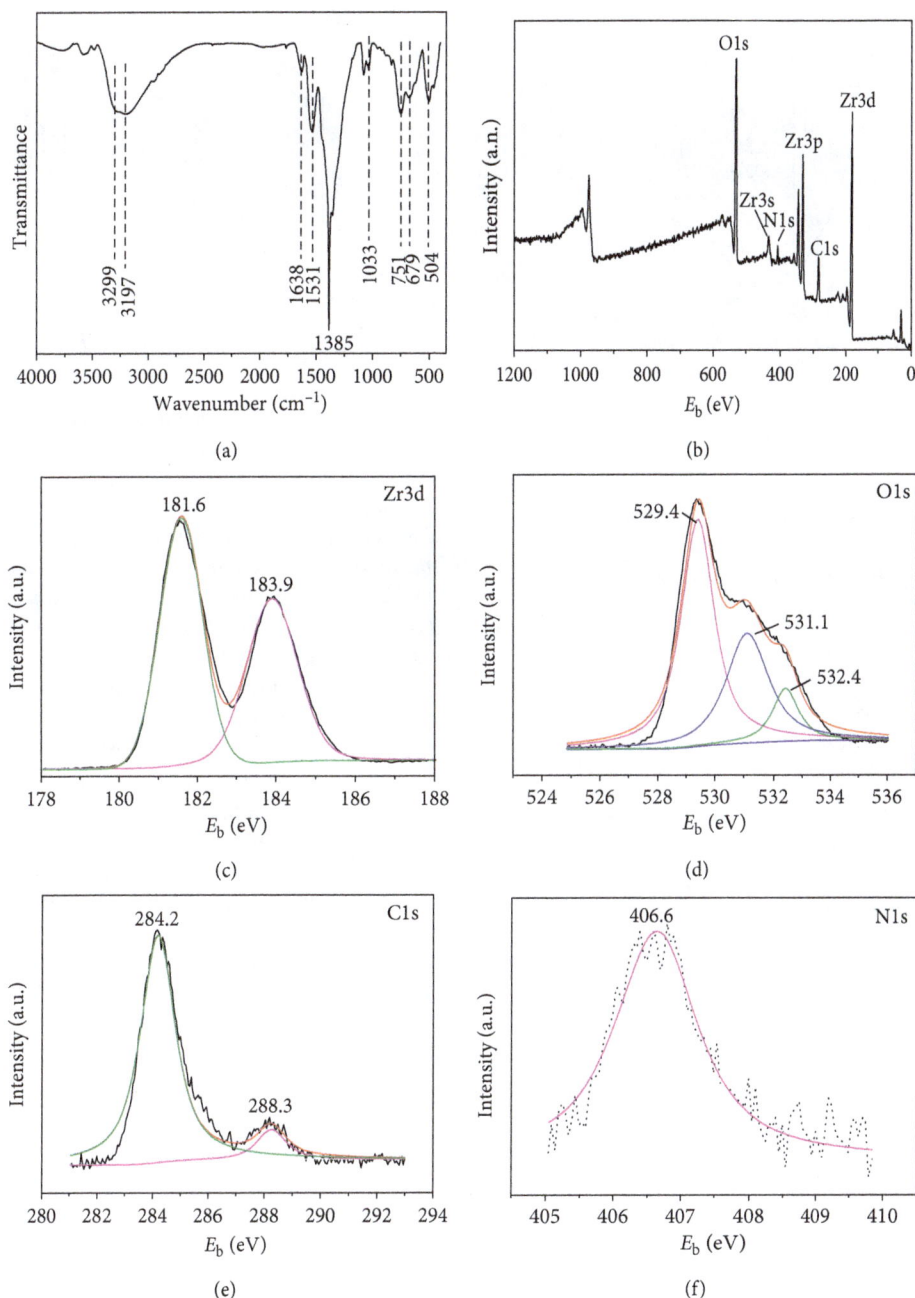

FIGURE 4: FTIR (a) and XPS (b–f) results showing the survey spectra of ZrO_2 products.

up to nearly 100% in 30 min, which shows an improved photocatalytic activity compared to the previous reports [24] (RhB degradation 100% in 40 min with flower-like ZrO_2). All products synthesized with different molar ratios of Zr $(NO_3)_4 \cdot 5H_2O$ to CH_3COONa showed a superior photocatalytic performance (Figure 6(b)).

The excellent photocatalytic activity of star-like ZrO_2 nanostructures can be attributed to several factors. Firstly, the star-like nanostructures may provide more adsorption sites and stronger surface adsorption ability to the RhB molecules, so the photocatalytic reaction can take place efficiently. Secondly, the improved surface functions may also contribute to the high photocatalytic activity of catalysts. To confirm this

suggestion, the as-prepared star-like ZrO_2 was calcined (C-ZrO_2) at 600°C for 2 h and used as the catalyst at the same condition. It was found that only 60% of the RhB degraded after irradiation for 30 min with the calcined ZrO_2 as shown in Figure 6(a). Compared with the as-prepared ZrO_2, the as-calcined ZrO_2 showed a decreased photocatalytic activity toward the degradation of RhB. Furthermore, the FTIR was used to characterize the surface-adsorbed groups of ZrO_2 (Figure 7). It can be seen that the intensity of OH (3246 cm^{-1} and 1632 cm^{-1}) and NO_3^- absorption bands (1384 cm^{-1}) of calcined ZrO_2 at 600°C was significantly reduced compared to that of as-prepared ZrO_2. Furthermore, the vibration absorption bands of acetate groups completely disappeared after

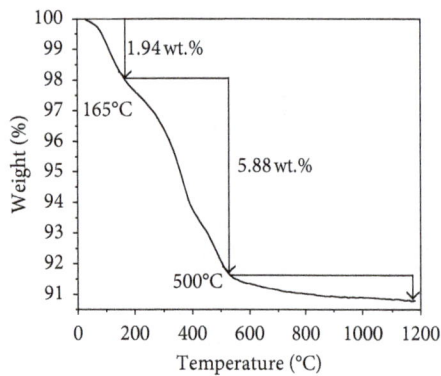

FIGURE 5: TG curve of ZrO$_2$ products obtained by reacting at 180°C for 6 h.

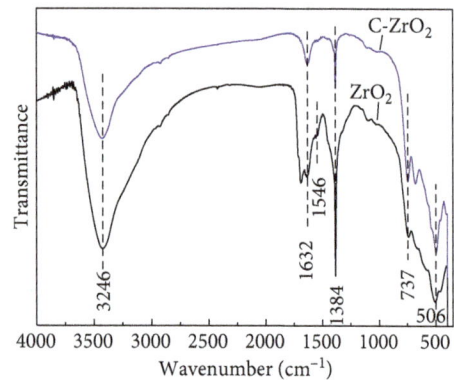

FIGURE 7: FTIR of the nanosized ZrO$_2$ and ZrO$_2$ calcined at 600°C.

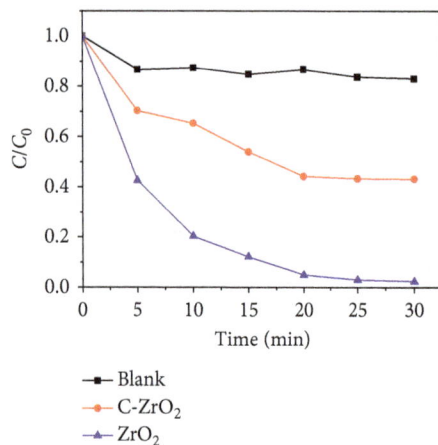

(a)

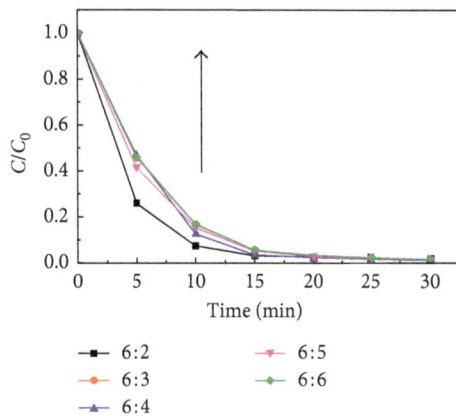

(b)

FIGURE 6: RhB concentration as a function of UV irradiation time over the product and calcined ZrO$_2$ (a) and the products with different molar ratios (b).

calcination. This indicates that the surface-adsorbed groups (hydroxyl or acetate groups) of the star-like ZrO$_2$ were destroyed during calcination and the surface chemistry of ZrO$_2$ changed. The acidity comes from hydroxyl groups since they capture light-induced holes [46], thereafter initiating the generation of strong active species (e.g., ˙OH), capable of

oxidizing adsorbed organic substrates [47]. Thus, the surface-adsorbed hydroxyl groups or water plays an important role in the high photocatalytic activity of star-like ZrO$_2$. Finally, to further probe the reason of the superior catalytic property of ZrO$_2$, the UV-Vis adsorption spectra of ZrO$_2$ and ZrO$_2$ calcined at 600°C for 2 h were measured (Figure 8(a)). It can be seen from Figure 8(a) that all star-like ZrO$_2$ showed a strong absorption in UV-Vis region with the maximum intensity at 300 nm, and an absorption band was also found at 348 nm. The absorption of the calcined ZrO$_2$ was, on the contrary, much weaker, and a considerable shift towards lower wavelength was found. For the obtained UV-Vis spectra, a bandgap was calculated using the Kubelka–Munk theory and Tauc method. The following equation was used to calculate the bandgap:

$$(Ah\nu)^2 = K\left(h\nu - E_g\right), \tag{2}$$

where A is the absorbance, K is the proportionality constant, and E_g is the bandgap energy. The plot of $(Ah\nu)^2$ versus $h\nu$ based on the direct transition is shown in Figure 8(b). The measured bandgap for the as-prepared ZrO$_2$ was found to be 3.56 eV, while the calcined ZrO$_2$ exhibited a wide bandgap of 5.22 eV. In our experiments, the bandgap of synthesized star-like ZrO$_2$ with various molar ratios of Zr(NO$_3$)$_4$·5H$_2$O to CH$_3$COONa was in the range of 3.50–3.85 eV, according to the results of UV-Vis spectra (Figures 9(a) and 9(b)). The narrow bandgap of the synthesized star-like ZrO$_2$ nanostructures could also relate to its surface functions and contribute to the superior photocatalytic activity.

The N$_2$ adsorption-desorption isotherm plots for the as-prepared ZrO$_2$ and the as-calcined ZrO$_2$ at 600°C are shown in Figure 10. Both samples exhibited type IV isotherms, indicating a typical mesoporous structure. The BET specific surface area of the as-prepared ZrO$_2$ was approximately 70.9 m^2/g (Figure 10(a)), which is higher than that of ZrO$_2$ calcined at 600°C. The ZrO$_2$ calcined at 600°C showed a smaller BET specific surface area of 26.5 m^2/g (Figure 10(b)), which indicates that the BET specific surface area of ZrO$_2$ decreased with calcining at 600°C. This may be another reason for the decreased photocatalytic activity towards the degradation of RhB by calcined ZrO$_2$.

As reported in the literature, when the solid material is irradiated with ultraviolet light, some holes are left in the

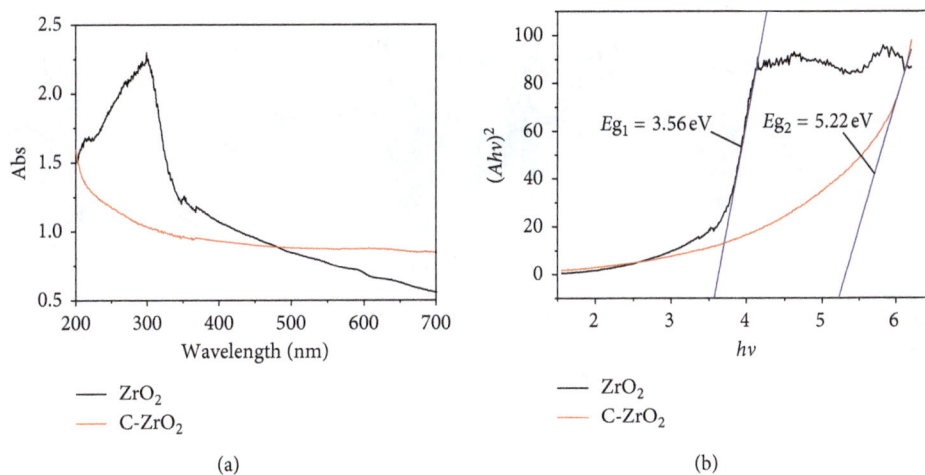

(a)

(b)

FIGURE 8: (a) UV-visible adsorption spectra of the nanosized ZrO_2 and the calcined ZrO_2; (b) plot of transformed Kubelka–Munk function versus light energy for different samples.

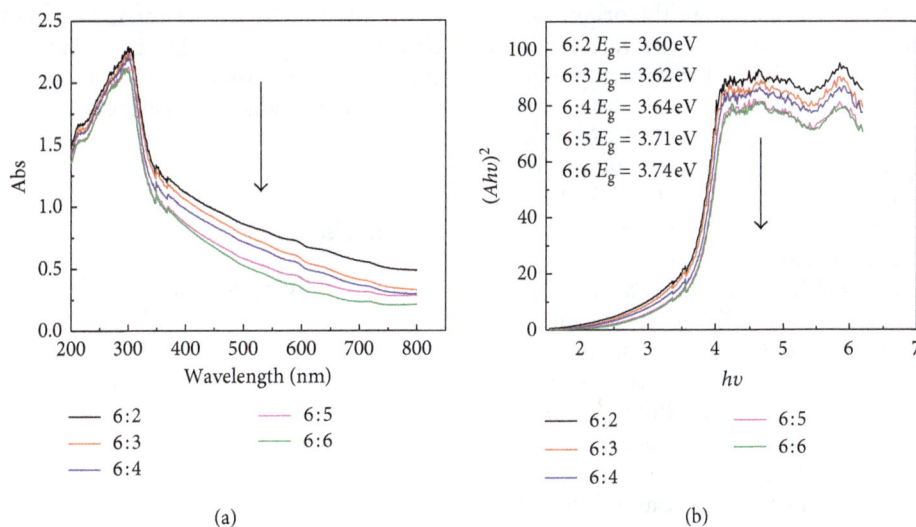

(a)

(b)

FIGURE 9: (a) UV-visible adsorption spectra of the star-like ZrO_2 with various molar ratios of $Zr(NO_3)_4 \cdot 5H_2O$ to CH_3COONa; (b) plot of transformed Kubelka–Munk function versus the energy of light for different samples.

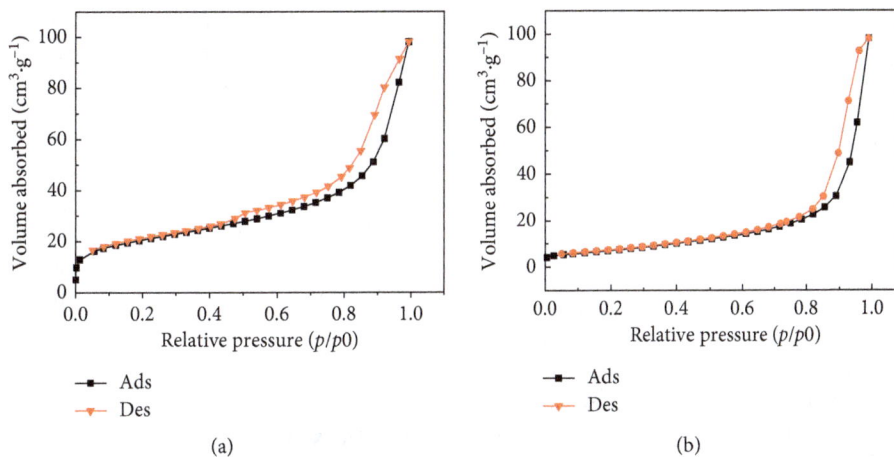

(a)

(b)

FIGURE 10: N_2 adsorption-desorption isotherms of (a) star-like ZrO_2 and (b) ZrO_2 calcined at 600°C.

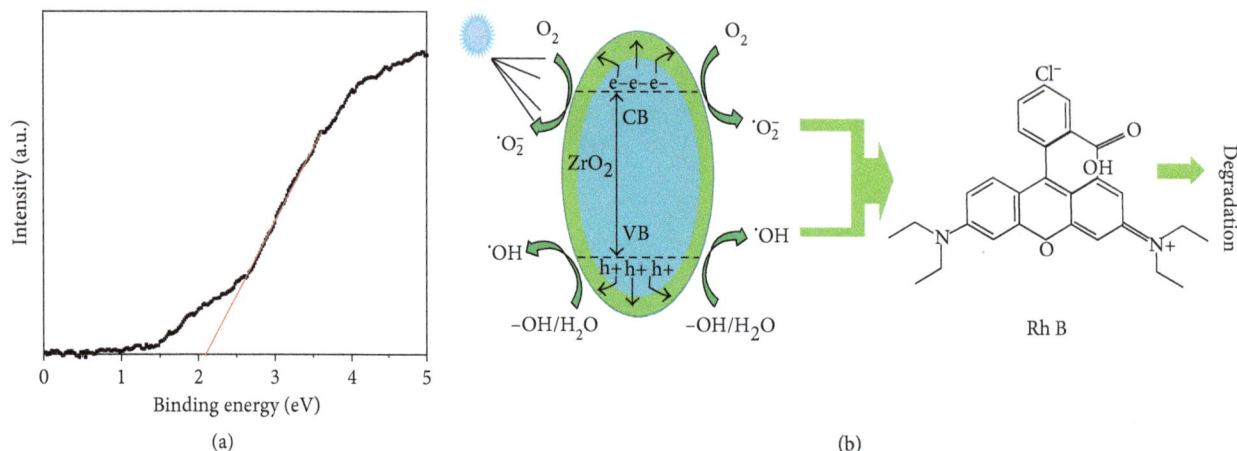

FIGURE 11: (a) Valence band (VB) XPS of the prepared ZrO_2 and (b) possible photocatalytic mechanism scheme with as-prepared ZrO_2 nanostructures.

valence band along with the process of electron transition from the valence band to the conduction band [48]. The photogenerated electrons and holes are the origin of the photocatalytic reaction. The results of the valence band (VB) XPS showed that the valence band energy of prepared ZrO_2 was 2.08 eV (Figure 11(a)). The optical bandgap energy of prepared ZrO_2 was 3.56 eV. According to the relationship $E_{VB} = E_{CB} + E_g$, the conduction band (CB) of prepared ZrO_2 would occur at −1.48 eV. Because the CB edge potential of ZrO_2 is more negative than E_{O_2/O_2^-} (−0.046 V) [49], the electrons in ZrO_2 can capture O_2 and reduce it to $\cdot O_2^-$, which could effectively suppress the electron-hole recombination rates [50]. Meanwhile, the holes are captured by OH groups or H_2O on the surface of ZrO_2 to produce hydroxyl radicals. Finally, the radicals formed, such as superoxide and hydroxyl, react with the RhB and degrade it completely. The abovementioned reactions take place on the surface of ZrO_2 with a high efficiency. The radicals produced react powerfully in the RhB solution and cause their degradation. The possible mechanism may be described as follows:

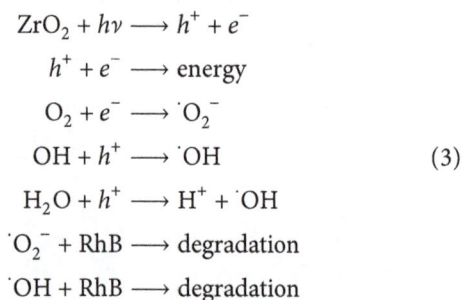

$$ZrO_2 + h\nu \longrightarrow h^+ + e^-$$
$$h^+ + e^- \longrightarrow energy$$
$$O_2 + e^- \longrightarrow \cdot O_2^-$$
$$OH + h^+ \longrightarrow \cdot OH \qquad (3)$$
$$H_2O + h^+ \longrightarrow H^+ + \cdot OH$$
$$\cdot O_2^- + RhB \longrightarrow degradation$$
$$\cdot OH + RhB \longrightarrow degradation$$

Sudrajat and Babel [43] studied the mechanism of nitrogen-doped ZrO_2-catalyzed degradation of rhodamine 6G. They found that $\cdot OH$ was the most dominant reactive species. The photogenerated h^+ also seems to play an important role in the dye degradation through direct attack of R6G molecules on the catalyst surface. $\cdot O_2^-$ is more easily produced through reduction of O_2 by the electron in the CB of N-ZrO_2 due to high CB potential of N-ZrO_2. Other references present a similar reaction mechanism.

Combining the above presented results with the literature reports, the possible photodegradation mechanism can be inferred (Figure 11(b)). In all the cases, the role of photogenerated electrons is negligible. This is an indication of an effective electron transfer from the catalyst surface to the adsorbed molecules to produce reactive species [43].

4. Conclusions

To sum up, the synthesis of star-like ZrO_2 nanostructures has been successfully carried out by using the hydrothermal method without any template and surfactant. The crystallite size of ZrO_2 increased with increasing reaction time, and the crystallite size was approximately 9.02 nm when the reaction time was 360 min. The FTIR and XPS spectra showed the surface of ZrO_2 was covered by acetate groups, hydroxyl, and adsorbed water. The bandgap of the as-synthesized star-like ZrO_2 with various molar ratios of $Zr(NO_3)_4 \cdot 5H_2O$ to CH_3COONa was in the range of 3.50–3.85 eV, according to the results of UV-Vis spectroscopy. The as-synthesized nano-ZrO_2 showed excellent photocatalytic activities in RhB degradation under UV irradiation which may be attributed to the surface functions, special morphology, and narrow bandgap of the star-like ZrO_2 nanostructures. The possible photodegradation mechanism was proposed, and potential applications of the synthesized star-like ZrO_2 have been considered. Reactive species$\cdot$OH and $\cdot O_2^-$ may play an important role in the RhB degradation. Overall, the star-like ZrO_2 nanostructures have been proven to be effective catalysts for the degradation of RhB under UV-Vis radiation and could be suitable candidates in environmental photocatalysis.

Conflicts of Interest

The authors declare that they have no conflicts of interest.

Acknowledgments

This research was supported by "the Fundamental Research Funds for the Central Universities" (Grant no. 2015 QNA07).

References

[1] G. Busca, *Heterogeneous Catalytic Materials: Solid State Chemistry, Surface Chemistry and Catalytic Behaviour*, Elsevier, New York, NY, USA, 2014.

[2] D. Ciuparu, A. Ensuque, G. Shafeev, and F. Bozon-Verduraz, "Synthesis and apparent bandgap of nanophase zirconia," *Journal of Materials Science*, vol. 19, no. 11, pp. 931–933, 2000.

[3] G. Poungchan, B. Ksapabutr, and M. Panapoy, "One-step synthesis of flower-like carbon-doped ZrO_2 for visible-light-responsive photocatalyst," *Materials Design*, vol. 89, pp. 137–145, 2016.

[4] A. R. Ardiyanti, A. Gutierrez, M. L. Honkela, A. O. I. Krause, and H. J. Heeres, "Hydrotreatment of wood-based pyrolysis oil using zirconia-supported mono- and bimetallic (Pt, Pd, Rh) catalysts," *Applied Catalysis A: Generation*, vol. 407, no. 1-2, pp. 56–66, 2011.

[5] S. Guerrero, P. Araya, and E. E. Wolf, "Methane oxidation on Pd supported on high area zirconia catalysts," *Applied Catalysis A: Generation*, vol. 298, pp. 243–253, 2006.

[6] S. Tekeli, "The solid solubility limit of Al_2O_3 and its effect on densification and microstructural evolution in cubic-zirconia used as an electrolyte for solid oxide fuel cell," *Materials Design*, vol. 28, no. 2, pp. 713–716, 2007.

[7] Y. W. Suh, J. W. Lee, and H. K. Rhee, "Synthesis of thermally stable tetragonal zirconia with large surface area and its catalytic activity in the skeletal isomerization of 1-butene," *Catalysis Letters*, vol. 90, no. 1-2, pp. 103–109, 2003.

[8] V. V. Brei, "Superacids based on zirconium dioxide," *Theoretical and Experimental Chemistry*, vol. 47, no. 3, pp. 165–175, 2005.

[9] S. Samantaray, B. G. Mishra, D. K. Pradhan, and G. Hota, "Solution combustion synthesis and physicochemical characterization of ZrO_2-MoO_3 nanocomposite oxides prepared using different fuels," *Ceramics International*, vol. 37, no. 8, pp. 3101–3108, 2011.

[10] Y. Lu, Z. Wang, S. Yuan, L. Shi, Y. Zhao, and W. Deng, "Microwave-hydrothermal synthesis and humidity sensing behavior of ZrO_2 nanorods," *RSC Advances*, vol. 3, no. 29, pp. 11707–11714, 2013.

[11] M. Habulan, "Deposition of sol-gel ZrO_2 films on stainless steel," *Evolution*, vol. 57, pp. 2904–2910, 2014.

[12] P. Murugavel, M. Kalaiselvam, A. R. Raju, and C. N. R. Rao, "Sub-micrometre spherical particles of TiO_2, ZrO_2 and PZT by nebulized spray pyrolysis of metal-organic precursors," *Journal of Materials Chemistry*, vol. 7, no. 8, pp. 1433–1438, 2015.

[13] T. S. Jeon, J. M. White, and D. L. Kwong, "Thermal stability of ultrathin ZrO_2 films prepared by chemical vapor deposition on Si(100)," *Applied Physics Letters*, vol. 78, no. 3, pp. 368–370, 2001.

[14] X. Dong, Z. Liu, J. Yan, J. Wang, and G. Hong, "Preparation of nanocrystalline ZrO_2 by reverse precipitation method," *Journal of Wuhan University of Technology*, vol. 20, no. 3, pp. 1–4, 2005.

[15] Z. Shu, X. Jiao, and D. Chen, "Hydrothermal synthesis and selective photocatalytic properties of tetragonal star-like ZrO_2 nanostructures," *CrystEngComm*, vol. 15, no. 21, pp. 4288–4294, 2013.

[16] E. Beltowska-Lehman, P. Indyka, A. Bigos, M. J. Szczerba, and M. Kot, "Ni-W/ZrO_2 nanocomposites obtained by ultrasonic DC electrodeposition," *Materials Design*, vol. 80, pp. 1–11, 2015.

[17] Q. X. Gao, X. F. Wang, X. C. Wu, Y. R. Tao, and J. J. Zhu, "Mesoporous zirconia nanobelts: preparation, characterization and applications in catalytical methane combustion," *Microporous and Mesoporous Materials*, vol. 143, no. 2-3, pp. 333–340, 2011.

[18] H. Q. Cao, X. Q. Qiu, B. Luo et al., "Synthesis and room-temperature ultraviolet photoluminescence properties of zirconia nanowires," *Advanced Functional Material*, vol. 14, no. 3, pp. 243–246, 2004.

[19] H. M. Abdellaal, "One-pot path for the synthesis of hollow zirconia sub-microspheres using hydrothermal approach," *Materials Letters*, vol. 212, pp. 218–220, 2018.

[20] J. L. Gole, S. M. Prokes, J. D. Stout, O. J. Glembocki, and R. Yang, "Unique properties of selectively formed zirconia nanostructures," *Advanced Materials*, vol. 18, no. 5, pp. 664–667, 2006.

[21] B. B. Nayak, S. K. Mohanty, M. Q. B. Takmeel, D. Pradhan, and A. Mondal, "Borohydride synthesis and stabilization of flake-like tetragonal zirconia nanocrystallites," *Materials Letters*, vol. 64, no. 17, pp. 1909–1911, 2010.

[22] A. M. Azad, "Fabrication of yttria-stabilized zirconia nanofibers by electrospinning," *Materials Letters*, vol. 60, no. 1, pp. 67–72, 2006.

[23] J. B. Joo, A. Vu, Q. Zhang, M. Dahl, M. Gu, and F. Zaera, "A sulfated ZrO_2 hollow nanostructure as an acid catalyst in the dehydration of fructose to 5-hydroxymethylfurfural," *ChemSusChem*, vol. 6, no. 10, pp. 2001–2008, 2013.

[24] Z. Shu, X. Jiao, and D. Chen, "Synthesis and photocatalytic properties of flower-like zirconia nanostructures," *CrystEngComm*, vol. 14, no. 3, pp. 1122–1127, 2012.

[25] X. Yang, X. Song, Y. Wei, W. Wei, L. Hou, and X. Fan, "Synthesis of spinous ZrO_2 core-shell microspheres with good hydrogen storage properties by the pollen bio-template route," *Scripta Materialia*, vol. 64, no. 12, pp. 1075–1078, 2011.

[26] T. F. Ma, J. Bai, and C. P. Li, "Facile synthesis of g-C_3N_4 wrapping on one-dimensional carbon fiber as a composite photocatalyst to degrade organic pollutants," *Vacuum*, vol. 145, pp. 47–54, 2017.

[27] P. Supriya, B. T. V. Srinivas, K. Chowdeswari, N. V. S. Naidu, and B. Sreedhar, "Biomimetic synthesis of gum acacia mediated Pd-ZnO and Pd-TiO_2-promising nanocatalysts for selective hydrogenation of nitroarenes," *Materials Chemistry and Physics*, vol. 204, pp. 27–36, 2017.

[28] Z. Liu, Q. Zhang, Y. Li, and H. Wang, "Solvothermal synthesis, photoluminescence and photocatalytic properties of pencil-like ZnO microrods," *Journal of Physics and Chemistry of Solids*, vol. 73, no. 5, pp. 651–655, 2012.

[29] N. F. Jaafar, A. A. Jalil, S. Triwahyono, M. N. M. Muhid, N. Sapawe, and M. A. H. Satar, "Photodecolorization of methyl orange over α-Fe_2O_3-supported HY catalysts: the effects of catalyst preparation and dealumination," *Chemical Engineering Journal*, vol. 191, pp. 112–122, 2012.

[30] J. Zhang, L. Li, and D. Liu, "Multi-layer and open three-dimensionally ordered macroporous TiO_2-ZrO_2 composite: diversified design and the comparison of multiple mode photocatalytic performance," *Materials Design*, vol. 86, pp. 818–828, 2015.

[31] A. Nezamzadeh-Ejhieh and M. Karimi-Shamsabadi, "Decolorization of a binary azo dyes mixture using CuO incorporated nanozeolite-X as a heterogeneous catalyst and solar irradiation," *Chemical Engineering Science*, vol. 228, pp. 631–641, 2013.

[32] C. Suciu, L. Gagea, A. C. Hoffmann, and M. Mocean, "Sol-gel production of zirconia nanoparticles with a new organic precursor," *Chemical Engineering Science*, vol. 61, no. 24, pp. 7831–7835, 2006.

[33] J. Zhao, X. Wang, L. Zhang, X. Hou, Y. Li, and C. Tang, "Degradation of methyl orange through synergistic effect of zirconia nanotubes and ultrasonic wave," *Journal of Hazardous Materials*, vol. 188, no. 1–3, pp. 231–234, 2011.

[34] L. Renuka, K. S. Anantharaju, S. C. Sharma et al., "Hollow microspheres Mg-doped ZrO_2 nanoparticles: green assisted synthesis and applications in photocatalysis and photoluminescence," *Journal of Alloys and Compounds*, vol. 672, pp. 609–622, 2016.

[35] L. Xu and H. K. Lee, "Zirconia hollow fiber: preparation, characterization, and microextraction application," *Analytic Chemistry*, vol. 79, no. 14, pp. 5241–5248, 2007.

[36] D. V. Pinjari, K. Prasad, and P. R. Gogate, "Intensification of synthesis of zirconium dioxide using ultrasound: effect of amplitude variation," *Chemical Engineering and Process*, vol. 74, pp. 178–186, 2013.

[37] A. Brisdon and K. Nakamoto, "Infrared and Raman spectra of inorganic and coordination compounds, part B, applications in coordination, organometallic, and bioinorganic chemistry," *Applied Organometallic Chemistry*, vol. 24, p. 489, 2010.

[38] A. Manikandan, L. J. Kennedy, and J. J. Vijaya, "Comparative investigation of zirconium oxide ($ZrO_{(2)}$) nano and microstructures for structural, optical and photocatalytic properties," *Journal of Colloid and Interface Science*, vol. 389, pp. 91–98, 2013.

[39] X. Fu., L. A. Clark, and Q. Yang, "Enhanced photocatalytic performance of titania-based binary metal oxides: TiO_2/SiO_2 and TiO_2/ZrO_2," *Environmental Science and Technology*, vol. 30, no. 2, pp. 647–653, 1996.

[40] J. Zhang, L. Li, Z. Xiao, D. Liu, S. Wang, and J. Zhang, "Hollow sphere TiO_2-ZrO_2 prepared by self-assembly with polystyrene colloidal template for both photocatalytic degradation and H_2 evolution from water splitting," *ACS Sustainable Chemistry and Engineering*, vol. 4, no. 4, pp. 2037–2046, 2016.

[41] A. V. Charanpahari, S. G. Ghugal, S. S. Umare, and S. Rajamma, "Mineralization of malachite green dye over visible light responsive bismuth doped TiO_2-ZrO_2 ferromagnetic nanocomposites," *New Journal of Chemistry*, vol. 39, no. 5, pp. 3629–3638, 2015.

[42] Z. Wang, Q. Long, L. Yuan, W. Zhang, X. Hu, and Y. Huang, "Functionalized N-doped interconnected carbon nanofibers as an anode material for sodium-ion storage with excellent performance," *Carbon*, vol. 55, pp. 328–334, 2013.

[43] H. Sudrajat and S. Babel, "Comparison and mechanism of photocatalytic activities of N-ZnO and N-ZrO_2 for the degradation of rhodamine 6G," *Environmental Science and Pollution Research*, vol. 23, no. 10, pp. 10177–10188, 2016.

[44] D. C. Perera, J. W. Hewage, and N. D. Silva, "Theoretical study of catalytic decomposition of acetic acid on MgO nanosurface," *Computational and Theoretical Chemistry*, vol. 1064, pp. 1–6, 2015.

[45] Y. Shanmugam, L. Fanyuan, C. Tsonghuei, and C. Yeh, "Thermal decomposition of metal nitrates in air and hydrogen environments," *Journal of Physical Chemistry B*, vol. 107, pp. 1044–1047, 2003.

[46] C. Anderson and A. J. Bard, "Improved photocatalytic activity and characterization of mixed TiO_2/SiO_2 and TiO_2/Al_2O_3 materials," *Journal of Physical Chemistry B*, vol. 101, no. 14, pp. 2611–2616, 1997.

[47] X. Chen, X. Wang, and X. Fu, "Hierarchical macro/mesoporous TiO_2/SiO_2 and TiO_2/ZrO_2 nanocomposites for environmental photocatalysis," *Energy & Environmental Science*, vol. 2, no. 8, pp. 872–877, 2009.

[48] L. Renuka, K. S. Anantharaju, S. C. Sharma et al., "A comparative study on the structural, optical, electrochemical and photocatalytic properties of ZrO_2 nanooxide synthesized by different routes," *Journal of Alloys and Compounds*, vol. 695, pp. 382–395, 2017.

[49] D. Wang, T. Kako, and J. Ye, "Efficient photocatalytic decomposition of acetaldehyde over a solid-solution perovskite $(Ag_{0.75}Sr_{0.25})(Nb_{0.75}Ti_{0.25})O_3$ under visible-light irradiation," *Journal of the American Chemical Society*, vol. 130, no. 9, pp. 2724–2725, 2008.

[50] X. Wang, L. Zhang, H. Lin et al., "Synthesis and characterization of a ZrO_2/g-C_3N_4 composite with enhanced visible-light photoactivity for rhodamine degradation," *RSC Advances*, vol. 4, no. 75, pp. 29–35, 2014.

20

Olive Stone Ash as Secondary Raw Material for Fired Clay Bricks

D. Eliche-Quesada,[1] M. A. Felipe-Sesé,[1,2] and A. Infantes-Molina[3]

[1]Department of Chemical, Environmental, and Materials Engineering, Advanced Polytechnic School of Jaén, University of Jaén, Campus Las Lagunillas s/n, 23071 Jaén, Spain
[2]International University of La Rioja, Avenida La Paz, No. 137, Logroño, 26002 La Rioja, Spain
[3]Department of Inorganic Chemistry, Crystallography and Mineralogy, Affiliate Unit of the ICP-CSIC, Faculty of Sciences, University of Málaga, Campus de Teatinos, 29071 Málaga, Spain

Correspondence should be addressed to D. Eliche-Quesada; deliche@ujaen.es

Academic Editor: Kaveh Edalati

This work evaluates the effect of incorporation of olive stone ash, as secondary raw material, on the properties of fired clay bricks. To this end, three compositions containing 10, 20, and 30 wt% olive stone ash in a mixture of clays (30 wt% red, 30 wt% yellow, and 40 wt% black clay) from Spain were prepared. The raw materials, clay and olive stone ash, were characterized by means of XRD, XRF, SEM-EDS, and TG-TDA analysis. The engineering properties of the press molded specimens fired at 900°C (4 h) such as linear shrinkage, bulk density, apparent porosity, water absorption, and compressive strength were evaluated. The results indicated that the incorporation of 10 wt% of olive stone ash produced bricks with suitable technological properties, with values of compressive strength of 41.9 MPa but with a reduced bulk density, by almost 4%. By contrast, the incorporation of 20 wt% and 30 wt% sharply increased the water absorption as a consequence of the large amount of open porosity and low mechanical strength presented by these formulations, which do not meet the standards for their use as face bricks. The bricks do not present environmental problems according to the leaching test.

1. Introduction

Among renewable energy sources, biomass plays a very important role in the new energy framework, since agricultural residues are produced in relatively large quantities all over the world. Spain, deficient in fossil energy resources, is very rich in biomass resources. According to the Food and Agriculture Organization of the United Nations (FAO), Spain is the main world producer of olives, with 4,577,000 tons in 2014, followed by Greece, Italy, Turkey, and Morocco [1]. In 2015, the production of olives in Spain increased, representing 7,344,820 tons with a cultivated surface area of 2,526,496 ha [2]. In particular, the olive sector in Andalusia, region of southern Spain, has grown over the years due to expansion and intensification of olive groves, which cover 1.5 million hectares of olives groves and depict 60% of the total cultivation area of the country in 2012 [3]. Therefore, the oil sector can be considered as the greater producer of biomass in Andalusia [4]. The main biomass employed for thermal use

in the Andalusian industrial sector is the products derived of its olive industry, as olive stone, pomace, and leaves [5].

The cultivation of the olive grove is dedicated to two uses: production of olive oil and table olives production. The most important waste generated in both industries is the olive stone. In Andalusia, about 360,000 t/year of crushed stone is generated: 0.083 tons of stones is generated per ton of olive (11.5%). Also in the table olive industry around 22,000 tons of whole stone [4] is generated. Between the different ways of exploitation of olive stone, its use as an adsorbent, after its transformation into active carbon; in the treatment of waste water of chemical and pharmaceutical industries [6–9]; as electrodes for Li-ion batteries [10, 11]; and for furfural production, plastic filled, abrasive and cosmetics, animal feed, or resin formation is found [12].

Recently, its use in the production of building bricks has been studied [13, 14]. Nonetheless, the main use is as fuel to produce electric energy or heat. This source of renewable energy, biomass, which has tremendous potential to mitigate the global warming, likewise contributes to the increase of the value of the residues and reduction of the environmental impact of the waste disposal. The stone is a fuel presenting excellent features: high density, average humidity of 15%, very uniform particle size, and a calorific power of 4,500 kcal/kg in a dry base. It is very suitable for thermal applications, both own mills and other industries, greenhouse, domestic and residential use, and municipal facilities as swimming pools, schools, and parks, due to the low emissions of particles in its combustion and odorless conditions.

The combustion of olive stone has been associated with the production of residues of combustion, such as ashes. Currently, the greater part of the combustion ashes of olive stone are either arranged in a landfill or recycled in agricultural fields. Although ashes are considered as nonhazardous industrial waste according to the Ministry of Environment and so specified in the European list of waste 10.01.01 and 10.01.03 codes [15], the ashes generated in the biomass combustion process carry an economic and environmental load.

On the other hand, the presence of volatile heavy metals in these ashes can also have negative environmental effects if managed and eliminated without care, due to the possible leaching in surface and underground water [16].

Nowadays, the ceramic and cement industry includes manufacturing processes that make the recovery of wastes possible, taking advantage of the calorific powder from their combustion or by incorporating them into the internal structure of materials, forming part of its own matrix and becoming an inert element especially viable [17, 18]. In this regard, in the last years a great number of researches are devoted to the incorporation of different combustion ashes into ceramic bricks in order to valorize a waste and improve the properties of the resultant material according to the standards. Palm oil fuel ash [19], rice husk ash [20, 21], sugarcane bagasse ash and rice husk ash [22], municipal solid waste incineration fly ash [23, 24], sewage sludge incinerator ash [25, 26], olive pomace bottom ash [27], olive pomace fly ash [28, 29], and wood ash [30] have been studied. However, the use of olive stone combustion ash to manufacture ceramic materials has not been studied yet.

Therefore, the objective of this work is to evaluate the use of olive stone ashes as secondary raw material in the manufacturing of ceramic materials. The raw materials, clay and stone ash, were characterized in terms of chemical composition, crystalline phases, and thermal behavior. The influence of olive stone ash proportion was therefore examined. Clay brick samples containing 0–30 wt% of waste were mixed, molded, and sintered at 900°C. Engineering properties such as bulk density, apparent porosity, water absorption, and compressive strength were studied as a function of ash content.

2. Experimental Procedures

2.1. Raw Materials. The olive stone ash was collected by Geolit Air Conditioning Enterprise, society promoted by Valoriza Energy, Inverjaen, Geolit, Agener, and Thermal Power Stations and Networks. This society located in the technological park of Geolit (located in Mengibar, Jaén, Spain) develops a centralized eco-efficient and innovative

air conditioning system, with important benefits for the environment. The olive stone ash samples were collected from the mechanical hopper in the station. The collected sample was mixed and homogenized using suitable coning and quartering procedures. Finally, the samples were sieved to particle diameter less than 100 μm.

The clay was supplied by a clay pit located in Bailen, Jaén (Spain). It was obtained by mixing three types of raw clay in the following percentages: 30 wt% red, 30 wt% yellow, and 40 wt% black clay. To obtain uniform particle size, the clay was crushed and ground to yield a powder with a particle size suitable to pass through a 500 μm sieve.

2.2. Characterization of Raw Materials. The surface of the olive stone ash and mixed clay was characterized using scanning electron microscopy and energy dispersive X-ray spectrometry SEM/EDX using the high-resolution transmission electron microscope JEOL SM 840. Samples were placed on an aluminium grate and coated with carbon using the ion sputtering device JEOL JFC 1100. Crystalline phases were analyzed by using X'Pert Pro MPD automated diffractometer (PANalytical) equipped with a Ge (111) primary monochromator, using monochromatic Cu Kα radiation ($\lambda = 1.5406$ Å) and X'Celerator detector. The chemical composition was determined by X-ray fluorescence (XRF) using the Philips Magix Pro (PW-2440). The thermal behavior was determined by thermogravimetric and differential thermal analysis (TG-DTA) with a Mettler Toledo 851e device in oxygen. The total content of carbon, hydrogen, nitrogen, and sulphur was determined by combustion of samples in O_2 atmosphere using the CHNS-O Thermo Finnigan Elementary Analyzer Flash EA 1112.

2.3. Preparation and Testing of Fired Olive Stone Ash-Clay Bricks. Three olive stone ash-clay mixtures were prepared by adding the proper amount of ashes in order to obtain mixtures containing 10, 20, and 30 wt% of ash. Samples were homogenized in a mortar. To enable comparative results, ten samples per series were prepared for testing. The necessary amount of water (8 wt% moisture) was added to the samples to obtain adequate plasticity and absence of defects, mainly cracks, during the semidry compression molding stage under 54.5 MPa of pressure, using uniaxial laboratory-type pressing Mega KCK-30 A. Waste-free mixtures were also made as references. Solid bricks with 30 × 10 mm cross sections and a length of 60 mm were obtained. Samples were fired in a laboratory furnace at a rate of 3°C/min up to 900°C for 4 h. Samples were then cooled to room temperature by natural convection inside the furnace. The shaped samples were designated as C for the bricks without olive stone ash and C-xOSA for the mixtures, where x denotes the ash content (%) in the clay matrix. Figure 1 depicts the scheme of the preparation process followed.

Linear shrinkage was obtained by measuring the length of samples before and after the firing stage by using a caliper with a precision of ±0.01 mm, according to ASTM standard C326 [31]. Water absorption values were determined from weight difference between the as-fired and water-saturated samples (immersed in boiling water for 2 h), according to

TABLE 1: Carbonate content, organic content, and CNHS analysis of raw materials.

Sample	Carbonate content[a] (%)	Organic matter content (%)	%C	%H	%N	%S
OSA	39.54 ± 0.89	11.83 ± 0.12	11.60 ± 0.01	0.40 ± 0.02	0.012 ± 0.0	0.0
Clay	7.36 ± 0.33	2.29 ± 0.09	2.25 ± 0.01	0.34 ± 0.004	0.05 ± 0.002	0.032 ± 0.008

[a] Determined according to ASTM D-2974.

TABLE 2: Chemical composition of OSA and clay.

Oxide content (%)	SiO_2	Al_2O_3	Fe_2O_3	CaO	MgO	MnO	K_2O	TiO_2	P_2O_5	SO_3	ZnO	SrO	ZrO_2	Cl	LOI
OSA	8.47	1.68	2.97	24.0	3.42	0.057	31.22	0.073	4.04	—	0.03	0148	—	0.074	23.8
Clay	54.4	12.4	4.58	8.76	2.46	0.03	3.37	0.60	0.11	0.68	0.03	0.027	0.033	—	12.5

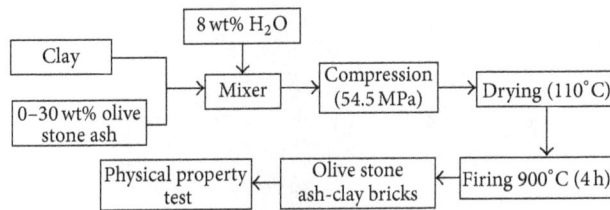

FIGURE 1: Flowchart of producing clay-olive stone ash bricks.

the ASTM standard C373 [32]. Bulk density was determined by the Archimedes method [32]. Water suction of a brick is defined as the volume of water absorbed during short partial immersion. Tests to determine water suction were implemented according to the standard procedure UNE-EN 772-1 [33].

Compressive strength of bricks is their bulk unit charge against breakage under axial compressive strength. For this trial, six fired samples were studied. Tests on compressive strength were performed according to the standard UNE-EN 772-1 [34] on MTS 810 Material Testing Systems laboratory press. The area of both bearing surfaces was measured and the average taken. All samples were submitted to a progressively increasing normal strength, with the load applied centered on the upper surface of the sample until breakage. The compressive strength of each sample was obtained by dividing the maximum load by the average surface of both bearing surfaces, expressed in MPa with 0.1 MPa accuracy.

Finally, leachability of heavy metals in the samples was studied using the toxicity characteristic leaching procedure (TCLP) according to EPA method 1311 (Environmental Protection Agency, Method 13-11, 1992) [35]. The concentrations in the filtrate were measured with an Inductively Coupled Plasma-Atomic Emission Spectrometer (ICP-AES Agilent 7500).

3. Results and Discussion

3.1. Raw Materials Characterization. The olive stone ash received had about 1.0% moisture content. In Figure 2, the SEM image of dried ash is shown. The particle morphology showed the presence of irregular, angled particles of 40–70 microns, some spherical particles, and clusters formed by sintering agglomeration during the thermal process. The

chemical composition of the olive stone ash determined by EDS indicated the presence of significant percentages of K, Ca, and Si. The spherical particles had a higher proportion of Ca.

The weight loss was investigated by calcination of the sample, grain size <100 μm, at 900°C for 3 h. The weight loss observed was 21.9%, indicating a high amount of unburned matter in the ash. This result indicated that the use of this material directly to form green bodies could lead to a significant volume reduction with the concomitant deformation or breakage of the bricks if the sintering process is not carried out at an adequate heating rate. The CNHS analysis of the olive ash (Table 1) showed that it was composed mainly of carbon (11.6%) and small quantities of hydrogen (0.40%) and nitrogen (0.012%). It did not contain sulphur. These data indicated that the olive stone ash contained large amounts of organic carbon pointing to incomplete combustion of the olive stone and therefore an inefficient fuel use [36, 37].

The particle size distribution of the olive stone ash showed an average particle size D_{50} of 59.4 μm. The ash presented a percentage of fine particles (<0.002 mm) of 1.31 vol%, a lime content (0.002–0.063 mm) of 47.1 vol%, and a sand content (0.063–2 mm) of 51.6 vol%. Therefore, the broad particle size caused a lack of homogeneity in the mixed, and so a further process to homogenize the grain size to a particle size of 100 μm was necessary.

The chemical composition of the raw materials, olive stone and clay, used in this study is shown in Table 2. If the ash is considered, it was mainly composed (>55 wt%) of K_2O (31.2 wt%) and CaO (24.0%). Oxides such as SiO_2, P_2O_5, MgO, Fe_2O_3, and Al_2O_3, in decreasing order of abundance, were present in a 1–10 wt% proportion, whereas SrO was a minor oxide (0.1–1 wt%). The chloride content was low, 0.07 wt%. High contents of fluxing oxides (K_2O) and auxiliary fluxing oxides (CaO, MgO, and Fe_2O_3) are desired to lower the firing process temperature for brick preparation. The total content of Si and Al (estimated as oxides) was lower than the values obtained in other bottom ashes due to their low level in the feeding fuel.

On the other hand, the clay was mainly composed of silica, aluminium, and calcium oxides (54.4 wt%, 12.4 wt%, and 8.8 wt%, resp.). The clay mixture also contained a small amount of fluxing agents, K_2O and MgO, accompanied by a higher amount of iron oxide (4.6 wt%). Clay and mainly olive stone ash presented high LOI values (12.5% for clay and 23.8%

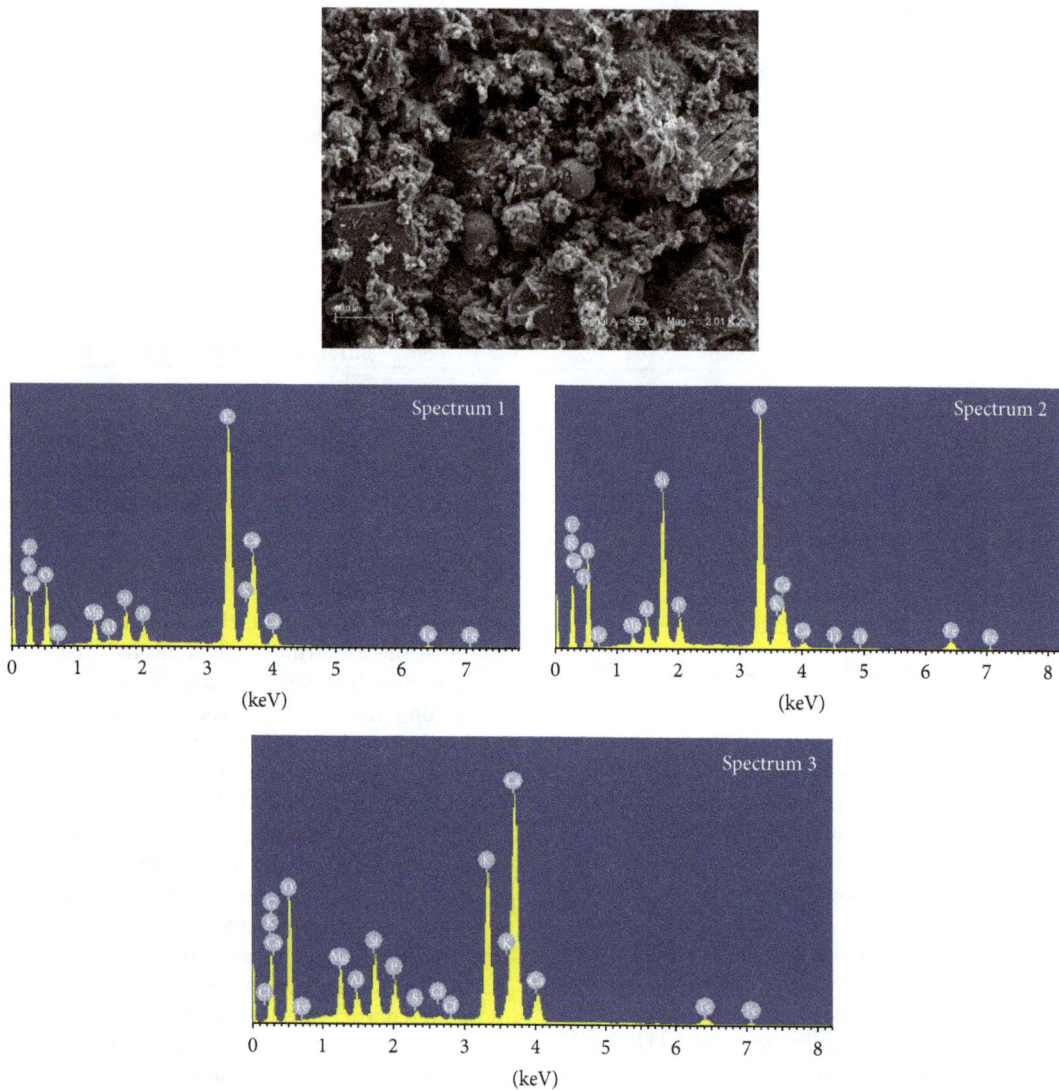

FIGURE 2: SEM/EDX of olive stone as used as raw material for clay bricks.

for OSA) which were associated with the organic matter (2.3% for clay and 11.8% for OSA) and carbonate contents (7.4% for clay and 39.5% for OSA) (Table 1). During firing, organic matter and carbonates act as pore-forming agents and carbonates also generate crystalline phases, which enhance mechanical strength [38].

The crystalline phases present in the raw materials were evaluated by means of XRD. Figure 3 lumped together both diffractograms. In the diffractogram corresponding to pure clay (Figure 3(a)), several contributions can be observed. The main diffraction peaks were related to the presence of quartz, calcite, and dolomite. Some tiny diffraction peaks also revealed the presence of feldspar and a small amount of gypsum and phyllosilicates. On the other hand, the diffractogram corresponding to olive stone ash was complex and a great number of diffraction peaks were noticeable. The phases present were mainly associated with carbonates of K and Ca. Thus the main phase was a carbonate of K and Ca hydrated, $K_2CO_3 \cdot (3/2)H_2O$ (PDF N°: 98-002-2257). As it will

be seen below, this sample loses 5 wt% of weight associated with moisture. The second phase observed was anhydrous carbonate of Ca and K, $K_2Ca(CO_3)_2$ (PDF N° 01-083-1921 and PDF N° 98-000-6177). Other tiny diffraction peaks matched with several silicates although the contribution of these peaks was low. XRD results agree with XRFS ones that indicated that Ca and K were the main elements present as well as a great amount of C derived from CNH analysis.

The thermal characterization of the raw materials is provided in Figure 4. The thermal decomposition profile of OSA showed a weight loss between 30 and 200°C with a weight loss of 5.6% associated with moisture. The organic matter and unburned elements, as residual carbon, decomposition occurred between 200 and 550°C [39] and a weight loss of 9.3% accompanied with a single strong exothermic peak centered at 425°C was present. Between 550 and 800°C, a weight loss of 5.1% was observed and several thermal effects were clearly visible. At 650°C, a slight endothermic peak associated with structural water release from hydroxide ions

TABLE 3: Technological properties of fired construction bricks made from clay and olive stone ash.

Sample	Waste content (wt%)	Linear shrinkage (%)	Loss on ignition (%)	Suction water (kg/m^2 min)
C	0	-0.518 ± 0.10	9.24 ± 0.09	2.32 ± 0.07
C-10OSA	10	-0.88 ± 0.12	12.97 ± 0.08	2.59 ± 0.22
C-20OSA	20	-1.88 ± 0.14	15.19 ± 0.08	3.06 ± 0.06
C-30OSA	30	-2.19 ± 0.15	17.19 ± 0.09	2.88 ± 0.08

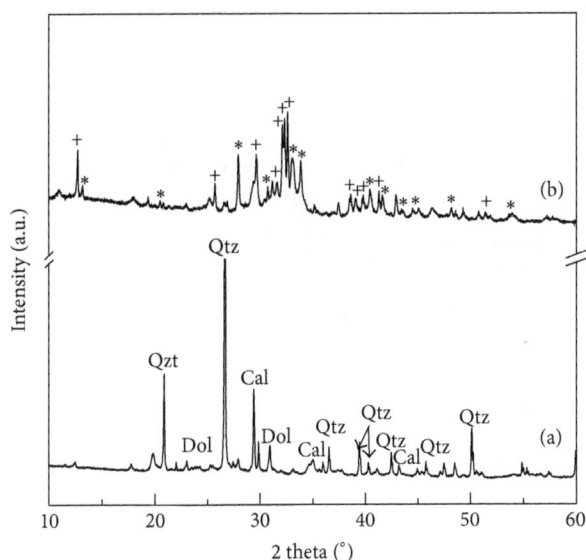

FIGURE 3: XRD patterns of (a) starting raw clay (Qtz: quartz, Cal: calcite, and Dol: dolomite) and (b) olive stone ash (+: $K_2CO_3\cdot(3/2)H_2O$; *: $K_2Ca(CO_3)_2$).

present in the ash was observed. The exothermic peak at 700°C was probably due to the combustion of unburned elements present in the ashes, whereas the endothermic effect at 750°C may be assigned to the decomposition of the carbonates in the ashes, emitting CO_2.

The TG-DTA analysis of the clay sample also showed several processes. Between 30 and 150°C, the moisture was removed (1.0 wt%). Between 100 and 650°C, the weight loss observed (2.75 wt%) was due to dehydroxylation reactions of clay minerals and organic matter decomposition. The main weight loss was obtained between 650 and 900°C (7.7 wt%) and associated with calcium carbonate decomposition. The DTA curve showed that dehydration, dehydroxylation, and decarbonation reactions were endothermic processes, with peaks centered at 85°C, 570°C, and 760°C, respectively. The decomposition of organic matter is an exothermic process, with peaks centered at about 375°C and 475°C.

3.2. Technological Properties of Fired Olive Stone Ash-Clay Bricks. In order to evaluate the application of clay bricks containing 10–30 wt% of olive stone ash, their technological properties were evaluated. During the firing process, changes in mass, dimensions, and color of the fired clay bricks occur. In this case, it was observed that by increasing the amount of ash in the brick the red color of the bricks was lighter.

No defects as cracks, bloating, or efflorescence were observed after firing at 900°C (Figure 5).

The changes in linear shrinkage for bricks after firing are shown in Table 3. Control bricks showed linear shrinkage of −0.5%, exhibiting an expansion behavior. The addition of olive stone ash increased the linear shrinkage up to −2.2% with the addition of 30 wt% of waste, which is a typical behavior of porous bricks, due to the relative high level content of organic matter and carbonates in the samples (Table 1). The OSA waste was formed by 8.5% of SiO_2 and 1.7% of Al_2O_3, only a 10.2% of waste weight was made of skeleton components. Flux materials (58.6%) and gaseous components (11.8% $CaCO_3$ and 39.5% organic matter) were found in higher proportions. When the temperature grew from room temperature to 900°C, the addition of OSA produced liquid phase at lower temperature and a decrease in temperature at sintering onset due to the higher content in flux materials. Also calcite decomposed at 800°C and organic matter at 550°C, generating CO_2. However, at firing temperature of 900°C, the liquid phase allowed gas trapping, an increase in gas pressure inside the closed pores, which tended to expand the pores [40]. As a result, bulk density decreased rapidly. Such behavior reflects that the effect of porosity formed in the material with high OSA content was greater than the effect provided by the melting capacity of the waste.

These values are considered to be within the safety limits for industrial production of fired clay bricks.

The loss of ignition (LOI) of the bricks after the firing process at 900°C was 9.24% for control bricks (Table 3). The control brick samples showed the lowest value and it is related to dehydroxylation reactions, loss of humidity, organic matter combustion, and carbonate decomposition. The olive stone ash is an inorganic-organic waste due to high amount of organic matter (Table 1). Therefore, by increasing OSA proportion, the loss on ignition of fired bricks increased from 12.97% up to 17.19% with additions of 10 and 30 wt% of waste, respectively.

Considering the bulk density, the addition of OSA decreased it (Figure 6), with values ranging from 1,767 to 1,553 kg/m^3, lower than the value for the control brick (1,839 kg/m^3). The addition of 10 wt% of ash resulted in a reduction of 3% achieving the maximum reduction after adding 30 wt% of OSA, 15.6%. There was an almost linear decrease in bulk density as the ash content increased. The reduction in bulk density is due to the combustion of the organic content and carbonates decomposition contained in the residue causing porosity in the clay body.

Another parameter evaluated was the water suction, which affects the quality and durability of the final materials.

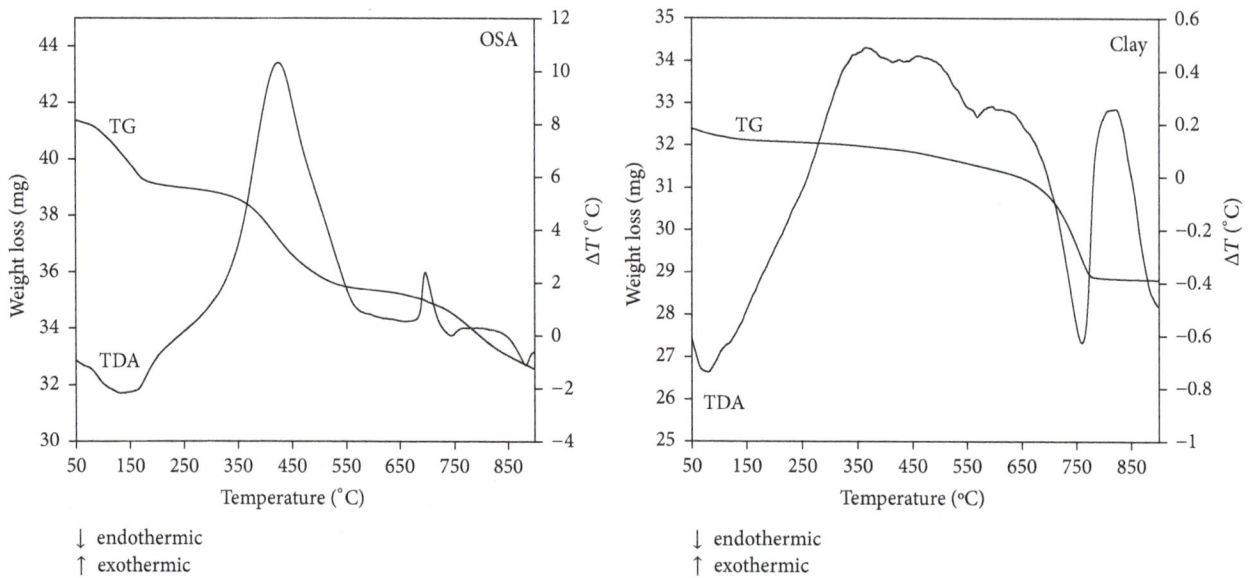

↓ endothermic
↑ exothermic

↓ endothermic
↑ exothermic

FIGURE 4: TG-DTA analysis of (a) olive stone ash and (b) raw clay.

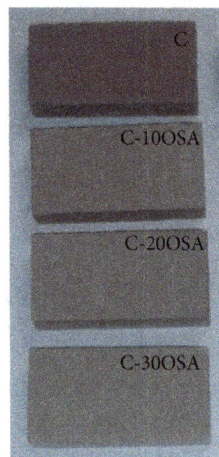

FIGURE 5: Bricks obtained by pressing of olive stone ash mixed with clay and fired at 900°C.

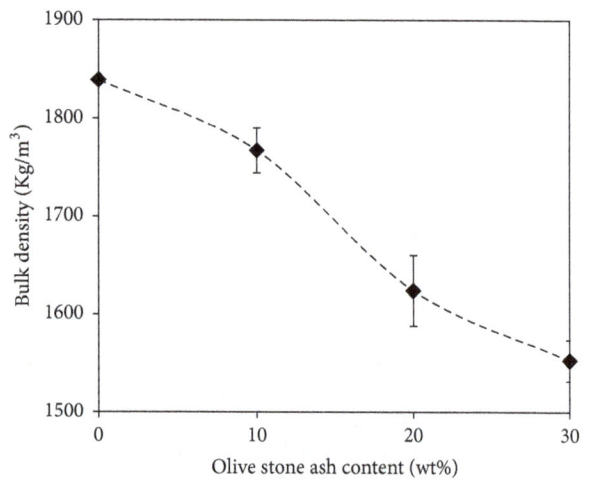

FIGURE 6: Bulk density of the fired bricks as function of olive stone ash addition.

This is a property of adherence among brick and mortar. The results of water suction as a function of OSA percentage added are included in Table 3. As seen from this table, the water suction of the clay showed a value of 2.32 kg/m² min and varied with the percentage of olive stone ash added. The maximum suction value, 3.06 kg/m² min, was achieved for the sample containing 20 wt% of ash. The water suction values were instead lower for C-10OSA and C-30OSA samples, 2.59 kg/m² min and 2.88 kg/m² min, respectively. Therefore, the incorporation of olive stone ash produced an increase in the interconnected surface porosity possibly due to pore growth, both in size and in number, due to the organic and carbonates content. In accordance with standard UNE bricks [33] water suction must be less than 4.5 kg/m² min because the brick removes water from the mortar causing inadequate curing. However, if the water suction value is higher than 0.15 kg/m² min, brief immersion in water of the brick is needed before its placement to avoid the dehydration of the mortar. All the bricks agree with regulations.

If the apparent porosity is considered, the addition of OSA produced a significant increase in the apparent porosity of the resultant bricks, increasing from 31.0% for the standard clay brick up to 36.9% for C-30OSA brick (Figure 7). The addition of 10 wt% of OSA gave rise to bricks with an apparent porosity of 34.8% increasing this property about 12% with regard to the control sample. The addition of higher amounts of OSA, 20 and 30 wt%, produced a higher increase (about 18%) in apparent porosity.

Water absorption is another key factor affecting the durability of brick and shows the same trend compared to apparent porosity. The water absorption is an indirect indicator of open porosity. The results of the water absorption

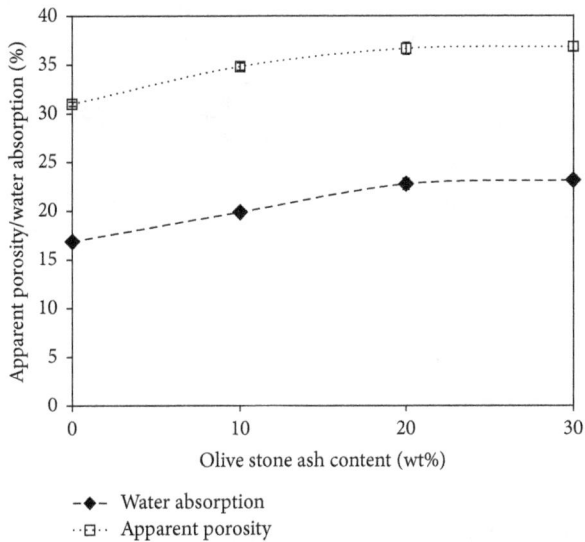

FIGURE 7: Apparent porosity and water absorption of the fired bricks as function of olive stone ash addition.

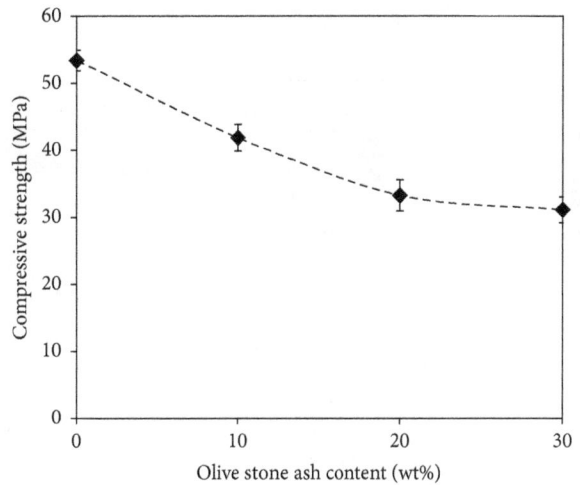

FIGURE 8: Compressive strength of the fired bricks as a function of olive stone ash addition.

tests (Figure 7) showed that the absorption capacity rose with OSA incorporation into the clay matrix. The water absorption of the control bricks presented the minimum value 16.9%. Instead, the values for C-10OSA, C-20OSA, and C-30OSA were 19.9%, 22.8%, and 23.1%, respectively, due to an increase of the open porosity as a consequence of organic matter and carbonates decomposed during the firing process. In any case, the maximum water absorption value is limited by standards ASTM C67-14 [41]. Depending on both, the main purpose of the brick and the environment, values must be below 17% for bricks exposed to severe weathering conditions and 22% for bricks exposed to moderate weathering conditions and no limit is set for negligible weathering resistance bricks. The addition of 20 wt% and 30 wt% of OSA resulted in bricks with very high water absorption values that do not fall within the standards of conventional bricks.

Compressive strength is the most critical index for building materials. According to ASTM C62-10 [42] and European Standard EN-772-1 [34], the compressive strength varies from 10 MPa for weather-resistant brick to 20 MPa in the case of severe weathering. Compressive strength of fired products is shown in Figure 8. Compressive strength rapidly decreased at higher ratio of olive stone ash in bricks, especially when the ratio is equal to or higher than 20 wt%. The highest value obtained corresponds to control bricks (53.3 MPa). The compressive strength values decreased a 21.6%, down to 41.9 MPa, for C-10OSA sample. In the case of C-20OSA and C-30OSA, the compressive strength was reduced up to 33.2 MPa and 31.1 MPa, respectively. These results are in accordance with data for bulk density, apparent porosity, and water absorption. Bulk density of bricks was reduced as the amount of olive stone ash increased, to reach 1,553 kg/m^3 compared to 1,839 kg/m^3 for control bricks. At high OSA additions (20 or 30 wt%), waste bricks of higher porosity and higher water absorption were produced. Open pores with irregular shape and microscopic imperfections may concentrate pressure and decrease the mechanical properties

of fired bricks [43]. In the present study, strength values of all samples containing ash were higher than 20 MPa.

In order to valorize the use of olive stone ash in obtaining fired clay bricks, the leachability of heavy metals was evaluated. Table 4 presents heavy metal concentrations measured in the extract obtained after the leaching test of the ceramic bricks with different quantities of olive stone ash by TCLP test EPA 1311 method [35]. The results show that only Cr was detected in small concentration, always lower than the control brick. This indicates that the amount of chrome was higher in the clay than in the olive stone ash. Concentrations of As, Cd, Co, Cr, Ni, Sn, and V increased as the proportion of olive stone ash did. However, it was observed that the use of olive stone ash gave rise to concentrations of heavy metals much lower than the limits established by EPA 658/2009 [35]. From these results, ceramic bricks fabricated with 10–30 wt% of olive stone ash can be classified as inert and nondangerous for the environment. Leaching tests showed a high degree of immobilization of heavy metals, highlighting that the incorporation of the different amounts (10–30 wt%) of olive stone ash in fired clay bricks production is an efficient method of immobilization. Hence, no environmental problems due to heavy metals disposal can be expected by the use of the OSA-clay fired bricks.

4. Conclusions

The valorization of olive stone ash (OSA) in the manufacturing of fired clay bricks is a sustainable way to reduce the harmful environmental impacts of these wastes. The OSA is a waste predominantly composed of potassium, calcium, and silicon but also contains organic matter. The addition of 10–30 wt% of OSA to a clay had a pronounced effect on the evolution of physical and mechanical properties of the resultant bricks fired at 900°C. Increasing additions of waste decreased the compressive strength and bulk density of bricks and increased simultaneously their apparent porosity and water absorption. The compressive strength decreased

TABLE 4: USEPA TCLP test results (ppm) and the maximum concentration of contaminants for toxicity characteristics of fired clay and fired clay-OSA bricks.

Component (ppm)	C	C-10OSA	C-20OSA	C-30OSA	USEPA regulated TCLP limits (ppm)
As	0.027	0.061	0.028	0.015	5
Ba	0.204	0.225	0.558	0.188	100
Cd	0.0003	0.00020	0.00012	0.00008	1
Co	0.010	0.0084	0.0013	0.0010	—
Cr	0.554	0.297	0.282	0.234	5
Cu	0.296	0.467	0.986	0.330	5
Ni	0.047	0.070	0.012	0.0004	—
Pb	0.0009	0.00043	0.087	0.00019	5
Sb	0.0010	0.0036	0.0098	0.0057	—
Se	0.013	0.018	0.0031	0.0032	1
Sn	0.00014	0.00018	0.00016	0.00012	—
V	0.161	0.351	0.200	0.115	—
Hg	0.00004	0.00009	0.00011	0.00009	0.2
Zn	0.058	0.012	0.005	0.006	300

from 53.4 MPa in control bricks without OSA to 31.1 MPa in bricks containing up to 30 wt% of waste. The decrease in the mechanical properties of OSA-fired clay bricks was due to a marked increase in open porosity as indicated by the water absorption data. The bulk density was between 4 and 16% lower than the bulk density of control bricks.

The clay brick containing 10 wt% of olive stone ash showed an appropriate balance among the physical and mechanical properties such as bulk density (1,767 kg/cm^3), apparent porosity (34.8%), water absorption (19.8%), and compressive strength (41.9 MPa). In order to decrease open porosity and to raise the amount of OSA waste in fired clay bricks further studies are required and should be focused on the sintering temperature.

The environmental behavior (TCLP test EPA 1311 method) of bricks showed that the leaching of all heavy metals was under the requested limits.

Competing Interests

The authors declare that they have no competing interests.

Acknowledgments

This work has been funded by the Project "Valuation of Various Types of Ash for the Obtaining of New Sustainable Ceramic Materials" (UJA2014/06/13), Own Plan University of Jaén, sponsored by Caja Rural of Jaén. Technical and human support provided by CICT of Universidad de Jaén (UJA, MINECO, Junta de Andalucía, FEDER) is gratefully acknowledged. A. Infantes-Molina thanks the Ministry of Economy and Competitiveness for a Ramón y Cajal contract (RyC2015-17870).

References

[1] Food and Agriculture Organization of the United Nations. Statistics Division, 2016, http://faostat3.fao.org/browse/Q/QC/E.

[2] Ministerio de Agricultura, Alimentación y Medio Ambiente. Gobierno de España. Estadísticas, 2016, http://www.magrama.gob.es/es/estadistica/temas/estadisticas-agrarias/agricultura/superficies-producciones-anuales-cultivos/.

[3] Ministerio de Agricultura, Alimentación y Medio Ambiente, Gobierno de España, Encuesta sobre Superficies y Rendimientos, 2013.

[4] Agencia Andaluza de la Energía (AAE), La biomasa en Andalucía, 2015, https://www.agenciaandaluzadelaenergia.es/sites/default/files/3_2_0161_15_la_biomasa_en_andalucia_octubre_2015.pdf.

[5] Junta de Andalucía Biomasa Forestal en Andalucía, 1. Modelos de existencias, crecimiento y producción, Coníferas, 2012, http://www.juntadeandalucia.es/medioambiente/portal_web/web/temas_ambientales/montes/usos_y_aprov/jornadas_biomasa/Publicaciones/biomasa1.pdf.

[6] M. C. Trujillo, M. A. Martín-Lara, A. B. Albadarin, C. Mangwandi, and M. Calero, "Simultaneous biosorption of methylene blue and trivalent chromium onto olive stone," *Desalination and Water Treatment*, vol. 57, no. 37, pp. 17400–17410, 2015.

[7] S. Larous and A. Meniai, "Adsorption of Diclofenac from aqueous solution using activated carbon prepared from olive stones," *International Journal of Hydrogen Energy*, vol. 41, no. 24, pp. 10380–10390, 2016.

[8] A. Erto, B. Tsyntsarski, M. Balsamo et al., "Synthesis of activated carbons by thermal treatments of agricultural wastes for CO_2 capture from flue gas," *Combustion Science and Technology*, vol. 188, no. 4-5, pp. 581–593, 2016.

[9] M. Balsamo, B. Tsyntsarski, A. Erto et al., "Dynamic studies on carbon dioxide capture using lignocellulosic based activated carbons," *Adsorption*, vol. 21, no. 8, pp. 633–643, 2015.

[10] N. Moreno, Á. Caballero, J. Morales, and E. Rodríguez-Castellón, "Improved performance of electrodes based on carbonized olive stones/S composites by impregnating with mesoporous TiO_2 for advanced Li—S batteries," *Journal of Power Sources*, vol. 313, pp. 21–29, 2016.

[11] N. Moreno, A. Caballero, L. Hernán, and J. Morales, "Lithium-sulfur batteries with activated carbons derived from olive stones," *Carbon*, vol. 70, pp. 241–248, 2014.

[12] G. Rodríguez, A. Lama, R. Rodríguez, A. Jiménez, R. Guillén, and J. Fernández-Bolaños, "Olive stone an attractive source of bioactive and valuable compounds," *Bioresource Technology*, vol. 99, no. 13, pp. 5261–5269, 2008.

[13] S. Arezki, N. Chelouah, and A. Tahakourt, "The effect of the addition of ground olive stones on the physical and mechanical properties of clay bricks," *Materiales de Construcción*, vol. 66, no. 322, article no. e082, 2016.

[14] S. Serrano, C. Barreneche, and L. F. Cabeza, "Use of by-products as additives in adobe bricks: mechanical properties characterisation," *Construction and Building Materials*, vol. 108, pp. 105–111, 2016.

[15] Decisión de la Comisión 2000/532/CE sobre lista de residuos, DO L nº226, 6.9.2000 modificada por Decisiones 2001/118/CE, 2001/119/CE y 2001/573/CE y rectificada por DO L nº112, 27.4.2002, http://eur-lex.europa.eu/LexUriServ/LexUriServ.do?uri=CELEX:32000D0532:ES:HTML.

[16] D. Vamvuka, "Comparative fixed/fluidized bed experiments for the thermal behaviour and environmental impact of olive kernel ash," *Renewable Energy*, vol. 34, no. 1, pp. 158–164, 2009.

[17] A. Ravaglioli, C. Fiori, and B. Fabbri, *Materie Prime Ceramiche. Argille, Materiali non Argillosi e Sottoprodotti Industriali*, Faenza Editrice S.P.A. Faenza, 1989.

[18] L. Sánchez-Muñoz and J. B. Carda Castelló, "Materiales residuales," in *De Materias Primas y Aditivos Cerámicos*, F. E. Iberica and S. L. Castellón, Eds., pp. 159–160, 2002.

[19] A. A. Kadir, N. A. Mohd Zahari, and N. Azizi Mardi, "Utilization of palm oil waste into fired clay brick," *Advances in Environmental Biology*, vol. 7, no. 12, pp. 3826–3834, 2013.

[20] C.-L. Hwang and T.-P. Huynh, "Investigation into the use of unground rice husk ash to produce eco-friendly construction bricks," *Construction and Building Materials*, vol. 93, pp. 335–341, 2015.

[21] S. Janbuala and T. Wasanapiarnpong, "Effect of rice husk and rice husk ash on properties of lightweight clay bricks," *Key Engineering Materials*, vol. 659, pp. 74–79, 2015.

[22] S. M. Kazmi, S. Abbas, M. A. Saleem, M. J. Munir, and A. Khitab, "Manufacturing of sustainable clay bricks: utilization of waste sugarcane bagasse and rice husk ashes," *Construction and Building Materials*, vol. 120, pp. 29–41, 2016.

[23] Z. Haiying, Z. Youcai, and Q. Jingyu, "Utilization of municipal solid waste incineration (MSWI) fly ash in ceramic brick: product characterization and environmental toxicity," *Waste Management*, vol. 31, no. 2, pp. 331–341, 2011.

[24] K. L. Lin, "Feasibility study of using brick made from municipal solid waste incinerator fly ash slag," *Journal of Hazardous Materials*, vol. 137, no. 3, pp. 1810–1816, 2006.

[25] M. Smol, J. Kulczycka, A. Henclik, K. Gorazda, and Z. Wzorek, "The possible use of sewage sludge ash (SSA) in the construction industry as a way towards a circular economy," *Journal of Cleaner Production*, vol. 95, pp. 45–54, 2015.

[26] P. Pavšič, A. Mladenovič, A. Mauko et al., "Sewage sludge/biomass ash based products for sustainable construction," *Journal of Cleaner Production*, vol. 67, pp. 117–124, 2014.

[27] D. Eliche-Quesada and J. Leite-Costa, "Use of bottom ash from olive pomace combustion in the production of eco-friendly fired clay bricks," *Waste Management*, vol. 48, pp. 323–333, 2016.

[28] C. Fernández-Pereira, J. A. De La Casa, A. Gómez-Barea, F. Arroyo, C. Leiva, and Y. Luna, "Application of biomass gasification fly ash for brick manufacturing," *Fuel*, vol. 90, no. 1, pp. 220–232, 2011.

[29] J. A. de la Casa and E. Castro, "Recycling of washed olive pomace ash for fired clay brick manufacturing," *Construction and Building Materials*, vol. 61, pp. 320–326, 2014.

[30] L. Pérez-Villarejo, D. Eliche-Quesada, F. J. Iglesias-Godino, C. Martínez-García, and F. A. Corpas-Iglesias, "Recycling of ash from biomass incinerator in clay matrix to produce ceramic bricks," *Journal of Environmental Management*, vol. 95, supplement, pp. S349–S354, 2012.

[31] "Test method for drying and firing shrinkage of ceramic whiteware clays," ASTM C 326, American Society for Testing and Materials, 1997.

[32] "Test method for water absorption, bulk density, apparent porosity, and apparent specific gravity of fired whiteware products," ASTM C373, American Society for Testing and Materials, 1994.

[33] "Methods of test for masonry units—part 11: determination of water absorption of aggregate concrete, manufactured stone and natural stone masonry units due to capillary action and the initial rate of water absorption of clay masonry units," UNE-EN 772-11, 2011.

[34] "Methods of test for masonry units—part 1: determination of compressive strength," UNE EN 772-1, 2011.

[35] U.S. Environmental Protection Agency, *Method 13–11 Toxicity Characteristics Leaching Procedure (TCLP)*, vol. 51, Federal Register, Washington, DC, USA, 1992.

[36] A. Demirbas, "Potential applications of renewable energy sources, biomass combustion problems in boiler power systems and combustion related environmental issues," *Progress in Energy and Combustion Science*, vol. 31, no. 2, pp. 171–192, 2005.

[37] O. Dahl, H. Nurmesniemi, R. Pöykiö, and G. Watkins, "Comparison of the characteristics of bottom ash and fly ash from a medium-size (32 MW) municipal district heating plant incinerating forest residues and peat in a fluidized-bed boiler," *Fuel Processing Technology*, vol. 90, no. 7-8, pp. 871–878, 2009.

[38] J. García-Ten, M. J. Orts, A. Saburit, and G. Silva, "Thermal conductivity of traditional ceramics: part II: Influence of mineralogical composition," *Ceramics International*, vol. 36, no. 7, pp. 2017–2024, 2010.

[39] I. J. Fernandes, D. Calheiro, A. G. Kieling et al., "Characterization of rice husk ash produced using different biomass combustion techniques for energy," *Fuel*, vol. 165, pp. 351–359, 2016.

[40] H. He, Q. Yue, Y. Su et al., "Preparation and mechanism of the sintered bricks produced from Yellow River silt and red mud," *Journal of Hazardous Materials*, vol. 203-204, pp. 53–61, 2012.

[41] "Standard test methods for sampling and testing brick and structural clay tile," ASTM C67-14, 2014.

[42] "Standard specification for building brick (solid masonry units made from clay or shale)," ASTM C62-10, ASTM International, West Conshohocken, Pa, USA, 2010.

[43] W. M. Carty and U. Senapati, "Porcelain—raw materials, processing, phase evolution, and mechanical behavior," *Journal of the American Ceramic Society*, vol. 81, no. 1, pp. 3–20, 1998.

21

Comprehensive Study on Elastic Moduli Prediction and Correlation of Glass and Glass Ceramic Derived from Waste Rice Husk

Chee Sun Lee,[1] **Khamirul Amin Matori,**[1,2] **Sidek Hj. Ab Aziz,**[1] **Halimah Mohamed Kamari,**[1]
Ismayadi Ismail,[2] **and Mohd Hafiz Mohd Zaid**[1,2]

[1]*Department of Physics, Faculty of Science, Universiti Putra Malaysia (UPM), 43400 Serdang, Selangor, Malaysia*
[2]*Materials Synthesis and Characterization Laboratory, Institute of Advanced Technology, Universiti Putra Malaysia (UPM), 43400 Serdang, Selangor, Malaysia*

Correspondence should be addressed to Mohd Hafiz Mohd Zaid; mhmzaid@gmail.com

Academic Editor: Gianluca Cicala

Zinc silicate (ZnO–SiO_2) systems were fabricated using zinc oxide (ZnO) and white rice husk ash (WRHA) with compositions of $(ZnO)_x(WRHA)_{1-x}$ ($x = 0.55$, 0.60, 0.65, and 0.70 wt.%) was symbolized by S1, S2, S3, and S4, respectively. The ZnO–SiO_2 samples were fabricated by applying the melt-quench method and the physical and elastic properties of the samples were investigated. Physical properties used in this study are density and molar volume while the theoretical elastic moduli of the samples produced were obtained using direct calculation of theoretical model compared with the experimental elastic moduli obtained by acquiring ultrasonic velocities using ultrasonic pulse-echo technique. Values of experimental elastic moduli including longitudinal modulus (L), shear modulus (S), Young's modulus (E), bulk modulus (K), and Poisson's ratio (σ) were compared with theoretical model calculated using Rocherulle's model. All the configurations of the elastic moduli obtained experimentally match very well with the configuration from Rocherulle's model but Poisson's ratio obtained experimentally differs from the values of Poisson's ratio obtained through Rocherulle's model.

1. Introduction

Rice is the main source of carbohydrates and consumption for the populations in Asia and the main by-product is the waste rice husk (RH) which is produced in a vast production of paddy field in Malaysia and also throughout the globe. Rice cultivation in Malaysia has resulted in 400,000 metric tonnes of RH annually as it has been considered valueless and the main problem arises once it comes to the problem of getting rid of it [1]. RH is obtained as a worthless by-product as it cannot fetch a good price in the open market and most of the RH milling factories are massively influenced by this problem and disposing RH ethically will become unprofitable for them [2, 3]. RH produced is generally discarded or burned and this has caused a serious threat to our mother earth and by developing new ways to reuse RH, it can be a useful by-product such as substitution for conventional silica in the glass industry [4, 5]. Next, substitution silica such as white rice husk ash (WRHA) can be derived from RH by a complete combustion [6]. RH serves as good substituents for conventional silica as it possesses high concentration of silica after complete combustion of 1000°C from WRHA and it has relatively lower cost compared to its conventional silica counterpart [7, 8]. RH glass system which is derived can be doped into lithium silicate glass system and it mimics the thermal behavior and crystalline phases of the glass and glass ceramic lithium silicate systems and it also increases the fracture toughness of both systems by 100% thus implying that RH will act as an adequate replacement for conventional silica [9]. RH glass ceramic can be used to derive nepheline–forsterite glass ceramic where its main source of silica is from WRHA [10].

Zinc silicate (ZnO–SiO_2) glass is an amorphous glass with high concentration of ZnO in its major composition and it grabs much attention from researches as it possesses various applicable uses in the range of glass ceramics and technical glasses [11]. Researchers can do a lot more to understand the applications of ZnO–SiO_2 glass and by variating SiO_2 and ZnO compositions, endless possibilities can be achieved. ZnO is one of the major compositions in the samples produced and incorporating it in the glass and glass ceramic system makes ZnO–SiO_2 system a remarkable one. ZnO based glass is one of the most interesting network modifiers in the list of all the network modifiers because of its influence in the optical, electrical, and magnetic properties of the glass produced causing the glass to be environment-friendly and of low cost which grabs the attention of researches and manufacturing industry [12, 13]. Silicate glass doped with high concentration of ZnO has lower elastic moduli because of the effect presented by ZnO which acts as a network modifier and its effect is what causes these types of glass to be outstanding compared to other types of glass system [14].

In this study, ZnO and WRHA will be manipulated to study the physical and elastic properties of ZnO–SiO_2 glass and glass ceramic system. Influence of ZnO on its elastic moduli and comparison of elastic moduli obtained experimentally and elastic moduli obtained theoretically were also discussed.

2. Experimental Details

RH species used is the *Oryza sativa* (Asian rice) which can be found in majority of Asian countries and RH from this study is acquired from a local rice factory at Tanjung Karang, Selangor, Malaysia. RH samples used were just milled a day before the study and they are kept in an indoor compound to minimize possible unwanted contaminations from the surroundings. RH then is examined to remove sands and small rocks which jeopardize the study. Next, RH is then washed for several times to get rid of possible dirt and contaminations as it is in contact with sand and it is drenched in water for 2 hours in a vast plastic basin to have the dirt and sands sunken to the bottom of the plastic basin. RH which is clean and clear without contaminations will float on the surface and it is removed to a large plastic sieve to dry the RH with an aluminum foil on top of it to minimize contaminations from the surroundings as it will be left in the lab for 24 hours to dry at room temperature. After 24 hours, RH will be subsequently dried in the oven for 2 hours at 100°C for 24 hours to obtain a fully dried RH. After obtaining the dry RH, 500 g of the dried RH is transferred to a ceramic crucible and distributed evenly so that the combustion of RH to WRHA can be performed evenly. RH will be burned at 1000°C for 2 hours at a heating rate of 10°C/min and the changes in color of the WRHA produced are observed. Ceramic crucible selected must be flat as it delivers a complete homogeneous burning of RH. The EDXRF measurement has been done to obtain the percentage of the oxide composition in the WRHA [16].

In this study WRHA is acquired to replace conventional silica in the formation of ZnO–SiO_2 glass and glass ceramic system chemical formula of $(ZnO)_x(WRHA)_{1-x}$ (x = 0.55, 0.60, 0.65, and 0.70 wt.%) was symbolized by S1, S2, S3, and S4, respectively, acquiring melt-quench technique. After

mixing the samples, the mixture is then milled for 24 hours to ensure a fully homogenous mixture. Usually, melting silicate glass system would require up to 1700°C as it belongs to type of glass with high melting point but in this study, where the silica source is being manipulated by switching it to WRHA to form ZnO–SiO_2 glass system, a new method is being done with the lowest temperature at which ZnO–SiO_2 system can exist at 1450°C with 2 hours. After all the samples inside the crucible were successfully melted, the crucible was removed with a platinum clamp and the samples are transferred to a stainless steel mold which has been preheated at 500°C and the samples then are annealed at 500°C for 5 hours while the leftover samples were transferred into the same stainless steel base which holds the mold and is annealed at the similar condition. Finally the samples were cut using a diamond blade cutter to a dimension of $(1.0 \times 2.0 \times 0.3)$ cm^3 and later the samples were grinded using several grinding papers in the range of 320–1000 grit to ensure a perfect smooth surface for ultrasonic measurement. Samples thickness of $(1.0 \times 2.0 \times 0.3)$ cm^3 is the optimum thickness because samples which are too thick or too thin may cause the decrease in ultrasonic wave intensity.

The compositions of the glass and glass ceramic samples are confirmed using EDXRF diffractometer (Shimadzu, EDX-720). Next, the densities of the samples were determined using the Archimedes method where it acquires the weight of the samples in air and distilled water. Based on the Archimedes method, the buoyant force produced when the sample is immersed in distilled water is the same with the magnitude of the weight of distilled water displaced by the volume of the samples. Hence, the density of samples, ρ_{sample}, can be expressed as

$$\rho_{sample} = \frac{\text{weight of sample in air}}{\text{weight of sample in air} - \text{weight of sample in distilled water}} \times \rho_{distilled\ water}, \tag{1}$$

where $\rho_{distilled\ water}$ = 1.000 g cm^{-3}. Distilled water was chosen as the immersed liquid as it is the cheap, inert, and low surface tension. Molar volume (V_m) of any substance is defined as the volume occupied by one mole of any material from chemical compound or chemical element at a specified temperature and pressure formula of V_m is equal to the total molar mass (M) divided by the mass density (ρ).

The molar volumes were measured in cubic centimeters per mole (cm^3 mol^{-1}) and for liquids and solids; they can be expressed as

$$V_m = \sum \frac{M_T}{\rho}, \tag{2}$$

where M_T is the total molecular weight of the multicomponent system given by

$$M_T = x_i Z_i, \tag{3}$$

where x_i is the mole fraction of the ith oxides, and Z_i is the molecular weight of the ith oxides. Ultrasonic velocities of

TABLE 1: Analysis of chemical composition of ZnO–SiO$_2$ samples.

Samples	ZnO	SiO$_2$	K$_2$O	Al$_2$O$_3$	CaO	MgO	Na$_2$O	Fe$_2$O$_3$	CuO
S1	53.83	43.02	0.84	0.75	0.59	0.26	0.24	0.25	0.22
S2	58.40	38.84	0.78	0.66	0.58	0.23	0.20	0.22	0.09
S3	65.01	32.46	0.74	0.61	0.56	0.21	0.19	0.20	0.02
S4	70.02	28.68	0.44	0.34	0.29	0.07	0.10	0.05	0.01

TABLE 2: Density, molecular weight, and molar volume of ZnO–SiO$_2$ samples.

Samples	S1	S2	S3	S4
Density (g cm^{-3})	2.94	3.08	3.44	3.66
M_T (g mol^{-1})	70.43	71.59	72.79	73.91
V_m (cm^3 mol^{-1})	23.97	23.21	21.17	20.18

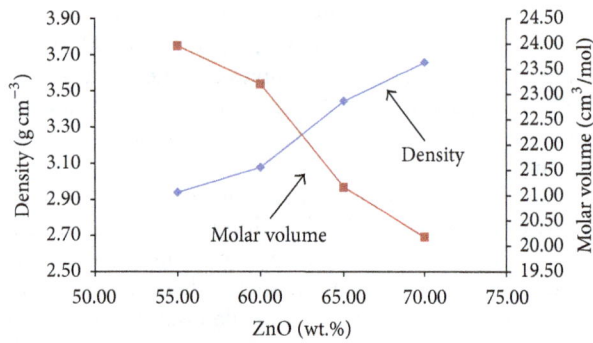

FIGURE 1: The density and molar volume versus wt.% of ZnO in ZnO–SiO$_2$ samples.

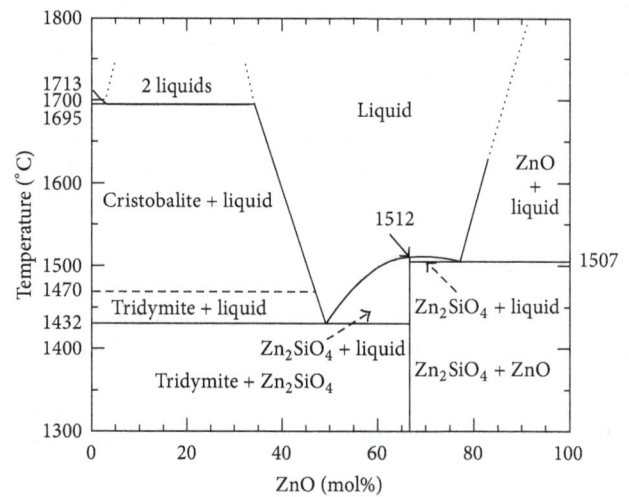

FIGURE 2: Binary system of ZnO–SiO$_2$ system [15].

the samples which consist of longitudinal velocity (V_L) and shear velocity (V_S) were determined using Ritec, Ram-5000 Snap System machine with frequency of 5 MHz. Longitudinal velocity of the samples was acquired using the longitudinal probe while the shear velocity of the samples was acquired using the shear probe. Adhesive agent used for the longitudinal probe is longitudinal gel while the shear probe is shear gel. Adhesive agent is used to minimize the refraction of wave in air gap while acquiring both ultrasonic velocities.

Ultrasonic technique is a method that is useful and flexible to measure the structure and mechanical properties of materials. Next, ultrasonic waves can be transmitted through the material and detect the mechanical properties and the defects found in the material. Generally, ultrasonic measurement can be used to measure the metal and nonmetal samples. Ultrasonic pulse-echo technique can also be known as the longitudinal modulus and shear modulus.

3. Results and Discussion

Four series of ZnO–SiO$_2$ samples were obtained and chemical composition of the WRHA was analyzed using EDXRF and the chemical composition of the WRHA and samples produced was tabulated in Table 1.

Figure 1 and Table 2 show the density, molecular weight, and molar volume for ZnO–SiO$_2$ samples and it shows that the density of the samples increases with the increase of

ZnO from 2.94 to 3.66 g cm^{-1}. Increase in density is largely due to substituting Si (28.08) with a higher atomic mass element of zinc which is (65.39). Besides, the increase in density with the increase of ZnO content occurs due to the increase in the crystallinity in the samples in which sample S1 is fully amorphous and sample S4 provides higher amounts of crystalline compared to the other samples [17, 18]. Increase in crystallinity is due to the increase in melting point of the ZnO–SiO$_2$ samples as its melting point increases as ZnO in the system increases as shown in Figure 2. Not only that, silica present in the system will also act as a nucleating agent for the samples to increase in crystalline [15].

Increase in density can also be explained by the increase in ZnO and the reduction of silica (SiO$_2$) which will increase its overall molecular weight of the samples as ZnO is much heavier relative to molecular mass of SiO$_2$ [19]. Increase in density of the ZnO–SiO$_2$ samples may also be caused by structural reorganization of the atoms thus causing a possible modification of the geometrical configuration.

Molar volume (V_m) in this study was calculated using (2) and, based on Figure 1 and Table 2, molar volumes of the ZnO–SiO$_2$ samples produced indicate a decrease from 23.97 to 20.18 cm^3 mol^{-1} when the amount of ZnO in the

TABLE 3: Density, molar volume, molecular weight, and ultrasonic velocity of ZnO–SiO$_2$ samples.

Samples	S1	S2	S3	S4
Density (g cm^{-3})	2.94	3.08	3.44	3.66
V_m (cm^3 mol^{-1})	23.97	23.21	21.17	20.18
M_T (g mol^{-1})	70.43	71.59	72.79	73.91
V_L (ms^{-1})	3857.03	3609.90	2794.69	2482.31
V_S (ms^{-1})	1951.95	1611.21	1185.46	966.27

samples increases. Molar volume is vastly dependent on the overall molecular weight of a particular composition and this decrease is mainly because of greater rise in density compared to the overall molecular weight. Generally, silicate glass and glass ceramic have an opposite trend to the density of its system when it comes to the molar volume and, in this study, SiO^{2+} (0.038 nm) which is greater in ionic radius is substituted with Zn^{2+} (0.074 nm) thus causing the increase in the interatomic spacing or bond length between the atoms [20]. Besides that, the overall reduction in the interatomic distances will result in the increase of compactness of the glass matrix [21].

Decrease in V_m can also be attributed to the addition of ZnO as it causes some kind of structural rearrangement of the atoms by entering the glass system and destroying the Si–O–Si bonds, thus creating ionic bonds between Zn^{2+} and single bonded oxygen atoms [20]. Next, there is also a possibility for the modification of the geometrical configuration upon swapping ZnO into the glass network and the structure of the studied glasses will not be expanded, which have more compact and low number of covalent bonds with high reduction in the number of bridging oxygens (BOs) [22]. Longitudinal velocity (V_L) and the shear velocity (V_S) of the samples were acquired and are presented in Table 3 and were plotted in Figure 3. Generally, V_L and V_S of majority of materials will increase as the density of the system increases but, for this study, V_L and V_S decrease as the density of the system increases. Similar trends were also observed for soda lime silicate and silicate glass system whereby the V_L and V_S of the samples decrease when the density in the sample increases [23]. Decrease in V_L and V_S of the samples is due to rise in amount of ZnO present in the samples which contains Zn^{2+} ions where its zinc atom had higher atomic radius compared to Si atom therefore encouraging the transformation of BOs to NBOs in the system.

Based on Table 3, values of V_S are much smaller compared to the value of its counterpart V_L and it can be considered half of the value of V_L at the same compositions. For example, sample S1 which has a value of V_S is 1951.95 ms^{-1} while its V_L is 3857.03 ms^{-1} which shows that its shear velocity is half of the longitudinal velocity and the trend is constant throughout all samples. Next, the values of V_L decrease from 3857.03 ms^{-1} to 2482.31 ms^{-1} while V_S decreases from 1951.95 ms^{-1} to 966.27 ms^{-1} as the amount of ZnO increases. Increase in the amount of ZnO in the samples promotes the formation of NBOs and it is because ZnO which acts as a network former and Zn^{2+} ions are the main result which causes the increase in NBOs. In the glass system, average coordination number of

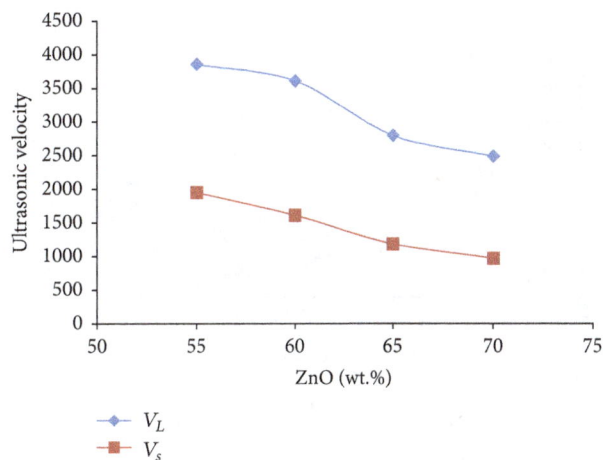

FIGURE 3: Longitudinal and shear velocity versus wt.% of ZnO in ZnO–SiO$_2$ samples.

SiO$_2$ gradually reduced to 3 from 4 as the values of ZnO rise and this causes the structural units of SiO$_4$ to be converted to SiO$_3$ initiating the rise of NBOs [24–26]. Decrease in both V_L and V_S with the rise of ZnO also indicates that the ultrasonic wave encounters more resistance in the ZnO–SiO$_2$ samples as the amount of ZnO increases [27, 28]. Not only that, rise of ZnO in the ZnO–SiO$_2$ samples also indicates that more Zn^{2+} ions exist in the samples and this caused overall more ions in samples to be opened up [29]. A result of lower ultrasonic velocity at higher content of ZnO indicates a major alteration in the structures of the samples.

Experimental values of longitudinal modulus (L), shear modulus (S), Young's modulus (E), bulk modulus (K), Poisson's ratio (σ), fractal bond connectivity (d), and microhardness (H) were calculated from the data obtained by ultrasonic velocities. The results of the elastic properties of the ZnO–SiO$_2$ samples are shown in Table 4. As seen, from Table 4, all the values of the elastic moduli decrease with the rise in ZnO.

Longitudinal modulus and shear modulus can be defined as an independent elastic constant which are produced by a small stress such as the elastic strain in an amorphous solid [30]. Longitudinal modulus is $L = \rho V_L^2$ and for pure shear modulus it is $S = \rho V_S^2$, where V_L and V_S, respectively, are the longitudinal and shear velocities of the samples [31]. The longitudinal modulus (L), shear modulus (S), Young's modulus (E), bulk modulus (K), Poisson's ratio (σ), and microhardness can be obtained using the ultrasonic velocities

TABLE 4: Experimental longitudinal modulus (L), shear modulus (S), Young's modulus (E), bulk modulus (K), Poisson's ratio (σ), fractal bond connectivity (d), and microhardness (H) of ZnO–SiO$_2$ samples.

Samples	L (GPa)	S (GPa)	E (GPa)	K (GPa)	σ	$d = 4G/K$	H (GPa)
S1	43.74	11.20	29.75	28.80	0.3279	1.5557	23.53
S2	40.10	7.99	21.98	29.45	0.3756	1.0851	10.58
S3	26.88	4.84	13.45	20.43	0.3903	0.9469	5.438
S4	22.54	3.42	9.64	17.99	0.4107	0.7596	2.971

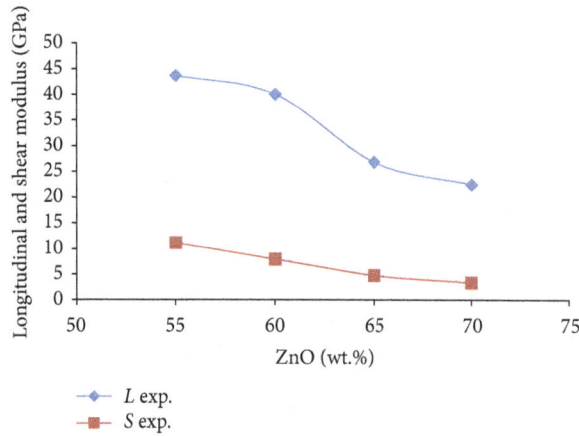

FIGURE 4: Longitudinal and shear modulus versus wt.% of ZnO in ZnO–SiO$_2$ samples.

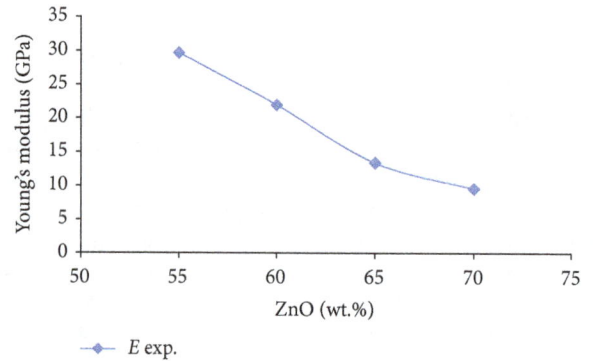

FIGURE 5: Young's modulus versus wt.% of ZnO in ZnO–SiO$_2$ samples.

by acquiring these equations from the following journals by Laoding et al. (2016) and Sidek et al. (2016) [32, 33]. Based on Table 4 and Figure 4 which illustrate the values and plot of longitudinal modulus and shear modulus of the ZnO–SiO$_2$ samples, it can be seen that as the amount of ZnO in the samples increases, the longitudinal modulus and shear modulus decrease. Decrease in longitudinal modulus from samples S1 to S4 is from 43.74 GPa to 22.54 GPa while decrease in shear modulus from samples S1 to S4 is from 11.20 GPa to 3.42 GPa. Longitudinal modulus for every sample is higher compared to its counterpart of shear modulus for every sample from S1 to S4 and this implies that the sample can withstand longitudinal stress compared to shear stress which indicates the samples are easier to be bent than elongated.

Young's modulus is defined as the ratio of stress against strain and it also represents the stiffness in a particular material which links to bonding strength between atoms in materials. Materials which have higher value of Young's modulus imply that that particular material is more stiffer and this Young's modulus is also affected by the dimensionality and connectivity of the system [33]. Not only that, samples that have higher value of Young's modulus can endure more stress compared to samples which possess lower Young's modulus and Figure 5 shows that as the amount of ZnO in the samples increases, Young's modulus of the samples decreases which indicates the increase in number of NBOs [34]. Decrease in the NBOs further links to the decrease in connectivity of the samples network which is caused by the decrease in Young's modulus. Lower Young's modulus as

ZnO increases shows that the ZnO–SiO$_2$ samples can reduce the speed of impacting mass thus causing a smaller stress compared to samples which have higher Young's modulus [34]. Results of Young's modulus can indicate that the samples can bear small stress acting against it even though its structure is less rigid and the decrease in Young's modulus indicates that the samples can work well with strain.

Bulk modulus (K) is known as the alteration in volume of a material when a small or moderate force is performed upon the specific surface. Figure 6 shows that as the amount of ZnO in the system increases, the bulk modulus of the samples increases and the bulk modulus of the samples is higher compared to Young's modulus which indicates that the samples work well with stress from multiple directions but not in one direction. Hence this clearly means that glass and glass ceramic acquired are usually hard if stress is applied from numerous directions simultaneously but would fracture if stress is applied at a single direction. Bergman and Kantor (1984) proposed the analysis of fractal bond connectivity with ratio of bulk to shear modulus and it can be simplified as $d = 4S/K$ where it gives the effective dimensionality of particular materials for any nonhomogeneous random mixture of fluid and a solid backbone near the percolation limit [35]. Value of $d = 4S/K$ acts as a parameter for network connectivity with a range of values obtained by the ratio [36]. Values of d differ for each kind of materials subjected on its dimension and $d = 1$ is for 1D chain and $d = 2$ and $d = 3$ for 2D and 3D chain structures correspondingly [37–39]. Results shown in Figure 7 and Table 4 indicate that the fractal bond connectivity of the ZnO–SiO$_2$ samples ranges from 0 to 1 which shows that ZnO–SiO$_2$ samples possess a 1D layer structure [40]. As ZnO increases in the samples, fractal bond

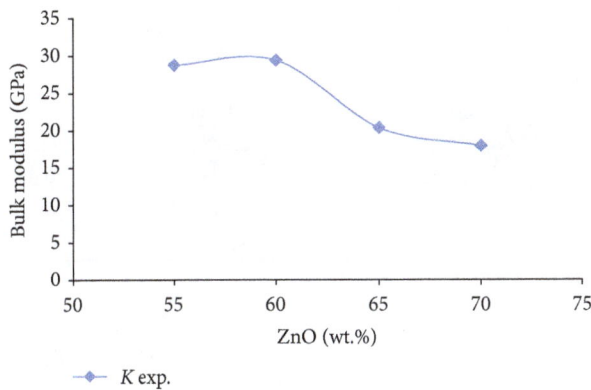

FIGURE 6: Bulk modulus versus wt.% of ZnO in ZnO–SiO$_2$ samples.

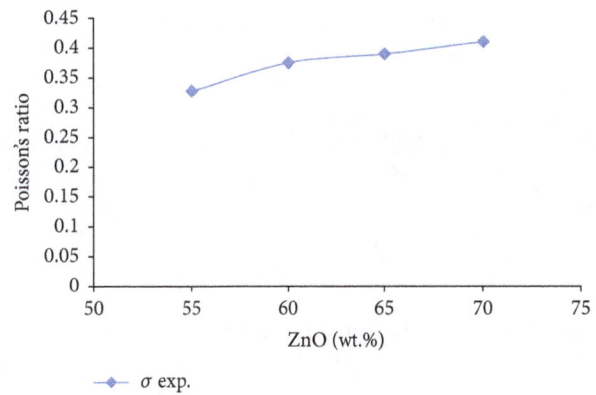

FIGURE 8: Poisson's versus wt.% of ZnO in ZnO–SiO$_2$ samples.

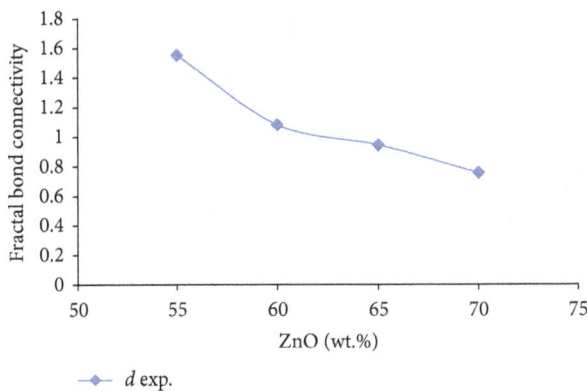

FIGURE 7: Fractal bond connectivity versus wt.% of ZnO in ZnO–SiO$_2$ samples.

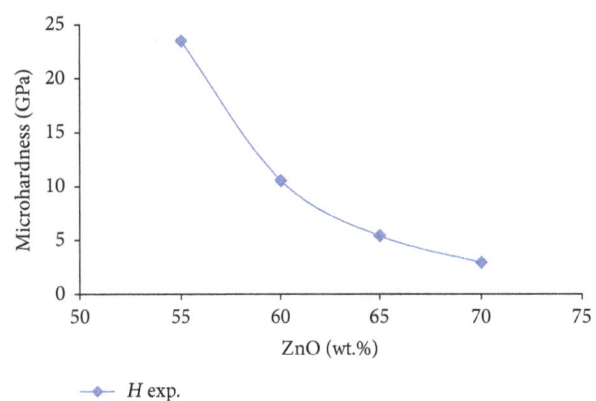

FIGURE 9: Microhardness versus wt.% of ZnO in ZnO–SiO$_2$ samples.

connectivity decreases and this will cause cross-links of the samples to be weakened and possibly break [23].

Poisson ratio is the measure of ratio of the shear strain to the longitudinal strain when tensile force is applied and it makes use of the degree of cross-link density of glass and glass ceramic network to find out the order of cross-link density [41]. Poisson's ratio measures the capability of a material to counter the alteration in volume when an incoming load acts on a certain material and the Poisson ratio (v) acquired can be in positive or negative [42]. In this study it is in the range of 0.327 to 0.410.

From Figure 8, it is observed that, as ZnO increases in the samples, results in higher values of Poisson's ratio show that the samples have a lower resistance towards lateral expansion when compressed compared to samples with lower amount of ZnO. Rise in Poisson's ratio also shows that the decrease in cross-link density and lateral strength with the increase in ZnO as network former does not have much significance in its cross-link density [43]. Relationship between Poisson's ratio and cross-link density was first discussed by Bridge et al. (1983) and its relationship is known as the number of bridging bonds per cations in the structure of the system where results of the cross-link density 0, 1, and 2 are associated with the values of Poisson's ratio obtained, 0.40, 0.30, and 0.15, correspondingly [44]. Average value of Poisson's ratio

acquired in these samples is 0.34 which suggests that the samples have 1D layer structure and values of this result correlate with the results from fractal bond connectivity which is around 1.

Microhardness (H) is known to be the total stress which is necessary to remove the free volume or distortion of the glass and glass ceramic network [45]. Based on Figure 9, the microhardness of the samples decreases as the amount of ZnO increases in the system and indicates a decrease in the connectivity of the glass and glass ceramic system. Application of hydrostatic pressure is necessary to decrease the free volume in the glass and glass ceramic system [46].

After completing experimental elastic moduli, it is always interesting to predict the results of elastic moduli without performing experimental methods. Theory of elastic prediction has been brought up by Makishima and Mackenzie (1973, 1975) on the prediction of elastic moduli of oxide glass by utilizing composition of oxides and packing density of chemical composition of the glass [47, 48].

Rocherulle et al. (1989) improved the packing density of the equation approximately 20 years later and incorporated A_mO_n as

$$C_i = \left(6.023 \times 10^{23}\right) \frac{4}{3}\pi\left(\frac{\rho}{M}\right)\left(mR_{\mathrm{A}}{}^3 + nR_{\mathrm{O}}{}^3\right), \quad (4)$$

TABLE 5: Theoretical calculated total packing density, elastic moduli, and Poisson's ratio of ZnO–SiO$_2$ samples.

Sample	C_t	Theoretical elastic moduli and Poisson's ratio				
		L_{cal} (GPa)	S_{cal} (GPa)	E_{cal} (GPa)	K_{cal} (GPa)	σ_{cal}
S1	0.5851	46.53	15.57	36.83	25.78	0.2626
S2	0.5819	44.96	15.10	35.68	24.83	0.2613
S3	0.5770	42.68	14.41	34.00	23.46	0.2592
S4	0.5731	40.93	13.88	32.71	22.42	0.2576

where R_A and R_O are Pauling's ionic radius of cation A and anions O, ρ and M are the density and effective molecular weight, and C_i is the new packing density [49]. This model offers a new derivation of elastic moduli and Poisson ratio's for the current system as

$$L_{cal} = \left[100 + \frac{4}{3} \left(\frac{300}{10.2C_t - 1} \right) \right] C_t^2 \sum_i G_i x_i,$$

$$S_{cal} = \left[\frac{300 C_t^2}{10.2 C_t - 1} \right] \sum_i G_i x_i,$$

$$E_{cal} = 83.6 C_t \sum_i G_i x_i,$$

$$K_{cal} = 100 C_t^2 \sum_i G_i x_i, \qquad (5)$$

$$\sigma_{cal} = 0.5 - \frac{1}{7.2 C_t},$$

$$G_t = \sum_i G_i x_i,$$

$$C_t = \sum_i C_i x_i,$$

where C_t, G_t, G_i, and x_i are the total packing density, total dissociation energy per unit volume, dissociation energy per unit volume, and the mole fraction of the oxide ith component correspondingly.

Table 5 illustrates the theoretical calculated total packing density, C_t, longitudinal modulus, L_{cal}, shear modulus, S_{cal}, Young's modulus, E_{cal}, bulk modulus, K_{cal}, and Poisson's ratio, σ_{cal}, based on Rocherulle's model. Based on the values of theoretical elastic moduli and Poisson's ratio of ZnO–SiO$_2$ samples obtained, it can be observed that the all the theoretical elastic moduli and Poisson's ratio decrease when composition of ZnO increases. Longitudinal modulus calculated decreases from 46.53 GPa to 40.93 GPa, shear modulus decreases from 15.57 GPa to 13.88 GPa, Young's modulus decreases from 36.83 GPa to 32.71 GPa, bulk modulus decreases from 25.78 GPa to 22.42 GPa, and Poisson's ratio decreases from 0.2626 to 0.2576. Decrease in elastic moduli and Poisson's ratio as ZnO increases suggests that ZnO would act as network modifier thus breaking BOs and converts it to NBOs which will decrease the connectivity and elastic moduli of the samples. Elastic moduli and Poisson's ratio predicted by using Rocherulle's model were illustrated in Figures 10 and 11. Comparisons of the experimental elastic moduli and Poisson's ratio and the theoretical elastic moduli and Poisson's

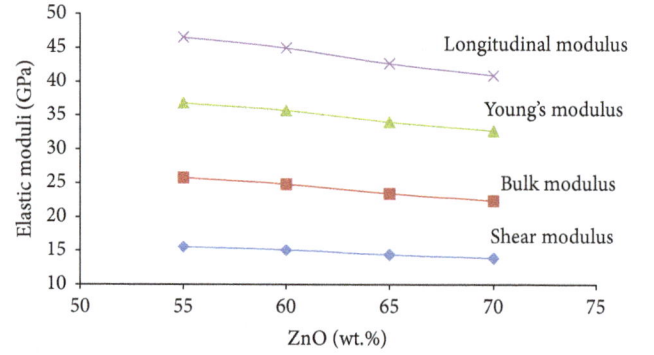

FIGURE 10: Elastic moduli versus wt.% of ZnO calculated using Rocherulle's model.

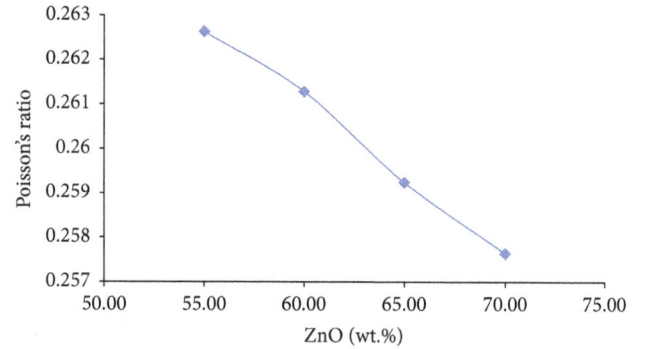

FIGURE 11: Poisson's ratio versus wt.% of ZnO calculated using Rocherulle's model.

ratio were illustrated in Figures 12–16. Experimental elastic moduli and theoretical elastic moduli have a great agreement in similar trend but vary in value except for Poisson's ratio.

4. Conclusion

(ZnO)$_x$(WRHA)$_{1-x}$ (x = 0.55, 0.60, 0.65 and 0.70 wt.%) was symbolized by S1, S2, S3, and S4, respectively, samples have been successfully fabricated, and the study of elastic moduli with prediction of its correlation with the variation of composition on physical and elastic characteristics of each sample was performed and analyzed. Physical analysis of the samples shows that as ZnO increases, the density of the overall samples decreases and this is very well associated with the formation of NBOs and its molar volume increases as ZnO increase which is caused by the substitution of ions of Zn^{2+} which is greater in interatomic spacing compared to SiO^{2+}.

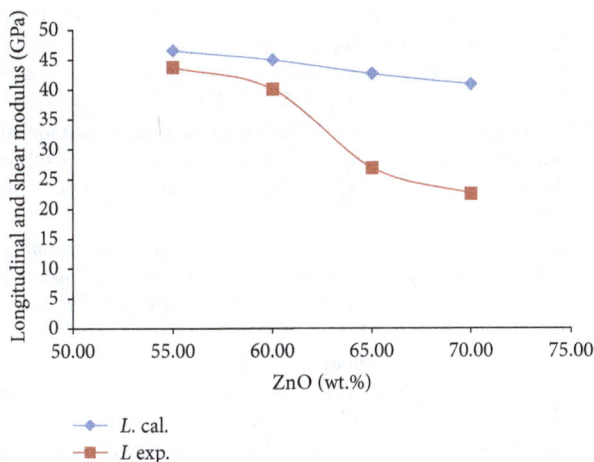

FIGURE 12: Experimental and theoretical longitudinal modulus versus wt.% of ZnO in ZnO–SiO$_2$ samples.

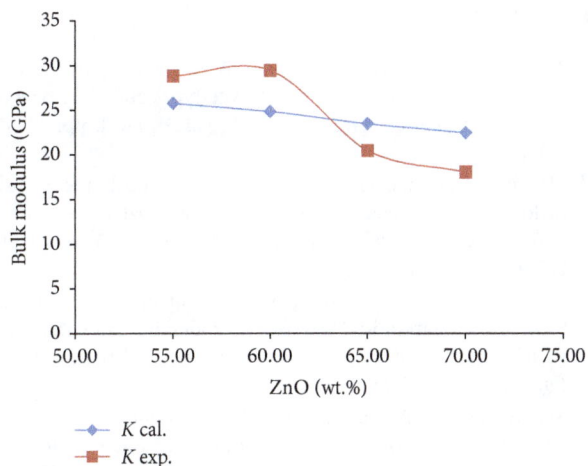

FIGURE 13: Experimental and theoretical shear modulus versus wt.% of ZnO in ZnO–SiO$_2$ samples.

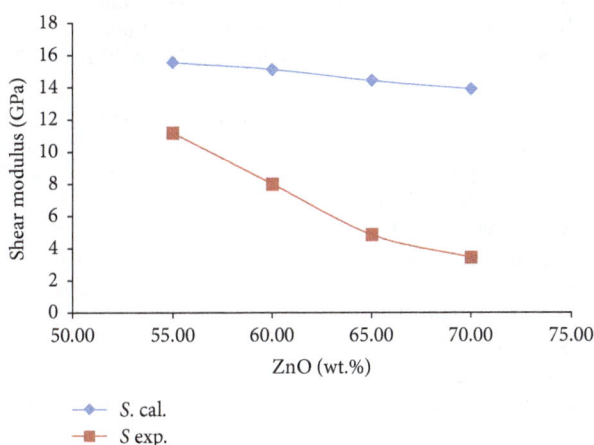

FIGURE 14: Experimental and theoretical Young's modulus versus wt.% of ZnO in ZnO–SiO$_2$ samples.

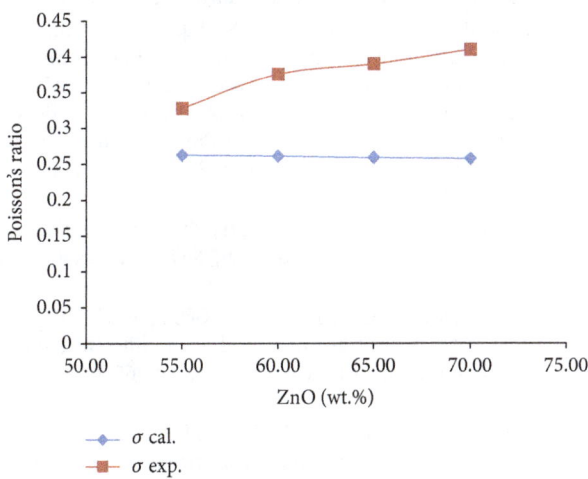

FIGURE 15: Experimental and theoretical bulk modulus versus wt.% of ZnO in ZnO–SiO$_2$ samples.

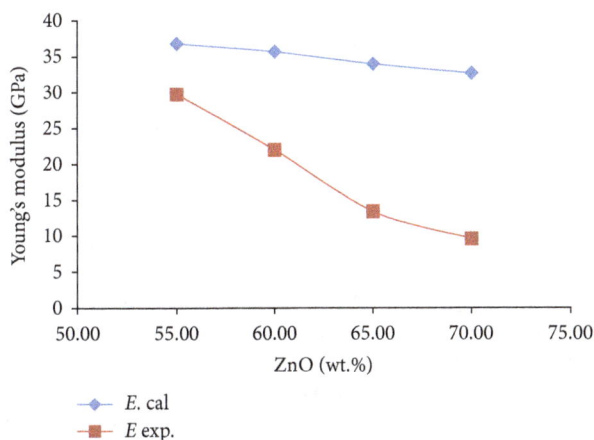

FIGURE 16: Experimental and theoretical Poisson's ratio versus wt.% of ZnO in ZnO–SiO$_2$ samples.

Last but not least, all the theoretical elastic moduli possess the same trend as the elastic moduli obtained experimentally suggesting that ZnO would act as network modifier thus breaking BOs and converts it to NBOs which will decrease the connectivity and elastic moduli of the samples.

Conflicts of Interest

The authors declare that there are no conflicts of interest regarding the publication of this paper.

Acknowledgments

The researchers gratefully acknowledge the financial support for this study from the Malaysian Ministry of Higher Education (MOHE) and Universiti Putra Malaysia through the Fundamental Research Grant Scheme (FRGS) and *Geran Putra Berimpak* research grant.

References

[1] A. M. Yusof, N. A. Nizam, and N. A. A. Rashid, "Hydrothermal conversion of rice husk ash to faujasite-types and NaA-type of zeolites," *Journal of Porous Materials*, vol. 17, no. 1, pp. 39–47, 2010.

[2] M. Ahmaruzzaman and V. K. Gupta, "Rice husk and its ash as low-cost adsorbents in water and wastewater treatment," *Industrial & Engineering Chemistry Research*, vol. 50, no. 24, pp. 13589–13613, 2011.

[3] O. Mohiuddin, A. Mohiuddin, M. Obaidullah, H. Ahmed, and S. Asumadu-Sarkodie, "Electricity production potential and social benefits from rice husk, a case study in Pakistan," *Cogent Engineering*, vol. 3, no. 1, Article ID 1177156, 2016.

[4] M. Arabani, S. A. Tahami, and M. Taghipoor, "Laboratory investigation of hot mix asphalt containing waste materials," *Road Materials and Pavement Design*, vol. 18, no. 3, pp. 713–729, 2017.

[5] S. Kamari and F. Ghorbani, "Synthesis of magMCM-41 with rice husk silica as cadmium sorbent from aqueous solutions: parameters' optimization by response surface methodology," *Environmental Technology (United Kingdom)*, pp. 1–18, 2016.

[6] S. A. Saad, M. F. Nuruddin, N. Shafiq, and M. Ali, "The effect of incineration temperature to the chemical and physical properties of ultrafine treated rice husk ash (UFTRHA) as supplementary cementing material (SCM)," in *Proceedings of the 4th International Conference on Process Engineering and Advanced Materials, ICPEAM 2016*, pp. 163–167, Kuala Lumpur, Malaysia, August 2016.

[7] B. O. Ayomanor and K. Vernon-Parry, "Potential synthesis of solar-grade silicon from rice husk ash," *Solid State Phenomena*, vol. 242, pp. 41–47, 2016.

[8] I. J. Fernandes, D. Calheiro, A. G. Kieling et al., "Characterization of rice husk ash produced using different biomass combustion techniques for energy," *Fuel*, vol. 165, pp. 351–359, 2016.

[9] F. A. Santos, C. dos Santos, D. Rodrigues Junior et al., "Lithium disilicate glass-ceramic obtained from rice husk-based silica," *Advances in Science and Technology*, vol. 63, no. 1, pp. 414–419, 2010.

[10] M. I. Martín, F. Andreola, L. Barbieri et al., "Crystallisation and microstructure of nepheline–forsterite glass-ceramics," *Ceramics International*, vol. 39, no. 3, pp. 2955–2966, 2013.

[11] M. H. M. Zaid, K. A. Matori, H. A. A. Sidek et al., "Fabrication and crystallization of ZnO-SLS glass derived willemite glass-ceramics as a potential material for optics applications," *Journal of Spectroscopy*, vol. 2016, Article ID 8084301, 2016.

[12] X. L. Duan, D. R. Yuan, D. Xu et al., "Preparation and characterization of Co^{2+}-doped $ZnO-Al_2O_3-SiO_2$ glass-ceramics by the sol–gel method," *Materials research bulletin*, vol. 38, no. 4, pp. 705–711, 2003.

[13] L. Yu and M. Nogami, "Local structure and photoluminescent characteristics of Eu^{3+} in $ZnO-SiO_2$ glasses," *Journal of Sol-Gel Science and Technology*, vol. 43, no. 3, pp. 355–360, 2007.

[14] K. A. Matori, M. H. M. Zaid, H. A. A. Sidek, M. K. Halimah, Z. A. Wahab, and M. G. M. Sabri, "Influence of ZnO on the ultrasonic velocity and elastic moduli of soda lime silicate glasses," *International Journal of Physical Sciences*, vol. 5, no. 14, pp. 2212–2216, 2010.

[15] E. N. Bunting, "Phase equilibria in the system $SiO_2–ZnO$," *Journal of the American Ceramic Society*, vol. 13, no. 1, pp. 5–10, 1930.

[16] A. J. A. Al-Nidawi, K. A. Matori, A. Zakaria, and M. H. M. Zaid, "Effect of MnO_2 doped on physical, structure and optical properties of zinc silicate glasses from waste rice husk ash," *Results in Physics*, vol. 7, pp. 955–961, 2017.

[17] S. Zhang and A. Stamboulis, "Effect of zinc substitution for calcium on the crystallisation of calcium fluoro-alumino-silicate glasses," *Journal of Non-Crystalline Solids*, vol. 432, pp. 300–306, 2016.

[18] L. Chen and Y. Dai, "Structure, physical properties, crystallization and sintering of iron-calcium-aluminosilicate glasses with different amounts of ZnO," *Journal of Non-Crystalline Solids*, vol. 452, pp. 45–49, 2016.

[19] M. H. M. Zaid, K. A. Matori, H. J. Quah et al., "Investigation on structural and optical properties of SLS–ZnO glasses prepared using a conventional melt quenching technique," *Journal of Materials Science: Materials in Electronics*, vol. 26, no. 6, pp. 3722–3729, 2015.

[20] K. A. Matori, M. H. M. Zaid, H. J. Quah, H. A. A. Sidek, Z. A. Wahab, and M. G. M. Sabri, "Studying the effect of ZnO on physical and elastic properties of $(ZnO)_x(P_2O_5)_{1-x}$ glasses using nondestructive ultrasonic method," *Advances in Materials Science and Engineering*, vol. 2015, Article ID 596361, 2015.

[21] S. Y. Marzouk and M. S. Gaafar, "Ultrasonic study on some borosilicate glasses doped with different transition metal oxides," *Solid State Communications*, vol. 144, no. 10-11, pp. 478–483, 2007.

[22] H. A. Saudi, H. A. Sallam, and K. Abdullah, "Borosilicate glass containing bismuth and zinc oxides as a hot cell material for gamma-ray shielding," *Physics and Materials Chemistry*, vol. 2, no. 1, pp. 20–24, 2014.

[23] M. H. M. Zaid, K. A. Matori, L. C. Wah et al., "Elastic moduli prediction and correlation in soda lime silicate glasses containing ZnO," *International Journal of Physical Sciences*, vol. 6, no. 6, pp. 1404–1410, 2011.

[24] N. A. Sharaf, R. A. Condrate Sr., and A. A. Ahmed, "FTIR spectral/structural investigation of the ion exchange/thermal treatment of silver ions into a silicate glass," *Materials Letters*, vol. 11, no. 3-4, pp. 115–118, 1991.

[25] P. F. Wang, Z. H. Li, J. Li, and Y. M. Zhu, "Effect of ZnO on the interfacial bonding between $Na_2O-B_2O_3-SiO_2$ vitrified bond and diamond," *Solid State Sciences*, vol. 11, no. 8, pp. 1427–1432, 2009.

[26] A. Bernasconi, M. Dapiaggi, A. Pavese, G. Agostini, M. Bernasconi, and D. T. Bowron, "Modeling the structure of complex aluminosilicate glasses: The effect of zinc addition," *Journal of Physical Chemistry B*, vol. 120, no. 9, pp. 2526–2537, 2016.

[27] Y. B. Saddeek, E. R. Shaaban, K. A. Aly, and I. M. Sayed, "Characterization of some lead vanadate glasses," *Journal of Alloys and Compounds*, vol. 478, no. 1-2, pp. 447–452, 2009.

[28] M. M. Umair and A. K. Yahya, "Elastic and structural changes of $xNa_2O-(35-x)V_2O_5-65TeO_2$ glass system with increasing sodium," *Materials Chemistry and Physics*, vol. 142, no. 2-3, pp. 549–555, 2013.

[29] K. A. Matori, M. I. Sayyed, H. A. A. Sidek, M. H. M. Zaid, and V. P. Singh, "Comprehensive study on physical, elastic and shielding properties of lead zinc phosphate glasses," *Journal of Non-Crystalline Solids*, vol. 457, pp. 97–103, 2017.

[30] C. Bootjomchai, "Comparative studies between theoretical and experimental of elastic properties and irradiation effects of soda lime glasses doped with neodymium oxide," *Radiation Physics and Chemistry*, vol. 110, pp. 96–104, 2015.

[31] C. Weigel, C. Le Losq, R. Vialla et al., "Elastic moduli of $XAlSiO_4$ aluminosilicate glasses: effects of charge-balancing cations," *Journal of Non-Crystalline Solids*, vol. 447, pp. 267–272, 2016.

[32] H. Laoding, H. Mohamed Kamari, A. Zakaria, A. H. Shaari, and I. Mansor, "Elastic properties of thulium doped zinc borotellurite glass," *Materials Science Forum*, vol. 863, pp. 70–74, 2016.

[33] H. A. A. Sidek, R. El-Mallawany, K. A. Matori, and M. K. Halimah, "Effect of PbO on the elastic behavior of $ZnO–P_2O_5$ glass systems," *Results in Physics*, vol. 6, pp. 449–455, 2016.

[34] R. A. Mccauley, A. K. De, and D. S. Carr, "Improved impact resistance in soda-lime-silica glasses through zinc oxide substitutions," *Journal of the American Ceramic Society*, vol. 64, no. 11, pp. 157-158, 1981.

[35] D. J. Bergman and Y. Kantor, "Critical properties of an elastic fractal," *Physical Review Letters*, vol. 53, no. 6, pp. 511–514, 1984.

[36] K. Shinozaki, T. Honma, and T. Komatsu, "Elastic properties and Vickers hardness of optically transparent glass-ceramics with fresnoite $Ba_2TiSi_2O_8$ nanocrystals," *Materials Research Bulletin*, vol. 46, no. 6, pp. 922–928, 2011.

[37] R. Bogue and R. J. Sladek, "Elasticity and thermal expansivity of $(AgI)_x(AgPO_3)_{1-x}$ glasses," *Physical Review B*, vol. 42, no. 8, pp. 5280–5288, 1990.

[38] G. A. Saunders, T. Brennan, M. Acet et al., "Elastic and nonlinear acoustic properties and thermal expansion of cerium metaphosphate glasses," *Journal of Non-Crystalline Solids*, vol. 282, no. 2-3, pp. 291–305, 2001.

[39] N. Ghribi, M. Dutreilh-Colas, J.-R. Duclère et al., "Structural, mechanical and optical investigations in the TeO_2-rich part of the TeO_2-GeO_2-ZnO ternary glass system," *Solid State Sciences*, vol. 40, pp. 20–30, 2015.

[40] A. Abd El-Moneim, "Correlation between acoustical and structural properties of glasses: Extension of Abd El-Moneim model for bioactive silica based glasses," *Materials Chemistry and Physics*, vol. 173, pp. 372–378, 2016.

[41] L. A. El Latif, "Ultrasonic study on the role of Na_2O on the structure of Na_2O-B_2O_3 and Na_2O-B_2O_3-SiO_2 glasses," *Journal of Pure and Applied Ultrasonic*, vol. 27, no. 2-3, pp. 80–91, 2005.

[42] G. N. Greaves, A. L. Greer, R. S. Lakes, and T. Rouxel, "Poisson's ratio and modern materials," *Nature Materials*, vol. 10, no. 11, pp. 823–837, 2011.

[43] N. B. Mohamed, A. K. Yahya, M. S. M. Deni, S. N. Mohamed, M. K. Halimah, and H. A. A. Sidek, "Effects of concurrent TeO_2 reduction and ZnO addition on elastic and structural properties of $(90 − x)TeO_2$-$10Nb_2O_5$-$(x)ZnO$ glass," *Journal of Non-Crystalline Solids*, vol. 356, no. 33-34, pp. 1626–1630, 2010.

[44] B. Bridge, N. D. Patel, and D. N. Waters, "On the elastic constants and structure of the pure inorganic oxide glasses," *Physica Status Solidi (A)*, vol. 77, no. 2, pp. 655–668, 1983.

[45] N. Sasmal, M. Garai, and B. Karmakar, "Influence of Ce, Nd, Sm and Gd oxides on the properties of alkaline-earth borosilicate glass sealant," *Journal of Asian Ceramic Societies*, vol. 4, no. 1, pp. 29–38, 2016.

[46] T. Rouxel, "Driving force for indentation cracking in glass: composition, pressure and temperature dependence," *Philosophical Transactions of the Royal Society A: Mathematical, Physical and Engineering Sciences*, vol. 373, no. 2038, Article ID 20140140, 2015.

[47] A. Makishima and J. D. Mackenzie, "Direct calculation of Young's moidulus of glass," *Journal of Non-Crystalline Solids*, vol. 12, no. 1, pp. 35–45, 1973.

[48] A. Makishima and J. D. Mackenzie, "Calculation of bulk modulus, shear modulus and Poisson's ratio of glass," *Journal of Non-Crystalline Solids*, vol. 17, no. 2, pp. 147–157, 1975.

[49] J. Rocherulle, C. Ecolivet, M. Poulain, P. Verdier, and Y. Laurent, "Elastic moduli of oxynitride glasses. Extension of Makishima and Mackenzie's theory," *Journal of Non-Crystalline Solids*, vol. 108, no. 2, pp. 187–193, 1989.

22

Joining of C$_f$/SiC Ceramic Matrix Composites

Keqiang Zhang,[1] **Lu Zhang,**[1] **Rujie He** ⓘ,[1] **Kaiyu Wang,**[2] **Kai Wei** ⓘ,[2] **and Bing Zhang**[3]

[1]*Institute of Advanced Structure Technology, Beijing Institute of Technology, Beijing 100081, China*
[2]*State Key Laboratory of Advanced Design and Manufacturing for Vehicle Body, Hunan University, Changsha 410082, China*
[3]*Bristol Composites Institute (ACCIS), University of Bristol, Queen's Building, University Walk, Bristol BS8 1TR, UK*

Correspondence should be addressed to Rujie He; herujie@bit.edu.cn

Academic Editor: Mikhael Bechelany

Carbon fiber-reinforced silicon carbide (C$_f$/SiC) ceramic matrix composites have promising engineering applications in many fields, and they are usually geometrically complex in shape and always need to join with other materials to form a certain engineering part. Up to date, various joining technologies of C$_f$/SiC composites are reported, including the joining of C$_f$/SiC-C$_f$/SiC and C$_f$/SiC-metal. In this paper, a systematic review of the joining of C$_f$/SiC composites is conducted, and the aim of this paper is to provide some reference for researchers working on this field.

1. Introduction

With the rapid development of high-tech in aerospace and other industry fields, the demands for new materials, which can work in extreme harsh working environment of high temperatures, are growing. The needs for better efficiency and higher thrust-to-weight ratio promote the development of advanced materials at high temperatures, such as superalloys [1–3], ceramics [4–6], composites [7–10], and so on. Among these advanced materials, ceramic matrix composites (CMCs) are drawn great attentions for their engineering applications under extreme conditions because they can maintain low density, high strength, wear resistance, oxidation resistance, thermal shock resistance, corrosion resistance, and some other functions together [11].

Carbon fiber-reinforced silicon carbide (C$_f$/SiC) ceramic matrix composites, one of the most famous CMCs, are becoming the most promising candidates for high-temperature structural applications (as illustrated in Figure 1), such as sharp leading edges, nose cones, aeronautic jet engines, thermal protection systems for reusable atmosphere reentry vehicles [12, 13], as well as optical components [14] and nuclear fusion/fission reactors [15, 16], owing to their relatively low density (~2 g/cm^3), high thermal conductivity (~67 W/(m·K)), high strength (300–800 MPa) [17–19], low

coefficient of thermal expansion (CTE, $3.0–3.1 \times 10^{-6} \cdot K^{-1}$), especially good stability and excellent oxidation and creep resistance at elevated temperatures [13, 21–23, 25]. In particular, C$_f$/SiC composites have shown significant improvements in fracture toughness and thermal shock resistance. These improvements in mechanical properties are dependent on the specific properties of the carbon fiber and the silicon carbide. According to the type of carbon fiber, it can be divided into 1D C$_f$/SiC, 2D C$_f$/SiC, 2.5D C$_f$/SiC, and 3D C$_f$/SiC and applied in different fields.

For aerospace applications, as reported by NASA, the X-37B and X-38 aircrafts employed a large number of C$_f$/SiC composites in their nose cone [25], leading edge wing and engine components [11, 26, 27, 29]. For nuclear applications, C$_f$/SiC composites are used as the cladding materials in pressurized water reactors and flow channel insert materials in thermonuclear fusion reactors [29, 30, 32]. In most cases, typically, C$_f$/SiC composite components are usually geometrically complex in shape and always need to join with other materials to form a certain engineering part. However, unfortunately, due to their poor machinability and toughness, C$_f$/SiC composites lack good processing performance like metal material and thus cannot be processed into complex-shaped components by forging, extrusion molding, and other traditional methods. It is very difficult to produce

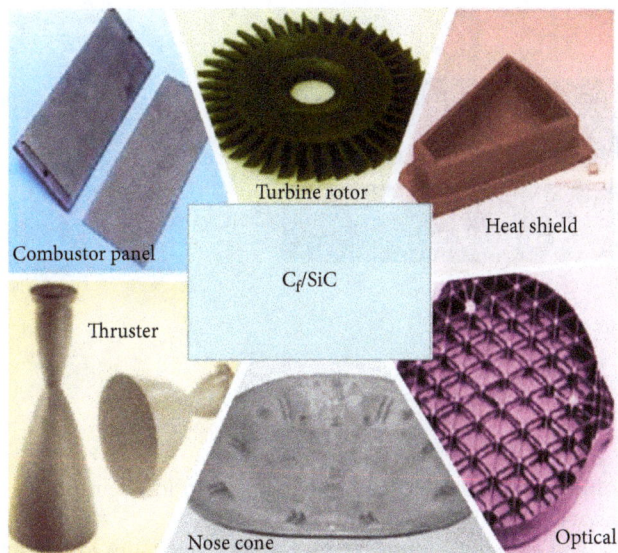

FIGURE 1: Various engineering applications of C$_f$/SiC composites.

large-size C$_f$/SiC composite components with complex shapes, resulting in they must be joined with themselves or other materials by appropriate joining technologies [32–34, 37]. There have been numerous reports on the joining of C$_f$/SiC composites in the past two decades, including self-joining of C$_f$/SiC composites [36, 37, 39], and joining of C$_f$/SiC composite to dissimilar materials, such as Ti [39, 40], Nb [41, 42, 45], Ni [44, 45, 48], TiAl alloys [47], and so on.

Up to date, various joining technologies of C$_f$/SiC composites are reported, including the joining of C$_f$/SiC-C$_f$/SiC and C$_f$/SiC-metal. Table 1 lists commonly used joining technologies, such as direct bonding of C$_f$/SiC-C$_f$/SiC, indirect bonding of C$_f$/SiC-C$_f$/SiC, brazing of C$_f$/SiC-metal, diffusion bonding of C$_f$/SiC-metal, online liquid infiltration of C$_f$/SiC-metal, ultrasonic-assisted joining of C$_f$/SiC-metal, and electric-assisted joining of C$_f$/SiC-metal. To the best knowledge of the authors, however, there has been no systematic summary of the joining of C$_f$/SiC composites. Therefore, we herein conduct a systematic review of the joining of C$_f$/SiC composites, and the aim of this paper is to provide some reference for researchers working on this field.

2. Self-Joining of C$_f$/SiC Composites

In some conditions, in order to obtain large size and complex-shaped C$_f$/SiC composite components, it is necessary that C$_f$/SiC composites should be joined with C$_f$/SiC composites themselves, named as "self-joining." There have been many reports about the self-joining of C$_f$/SiC composites in the last decades, usually including direct bonding and indirect bonding method.

2.1. Direct Bonding. Direct bonding method is a self-joining of C$_f$/SiC composites by solid-phase diffusion without any other materials (Figure 2(a)). As reported in previous papers,

TABLE 1: Commonly used joining methods of C$_f$/SiC composites.

Joining materials	Joining methods
C$_f$/SiC-C$_f$/SiC	Direct bonding
	Indirect bonding
C$_f$/SiC-metal	Brazing
	Diffusion bonding
	Online liquid infiltration
	Ultrasonic-assisted joining
	Electric-assisted joining

the main procedure of direct bonding usually includes three procedures: (1) plastic deformation, (2) diffusion, and (3) creep. Plastic deformation occurs on the interface because of heat and pressure; diffusion includes surface diffusion, bulk diffusion, grain boundary diffusion, and interfacial diffusion to achieve C$_f$/SiC bonding. Creep refers to the permanent movement or deformation of metal.

However, the bonding strength of the directly joined C$_f$/SiC composite is usually very low because a strong bonding of C$_f$/SiC composite is difficult to obtain without any other transition phases and because the diffusion between C$_f$/SiC composites is not easy owing to the strong covalent bond and the poor deformation ability of the SiC in the composites. Rizzo et al. [48] reported that a CVD-SiC coated C$_f$/SiC composite was directly joined to its counterparts using spark plasma sintering (SPS) technology. The results showed that the cracks in the CVD-SiC coating were visible among the interface and propagated from the SiC coating through the joint area (as is shown in Figure 3), due to the CTE mismatch between SiC coating and C$_f$/SiC substrate (as is shown in Table 2), and the apparent shear strength was as low as 5.6 MPa.

Therefore, direct bonding method is merely used owing to the low bonding strength. However, it is still very promising for direct bonding method of C$_f$/SiC composites, especially for extreme applications where it demands to avoid a second material.

2.2. Indirect Bonding. It is well known that it is very difficult to form diffusion between C$_f$/SiC composites owing to the strong covalent bond and the poor deformation ability of the SiC in C$_f$/SiC composites, thus resulting in a weak bonding strength of direct bonding joint (Figure 2(b)). Therefore, second-phase materials with plastic deformability, such as Ag-Cu-Ti [30, 49], Ti-Zr-Be [50], Ni [31, 51, 52, 55], calcia-alumina (CA) glass-ceramic [54], Ti$_3$SiC$_2$ [16, 36, 55, 56] and Si resin [57], and MoSi$_2$ [58], were widely reported to be used for the joining of C$_f$/SiC composites. These kinds of joining are known as indirect bonding method, always including using metal filler or nonmetal filler.

2.2.1. Metal Fillers. This method means that C$_f$/SiC composites are bonded with C$_f$/SiC composites using metal fillers, such as pure metal or alloys. Table 3 lists some typical reports on the self-joining of C$_f$/SiC composites using metal fillers.

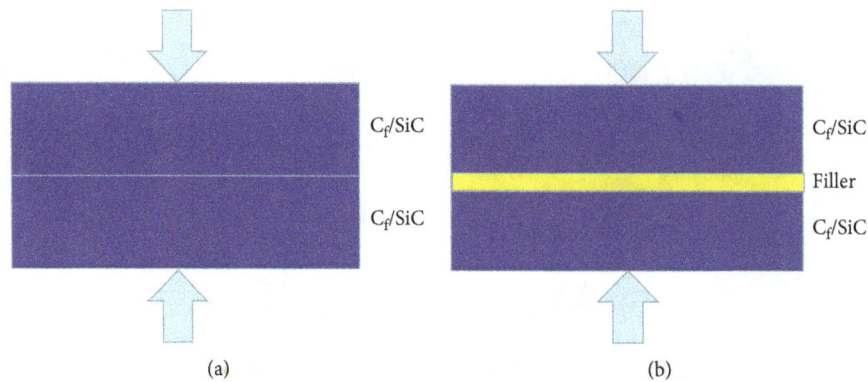

FIGURE 2: The diagram of (a) direct bonding and (b) indirect bonding.

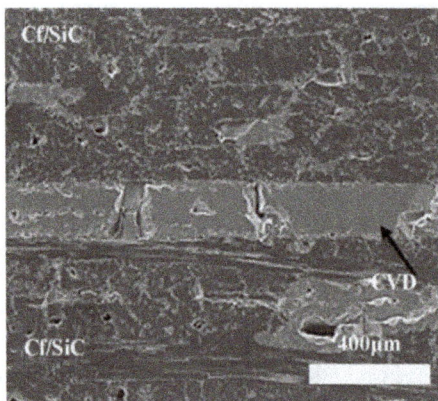

FIGURE 3: Scanning electron microscopy of polished cross sections of C_f/SiC specimens joined by SPS: direct bonding [48].

TABLE 2: CTE of typical materials (room temperature).

Materials	CTE ($\times 10^{-6} \cdot K^{-1}$)
C_f/SiC	3.0–3.1
Ti	8
Al	23.5
Cu	16.5
Ni	13
Ag	19.5
Nb	7.2
Mo	5.2
W	4.43
Zr	5.2
Co	6.8
Ta	6.7
C	1.5
TiAl	10.8
CrMo	12.5
SiC	4.8
Ti_3SiC_2	9.1
TiC	7.4

The low-temperature active filler is a relatively mature technology and widely used in C_f/SiC composites; however, the joint phase such as Ag, Cu, and other metals, usually with low melting point, leads to poor high-temperature strength and oxidation resistance. Therefore, C_f/SiC composites joined by metal fillers can only be used in low-temperature environment (<500°C).

Liu et al. [30] reported the C_f/SiC composites joined by ternary Ag-35.25 wt% Cu-1.75 wt% Ti and demonstrated that the mechanical strength decreased with the increase in temperature owing to the softening of filler. The flexural strength decreased to 46% and 26% at 300 and 500°C compared with that at room temperature, respectively. Stefano et al. [48] fabricated C_f/SiC-Ti-C_f/SiC sandwich by SPS and used pure Ti foils as filler. They also found that a Ti-Si-C-based phase (Ti_3SiC_2, as is shown in Figure 4) was the main reaction product, usually induced to strength decrease.

High-temperature metal fillers, such as Ni and its alloys, are reported and found to greatly improve the high-temperature resistance of the joint [61, 62]. Cheng [51, 52, 55] developed a novel joining process to join the 2D/3D C_f/SiC composites. Porous C_f/SiC composites were fabricated through chemical vapor infiltration (CVI) process, and Ni alloy was used to join the C_f/SiC composites together. Figure 5 shows the diagram of this joining process. Because the Ni alloy had a favorable wettability with C_f/SiC composites, melted Ni alloy easily infiltrated into the pores among C_f/SiC composites. Hence, the contact surface between Ni alloy and C_f/SiC composites matrix was greatly increased, thereby improved the bonding strength. Besides, Ni alloy had a higher melting point; hence, the joint was expected to be used at high temperatures (>1000°C).

Table 4 lists some typical reports on the self-joining of C_f/SiC composites with high-temperature fillers (Ni alloy). As a nontraditional joining method, the self-joining process using Ni alloy is usually carried out during composite preparation procedure, and the damage is minimal. And after the joining process, an afterward CVD process is conducted, which not only densify the porous composites but also provides antioxidation coating for the matrix and the joint.

2.2.2. Nonmetal Fillers. Nonmetal fillers, such as MAX ceramic [16, 36, 55, 56, 63], ceramic precursors [64], Si resin [57], and $MoSi_2$ [58], are also reported to be used in the self-joining of C_f/SiC composites (as listed in Table 5).

MAX phase ceramics are reported to exhibit not only high-temperature performance, thermal shock resistance,

TABLE 3: Self-joining of C_f/SiC composites using metal filler (SS = shear strength).

CMCs	Basic information	Metal filler	Process parameters	Bend strength (MPa)	Ref.
C_f/SiC	CVD-SiC	Ti	1700°C, 3 min, 60 MPa, vacuum	24.6 (SS)	[48]
C_f/SiC	3D, 10.0 vol.%, PIP	Cu-Au-Pd-V	1170°C, 10 min, 1.5×10^{-3} Pa	135	[59]
C_f/SiC	3D	Pd-Co-V	1250°C, 20 min, $3.0–7.0 \times 10^{-3}$ Pa	—	[60]
C_f/SiC	3D, PIP, 10.0 vol.%	Cu-Pd-V	1170°C, 10 min, $3.0–7.0 \times 10^{-3}$ Pa	128	[61]

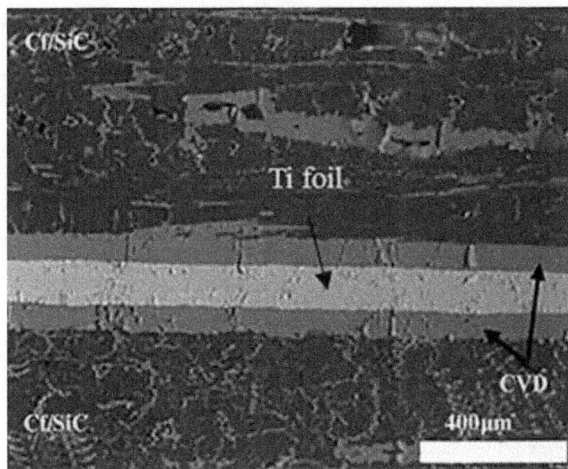

(a)

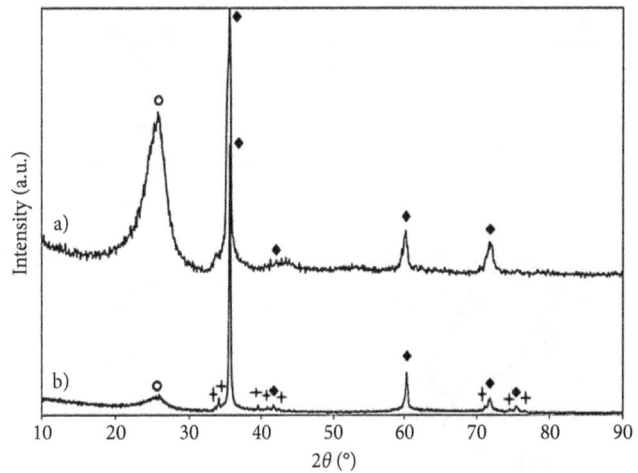

♦ SiC
+ Ti_3SiC_2
○ C

(b)

FIGURE 4: (a) Backscattered electron images of polished cross sections and (b) micro-XRD on the fracture surfaces of C_f/SiC joined by SPS with Ti foil [48].

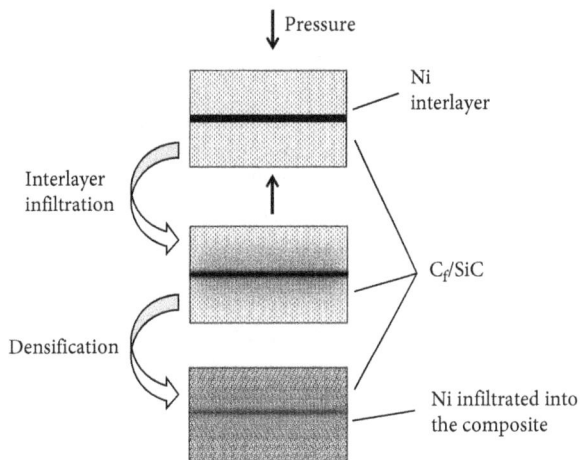

FIGURE 5: The diagram of self-joining using Ni alloy.

and wear resistance but also a good plastic deformation capacity. Among various MAX phase ceramics, Ti_3SiC_2 presents a suitable wettability and CTE toward C_f/SiC composites matrix (as is shown in Table 2) and is thus believed to be a promising candidate for the self-joining of C_f/SiC composites [16, 55, 56]. Dong et al. [55] used Ti_3SiC_2 as the nonmetal filler to join C_f/SiC composite together through hot pressing. The shear strength of the joint was reported as high as 110.4 MPa (56.7% of the C_f/SiC composite matrix). Chemical reactions took place at the interface between Ti_3SiC_2 and C_f/SiC, and residual thermal stress was investigated. The phase compositions of the fracture surfaces for the C_f/SiC joints joined at various temperatures were also analyzed by XRD (as is shown in Figure 6). In addition, the fracture behavior of joining interface and brazing application was explored in previous articles [9, 65, 66, 69]. Interfacial reactions can affect the formation of a joint from the onset of bonding through the development of equilibrated microstructure and to the optimization of the mechanical properties. It has been demonstrated that an adequate joining interface could lead to improvements of the composite wettability by C_f/SiC [39].

Besides, ceramic precursors are also used as nonmetal fillers for the self-joining of C_f/SiC composites. The ceramic precursor is transformed into amorphous ceramic at a certain temperature, and the composition and structure of the precursor are similar to those of the composite matrix. At the same time, the pyrolysis products are directly bonded with the composite matrix by chemical bonds. The thermodynamic properties of the joining layer obtained by this method are similar to those of the matrix [64]. And it has good compatibility with the composite matrix. Therefore,

TABLE 4: Self-joining of C$_f$/SiC composites with high-temperature fillers.

CMCs	Basic information	Joining material	Process parameters	Bend strength (MPa)	Ref.
C$_f$/SiC	3D, CVI	Ni alloy	1300°C, 45 min, 20 MPa, vacuum	260.3	[52]
C$_f$/SiC	2D, CVI	Ni alloy	1300°C, 45 min, 20 MPa, vacuum	60	[51]
C$_f$/SiC	2D, CVI	Ni alloy	1300°C, 15 MPa, vacuum	58	[53]

TABLE 5: Self-joining of C$_f$/SiC composites using inorganic filler (BS = bend strength).

CMCs	Basic information	Joining material	Process parameters	Shear strength (MPa)	Ref.
C$_f$/SiC	2D, CVI, 2.05 g/cm^3, 40 MPa	Ti$_3$SiC$_2$	1600°C, 30 min, 20–40 MPa, Ar	110.4 (BS)	[55]
C$_f$/SiC	3D, CVI	PSZ	1300°C, N$_2$	29.6	[64]
C$_f$/SiC	3D, PIP, 1.9 g/cm^3	Si resin	1400°C, 5 h, Ar	3.51	[57]
C$_f$/SiC	2D, CVI, 1.7–2.2 g/cm^3	MoSi$_2$/Si	1450°C, 5 min, Ar	—	[58]

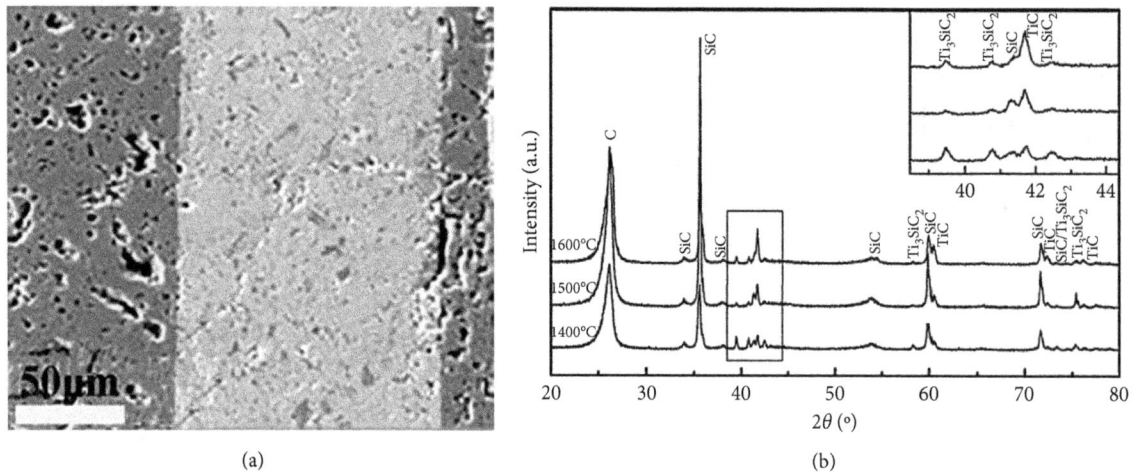

(a)   (b)

FIGURE 6: (a) The backscattered electron images and (b) XRD patterns of the fracture surfaces of the C$_f$/SiC joints [55].

the joint exhibits good mechanical strength. Previous reports showed that C$_f$/SiC composite was joined using Si-O-C ceramic precursor as filler [57].

Si resin is transformed into Si-O-C ceramic at low temperature; the Si-O-C ceramics infiltrating into the substrate improve the filler contact with the substrate closely and increase the connection area. Moreover, the Si-O-C ceramics infiltrating into the pits can form tiny "pins," thus increasing the shear strength of the joints. Gianchandani et al. [58] reported that a MoSi$_2$/Si composite obtained in situ by reaction of silicon and molybdenum at 1450°C in Ar flow is proposed as pressure-less joining material for C$_f$/SiC composites.

To sum up, we can know that the application of non-metal fillers method due to the phase consistency of joining material and matrix was similar, which not only avoid the CTE mismatch between the joining material and the matrix (CTE of typical materials is shown in Table 2) but also inhibit the adverse reactions of interface. It will be a very promising method in the future.

3. Joining of C$_f$/SiC Composites to Metals

In order to obtain large size and complex-shaped components, the joining of C$_f$/SiC composites to metals such as Ti

[40], Nb [42, 68, 69], Ni [70], and TiAl alloys [30, 46, 71] is necessary. Due to the differences in physical, chemical and mechanical properties between C$_f$/SiC composites and metals, there are several problems for the joining of C$_f$/SiC composites to metals: firstly, the chemical bonds of C$_f$/SiC composites are ionic bond and covalent bond and the valence state is stable, whereas metals mostly are metal bond and therefore it is difficult to wet the surface of C$_f$/SiC composites by metal [40]. Secondly, the CTE mismatch between metals and C$_f$/SiC composites is very large, which will produce residual stress at the joint interface; hence, cracks, pores, and other defects exist after cooling [41, 72, 73]. At last, a variety of chemical reactions occur in the interface, resulting in brittle compounds with high hardness, which usually is the reason for the brittle fracture of the joint during working [30].

At present, there are many technologies solving the above problems during the joining process. Brazing and diffusion bonding are the most commonly used methods. In addition, online liquid infiltration joining, ultrasonic-assisted joining, and electric-assisted field joining are also reported.

3.1. Brazing. Brazing is one of the earliest and most commonly used methods for joining CMCs to metals

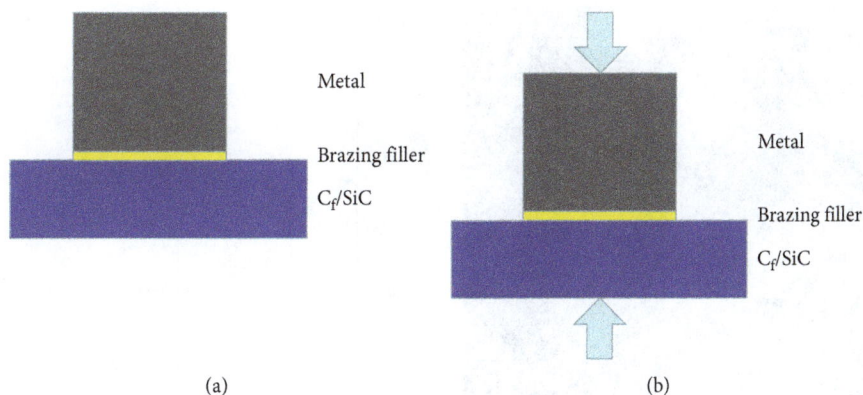

FIGURE 7: The schematic diagram of (a) brazing method and (b) diffusion bonding method.

(Figure 7(a)). It is divided into two kinds as follows: (1) metallizing the C_f/SiC composite surface and then brazing with ordinary brazing filler metals, usually known as indirect brazing, and (2) wetting CMCs surface directly using active metal, known as reactive brazing. Compared with indirect brazing, the scopes of application of reactive brazing are more extensive. Usually, metals and alloys with lower melting points are selected as the brazing fillers, and then the joint is heated to a certain temperature, which is higher than the melting point of brazing filler, and then brazing is conducted [74].

3.1.1. Low-Temperature Fillers.

Low-temperature filler is a kind of metal with low melting point, such as Ag and Cu, which can form brazing filler at lower temperature to realize the joining of metal. Due to the low joining temperature, the damage is low.

Brazing method is simple and convenient; however, the brazing filler is mainly active metal elements, so it is necessary to protect the active metal elements from oxidation. Once the active element is oxidized, it is difficult to react with C_f/SiC composites and to form a reliable joint; consequently, the joint strength is low. Therefore, brazing method is generally carried out in vacuum conditions or inert protective gases [39, 75]. Feng et al. [72, 76] investigated the microstructural evolution and joint strength of between TiAl alloys and C_f/SiC composite via vacuum brazing using Ag-Cu and Ag-Cu-Ti fillers. The diffusion of Al and Ti from TiAl to the matrix had an important effect on the structure and strength of joints. When active element Ti diffused into C_f/SiC composite, the formation of $AlCu_2Ti$ and Ag solid solution was detected with the dissolved Ti and Al; moreover, Ti_5Si_3 phase and TiC also formed adjacent to the composite (as is shown in Figure 8). The maximum shear strength achieved 85 MPa with the thickness of TiC layer of 4–5 μm. The fracture of the joint went through the TiC layer adjacent to its interface with the Ag solid solution and TiC bond layer.

As is shown in Table 6, Ag-Cu, Ag-Cu-Ti, and others are low-temperature fillers (900°C) and have low yield strength and good deformation ability, which is helpful to alleviate the residual stress of the joint, thus increasing the shear strength of the joints.

3.1.2. High-Temperature Fillers.

Ag-Cu-Ti alloys have good plastic deformation behaviors (as is shown in Table 7); nevertheless, they always have low melting points and can only be used in low-temperature environments (<800°C). Once the temperature increased, the strength of the joint drops sharply. Therefore, it is necessary to develop suitable high-temperature brazing filler for high-temperature conditions.

Huang et al. [46] joined C_f/SiC composite to TC4 alloy using (Ti-Zr-Cu-Ni) and W powder as brazing fillers. Ti and Zr elements reacted with C, Cu, and Ni in the interlayer. As elements diffused to each other, a reaction layer was formed between the C_f/SiC composite and TC4 alloy. The brazing parameters had a significant effect on the interfacial reaction between C_f/SiC composite and joining material, which affected the shear strength of the joints. A continuous reaction layer adjacent to C_f/SiC composite and a diffusion layer near TC4 alloy can be clearly observed (Figures 9 and 10). The addition of appropriate W powder helped to relieve residual stress and improved the strength of the joints. The shear strength of the joint was 166 MPa and 96 MPa at room temperature and 800°C, respectively. Therefore, the joint can be used under high temperature.

However, the effect of W powder on the residual stress was small and the residual stress was still high. Ti-Zr-Cu-Ni alloy and pure Ti metals were used as joining materials [30]; the molten Ti-Zr-Cu-Ni reacted with solid Ti in the liquid-solid reaction to form an in situ alloy. The effects of Ti contents on the strength of joints were explored. With the increase in the Ti content, more tearing ridges appeared in the fracture surfaces, which indicated that the fracture possessed more plasticity. When the Ti content reached up to 40%, the shear strength of the joint reached up to 283 MPa, which was 79% higher than using Ti-Zr-Cu-Ni alone. The main reason was that the metal Ti had better plasticity, and the proper addition was beneficial for improving the interfacial reaction between C_f/SiC composite and Ti-6Al-4V alloy.

There are many research studies using brazing method for joining C_f/SiC to metals as listed in Table 7. The low-expansion material (W), the soft metal (Ni), and the high-temperature metal (Mo) as the reinforcing phase are added into the brazing filler, so that the CTE of the brazing filler is

(a)

(b)

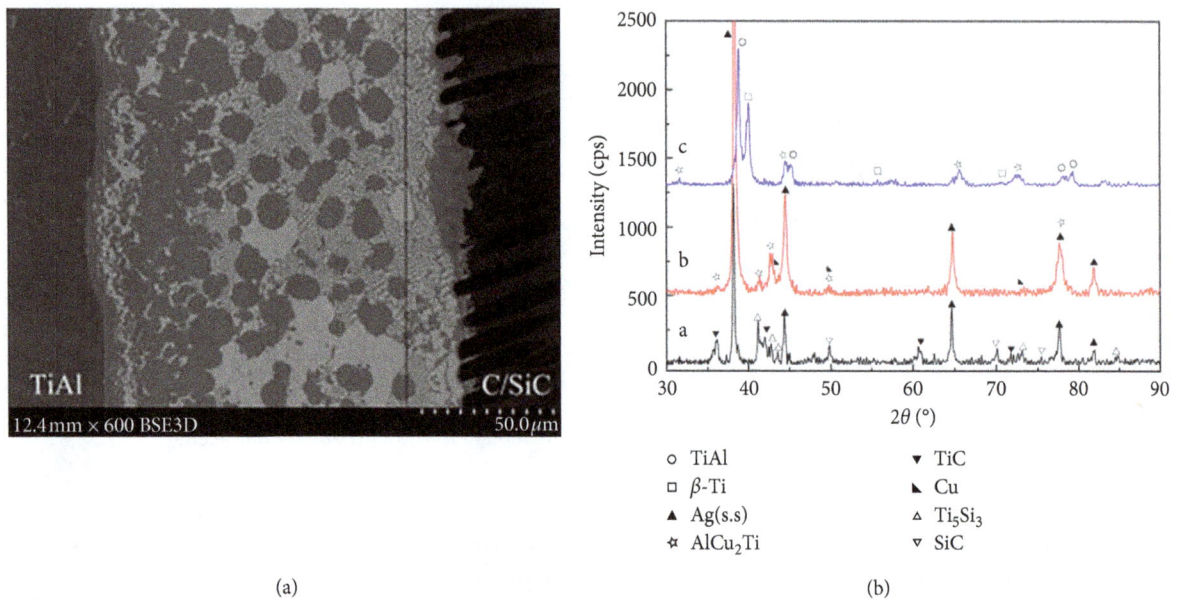

FIGURE 8: (a) Microstructures and (b) XRD patterns of the joint from C/SiC composites to TiAl brazed at 900°C for 10 min [72].

TABLE 6: Brazing of C_f/SiC composites to metals with low-temperature fillers.

CMCs	Basic information	Metal	Brazing material	Process parameters	Shear strength (MPa)	Ref.
C_f/SiC	3D, PIP, 1.86 g/cm^3	Ti Al alloys	Ag-Cu	900°C, 10 min, 5×10^{-3} Pa	85	[72]
C_f/SiC	3D, CVI	TC$_4$ alloys	Ag-Cu-Ti	900°C, 5 min, 10^{-4} Pa	102	[77]
C_f/SiC	—	Nb alloys	Ag-Cu-Ti	930°C, 15 min, 2×10^{-3} Pa	—	[68]
C_f/SiC	3D, 1.8 g/cm^3, 10–15%	Ti alloys	C_f/Ag-Cu-Ti	900°C, 30 min, 2.2×10^{-3} MPa, 6×10^{-3} Pa	84	[40]

TABLE 7: Brazing of C_f/SiC composites to metals with high-temperature fillers.

CMCs	Basic information	Metal	Brazing material	Process parameters	Shear strength (MPa)	Ref.
C_f/SiC	3D, 2.0–2.1 g/cm^3, 10–15 vol.%, 400 MPa	Ti alloys	Ag-Cu-Ti + 15 vol.%W	900°C, 5 min, 2.2×10^{-3} MPa, 6×10^{-3} Pa	180	[38]
C_f/SiC	3D, PIP, 1.7–1.8 g/cm^3, 10–15 vol.%, 400 MPa	TC$_4$ alloys	Ti-Zr-Cu-Ni + 15 vol.% W	930°C, 20 min, 6×10^{-3} Pa	166	[46]
C_f/SiC	3D	42CrMo	Ag-Cu-Ti + 5 vol.% Mo	900°C, 10 min, vacuum	587 (BS)	[78]
C_f/SiC	3D, PIP	Nb	Ti-Cu-Ni-Zr	930°C, 10 min, 5×10^{-3} Pa	124	[69]
C_f/SiC	3D	Ti-6Al-4V	Ti + (Ti-Cu-Ni-Zr)	940°C, 20 min, 5×10^{-3} Pa	283	[47]
C_f/SiC	3D, PIP, 1.98 g/cm^3, 21.5 vol.%	Nb-1Zr	Ti-Co-Nb	1280°C, 10 min, $1.0–3.0 \times 10^{-3}$ Pa	242	[43]
C_f/SiC	3D, PIP, 1.86 g/cm^3, 11.7 vol.%	Ti Al alloys	TiH$_2$-Ni-B	1180°C, 10 min, 5×10^{-3} Pa	105	[71]

reduced and the residual stress of the joint is facilitated. However, there are still some shortcomings for brazing process, such as the interface reaction is intense, to produce brittle compounds, which requires the appropriate adjustment of brazing filler and process parameters. More importantly, avoiding bad excessive interface reaction and accessing to excellent mechanical properties of joints are essential.

3.2. Diffusion Bonding.

In mid-1950s, the former Soviet Union scientists proposed a diffusion bonding method which was widely used to join ceramic to metals, including the joint of C_f/SiC composites to metals (as shown in Figure 7(b)). C_f/SiC composites and metals are contacted with each other under high temperatures, vacuum or inert atmospheres and pressures, and the plastic formation of connected surfaces is close to each other. After a certain period of soaking time, the intermolecular diffusion and chemical reaction are realized. During the diffusion bonding process, the interface is bonded by plastic deformation, diffusion, and creep mechanism. The joining temperature is high, the CTE and elastic modulus of the composites and

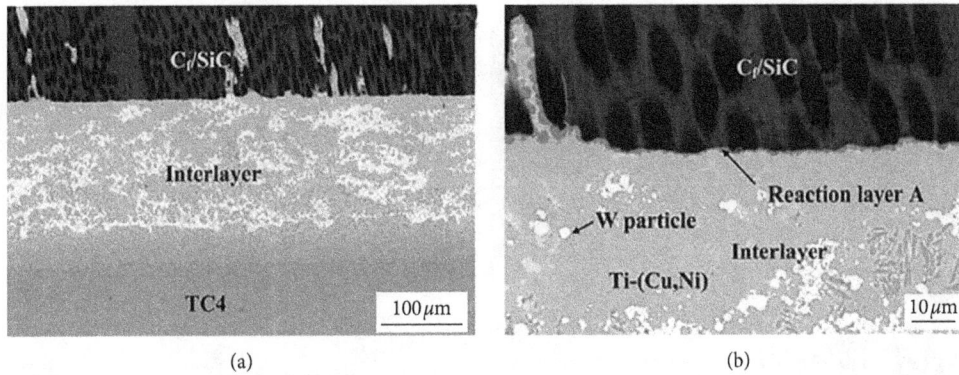

FIGURE 9: BSE images of the joint: (a) micrograph of the joint; (b) interface between C_f/SiC composite and interlayer [46].

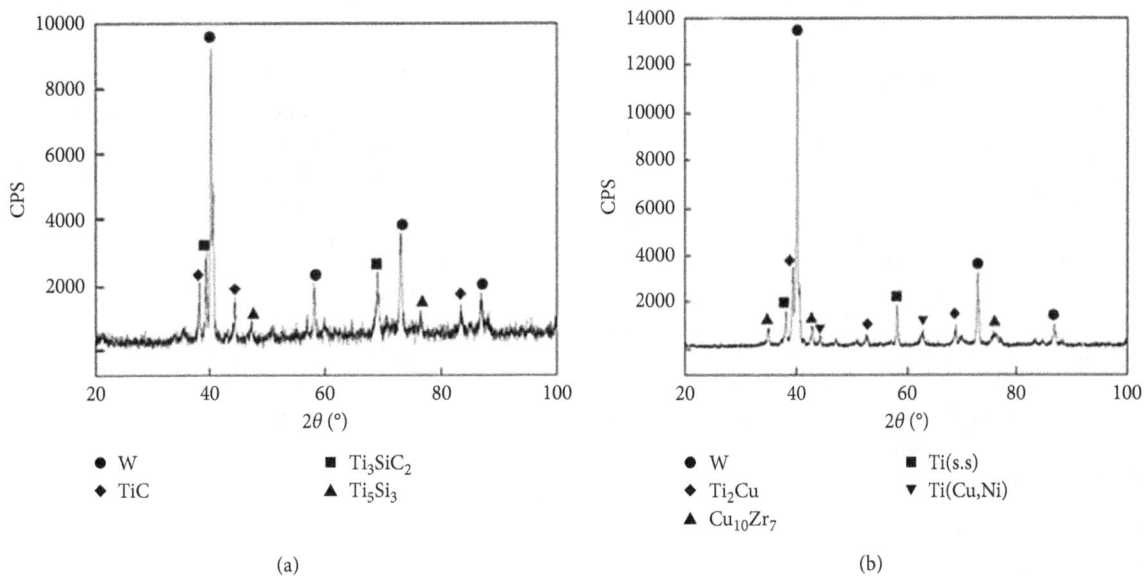

FIGURE 10: XRD pattern of the joint: (a) interface between C_f/SiC composite and interlayer; (b) interlayer [46].

metals are mismatch, and it is easy to induce high residual stress. Due to sharp structural transition near the interface and the lack of a buffer layer to relax the stress, the residual stress is high enough to lead to a lower joint strength.

Simply, diffusion bonding method is a solid-state bonding process, which has been demonstrated as a viable method to overcome the problems encountered in welding. There are many reports on the diffusion bonding C_f/SiC composites to metals. In order to join 3D/2D C_f/SiC composite to Nb alloy, Xiong et al. [41, 42] used Ti-Cu foil as the joining material to join C_f/SiC composite to Nb alloy through a two-stage joining process: solid-phase diffusion bonding and transient liquid-phase diffusion bonding. It was found that the Ti-Cu liquid eutectic alloy was formed by the reaction of Ti and Cu, not only infiltrated into open pores and microcracks as a nail but also reacted with ceramic coating. The remaining Cu was deformed by own plastic deformation and released the residual stress. In addition, the liquid layer formed by interlayer in the TLP-DB process had good wettability to C_f/SiC composite and can infiltrate into

C_f/SiC composite matrix and encapsulated C_f between the interlayer and C_f/SiC interface region. These processes were very beneficial for the mechanical strength of the joint. The shear strength of the joint between 2D C_f/SiC composite and Nb alloy was 14.1 MPa, and the shear strength of the joint between 3D C_f/SiC composite and Nb alloy reached up to 34.1 MPa. To our best knowledge, there were mainly two factors leading to a low shear strength of the joint between 2D C_f/SiC composite and Nb alloy: the CTE mismatch between 2D C_f/SiC composite and Nb alloy was larger compared with 3D C_f/SiC composite and Nb, resulting in a large residual stress, and the fiber direction among 2D C_f/SiC composite was parallel to the joining interface, whereas the fiber direction among 3D C_f/SiC composite was perpendicular to the joining interface. When the fiber was perpendicular to the joining interface, "nail effect" formed between reaction layer and C_f and shared more load than other regions in fracture test (as is shown in Figure 11). These results demonstrated that the direction of fiber was directly related to the interface structure of the joint, which

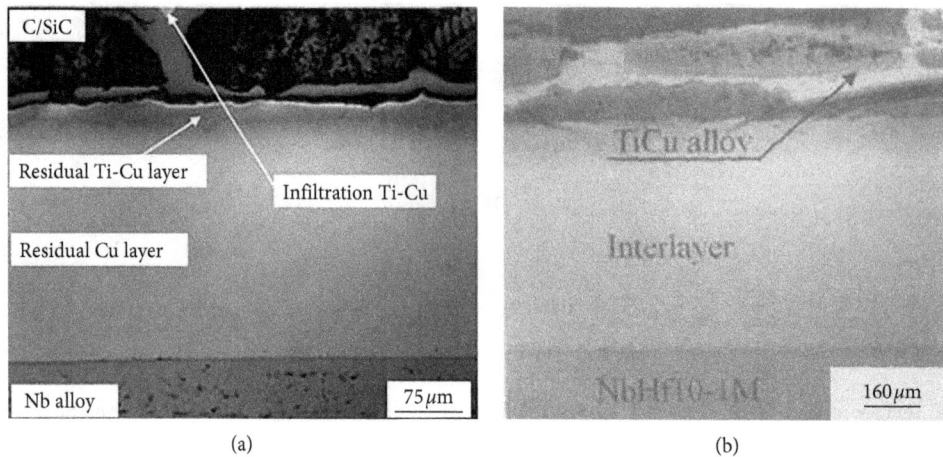

(a)   (b)

FIGURE 11: SEM micrograph of the joint: (a) 3D C$_f$/SiC and (b) 2D C$_f$/SiC [41, 42].

in turn affected the shear strength of joint. In this kind of research work, the influence of fiber must be considered; however, this interesting topic has not yet been studied systematically.

In addition, the reactions between joining material and composite matrix have been recently recognized as critical factors for determining the strength of the joint. C$_f$/SiC composite and Ti-6Al-4V alloy were joined by Ban et al. [79] with the mixed powder of Cu, Ti, and graphite under vacuum environment. In situ synthetic TiC that reduced the thermal stress significantly was synthesized by interdiffusing of C element in the graphite particle and Ti element in the liquid bonding layer (as is shown in Figure 12). The positive effect of TiC on joint strength was also described in other papers [56, 72, 80]. Table 8 summarizes the data of diffusion bonded joining. The utility model has the advantages of high strength, stable joint quality, and good corrosion resistance, especially for the joining of C$_f$/SiC composites and metals for high-temperature and corrosion-resistance application.

3.3. Online Liquid Infiltration Joining.
Online liquid infiltration joining is a novel technology, which is applied to the joining of fiber-reinforced ceramic matrix composites. C$_f$/SiC composites are usually porous both for CVI and PIP processing. An online liquid infiltration joining method that is suitable for the composites was reported. The porosity of C$_f$/SiC composites was controlled and then the compact process was carried out after the joining has finished, which reduced the damnification of joints as much as possible. The wettability between the joining material and C$_f$/SiC composite was improved; moreover, the joining material could be melted and infiltrated into the C$_f$/SiC matrix, which increased the joining area and reinforced the joint strength [81, 82]. In addition, a root-like morphology was formed in C$_f$/SiC composite substrate, which could greatly enhance the reliability of joint [83].

The only paper that attempts to join C$_f$/SiC composite to metal via online liquid infiltration joining was presented in 2004 [84]. The authors joined 2D/3D C$_f$/SiC composites to Nb with Ni-based filler by the online liquid infiltration

joining method (as shown in Figure 13). The joint between 2D C$_f$/SiC composite and Nb was failure and separated during the cooling. However, the favorable joint between 3D C$_f$/SiC composite and Nb was obtained. Approaches such as reactive brazed [68, 69] and diffusion bonding [41, 42] have also been successfully used to join C$_f$/SiC composites to Nb alloy. Unfortunately, the bonding processes above were usually conducted after the preparation of the composite matrix, which damaged the strength of the matrix. Online liquid infiltration joining, which is completed in the preparation process, is different from the above methods. Afterward, chemical vapor deposition (CVD) process not only complete the preparation of materials but also can provide antioxidation coating for the matrix and the joint, reflects the joining, preparation, and processing integration [51, 85].

3.4. Ultrasonic-Assisted Joining.
Ultrasonic-assisted joining is employed to join aluminum alloy structural parts at first. Afterward, ultrasonic is used for copper and alloy, gradually widely used in CMCs and metals, as shown in Figure 14 [86]. Since ultrasound exists as an energy form, it produces some unique ultrasonic effects when it propagates in the medium. The ultrasonic-assisted joining utilizes ultrasonic vibrations to interact the contact area of the CMCs with the metal. The ultrasonic effect causes the liquid joining material to spread on the surface of the matrix and form a joint with the metal [87]. In 1990s, ultrasonic-assisted joining technology facilitated the wetting of materials with poor wetting properties such as ceramics, glass, and stainless steel [88–90, 94]. The liquid-connecting materials spread and moisten, through the ultrasonic wave effect that from the vibrations of ultrasonic, the surface of the CMCs and metal to achieve good connection. Moreover, it is worth mentioning that ultrasonic-assisted joining technology can improve the wettability of connection materials on the surface of matrixes such as ceramics, glass, and stainless steel. Therefore, this technology has been widely applied in many fields.

The joining of SiC and Ti-6Al-4V alloy via ultrasonic-assisted joining was conducted by Chen et al. [91, 92]. SiC

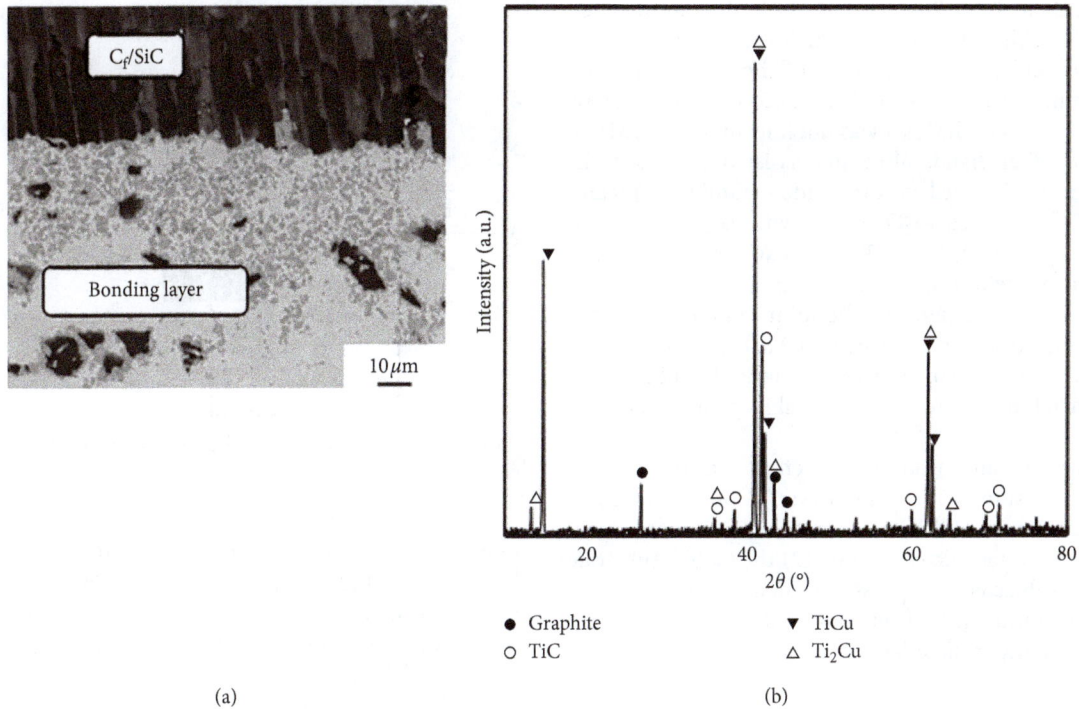

(a)

(b)

FIGURE 12: (a) Micrographs and (b) XRD pattern of the interface [79].

TABLE 8: Diffusion bonding of Cf/SiC composites to metals.

CMCs	Basic information	Metal	Joining material	Process parameter	Shear strength (MPa)	Ref.
Cf/SiC	3D, CVI, 2.1 g/cm³	Nb alloy	Ti-Cu bi-foil	800°C, 30 min, 6 MPa; 1020°C, 60 min, 0.05 MPa, 3.2×10^{-3} Pa	34.1	[41]
Cf/SiC	2D, CVI, 16 vol.%	Nb alloy	Ti-Cu-Cu	850°C, 40 min, 8 MPa; 980°C, 30 min, 0.05 MPa, 3.2×10^{-3} Pa	14.1	[42]
Cf/SiC	3D, 2.0–2.1 g/cm³, 10–15 vol.%, 400 MPa	TC₄ alloy	Cu-Ti-C	900–950°C, 5–30 min, 6.0×10^{-3} Pa	—	[79]
Cf/SiC	3D, 15 vol.%, 500 MPa	Ni alloy	Zr/Ta	1050°C, 10 min, 40.8 MPa, 10^{-2} Pa	110.89 (BS)	[70]

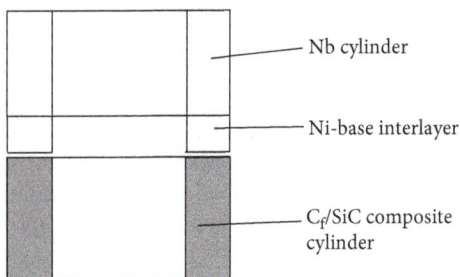

FIGURE 13: The diagram of specimens before being joined.

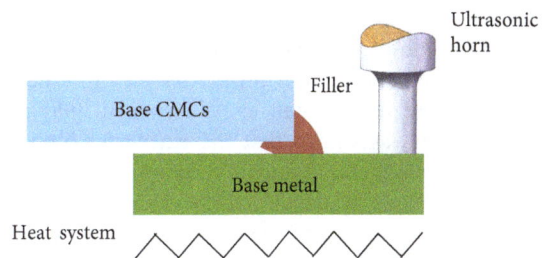

FIGURE 14: The diagram of capillary process that filler gets into clearance by action of ultrasonic vibration.

was employed by a joining material with an Al-12Si alloy at low temperature (620°C), and the shear strength of the joint was 84–94 MPa. In their study, the oxide layer of the matrix was broken by the ultrasonic, and the joining material can form a good interface between SiC and Ti-6Al-4V alloy. However, cracks were observed in SiC material and the

propagation direction was parallel to joint. The main reason was that the nonuniform shrinkage of material at the joint and residual stress, which leads to crack formation in the SiC substrate, was produced during the cooling process. They obtained an integrated joint when using the novel joining material in their study. SiC and Ti-6Al-4V alloy were joined with AlSnSiZnMg mixed metal, which reduced the joining temperature and the residual stress of the joint, inhibiting the occurrence of cracks and other defects. Unfortunately, the shear strength of the joint was not improved (77.8 MPa).

On the other hand, ultrasonic-assisted joining technology can also be used to join oxide ceramics to metals. Naka et al. [88] joined Al_2O_3 to Cu with Zn, Zn-5Al, and $Zn_{100-x}(Al_{0.6} + Cu_{0.4})_x$ ($x = 0$–30) as the joining materials. It was shown that with the time and joining temperature increased, the shear strength of the joint with Zn-6A1-4Cu filler was improved and reached ~62 MPa. In the above literatures, some of them were reported that the ultrasound was beneficial to improve the wettability of Al_2O_3 and metals.

The mechanism of ultrasonic effect on the joining process can be summarized as follows: (1) the macroscopic bubbles between the filled metal and the ceramic were removed by the ultrasonic cavitation; (2) the C_f/SiC substrate surface was subjected to high-speed impact of atoms under ultrasonic vibration; (3) the ultrasonic vibration and friction between the joining material and metal.

3.5. Electric-Assisted Field Joining. Although diffusion bonding is widely used to join CMCs and metals, generally, it requires high temperatures, high pressure, vacuum or inert atmosphere, and long joining time [93, 94, 98]. Electric-assisted field joining is an effective way to solve these above problems, as shown in Figure 15. Since the joining between CMCs and metal was realized by chemical reaction, interfacial structure formed by reaction determines the mechanical properties of the joint. Better joint can be obtained using the electric-assisted field method.

The interface between CMCs and metal were polarized under electrostatic field. On the one hand, it promotes atomic migration and vacancy diffusion. On the other hand, it accelerates the interface reaction, which reduces the joining temperature, the pressure, and the residual stress. Moreover, the interface reaction is easy to control, and joining time is very short [96–98, 101].

Initially, the electric-assisted field joining is mainly employed for joining ceramics to metals [100]. The interface composition and mechanical properties of joints between SiC and Ti were investigated by Wang et al. [98] in the electric field. It was shown that the external electric field reduced the joining temperature and time and improved the shear strength. It is important that the external electric field can improve the diffusion rate of interface atoms. Moreover, it promoted the interface reaction and improved the joining efficiency.

Owing to its simplicity and efficiency, electric-assisted field joining became a useful method employed for joining C_f/C composites [101] and C_f/SiC composites [48]. C_f/C composites were firstly joined by combining electric field-assisted sintering technology and using a Ti_3SiC_2 tape film as the interlayer [101]. In their work, the interdiffusion speed between the interlayer and the metal was accelerated by an electric field and the joining time was only 12 min. To our

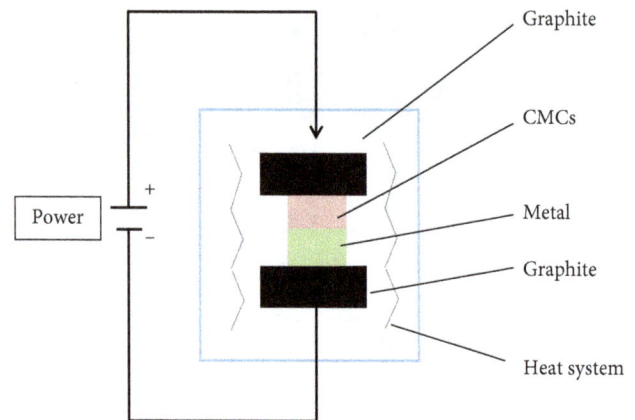

FIGURE 15: The diagram of the experimental equipment under electric field.

knowledge, the Ti_3SiC_2 exhibited pseudoplastic at 1300°C or higher [102, 103]. Therefore, Ti_3SiC_2 infiltrated into the composite matrix and a "nail" that clamps the matrix was observed, as shown in Figure 16, which improved shear strength of the joint. In the joining process, two key factors affected the strength of the joint: (1) the interdiffusion between the joining material and the matrix was promoted by electric field and (2) Ti_3SiC_2 showed good plastic deformation ability in the electric field.

Atoms spread to the interface under electric field. It is necessary to pass through the potential through the gap, the original position occupied by their own formed a new space. The energy of the atoms across the barrier was provided by the electric field. At the same time, the atoms across the barrier potential energy are reduced by the electric field. Combined with these two effects, the diffusion activation of atoms can be greatly reduced, thereby increasing the diffusion rate of solute atoms and obtaining a uniform structure [104]. Therefore, the electric-assisted fields joining method has drawn great attention and is expected to become an important way for the joining of C_f/SiC composites in the future.

4. Summary

With the rapid development of high-tech in aerospace and other industry fields, carbon fiber-reinforced silicon carbide (C_f/SiC) ceramic matrix composites, one of the most famous CMCs, are becoming the most promising candidates for high-temperature structural applications. In most cases, typically, it is very difficult to produce large-size C_f/SiC composite components with complex shapes, resulting in that they must be joined with themselves or other materials by appropriate joining technologies. At present, various joining technologies of C_f/SiC composites are reported, including the joining of C_f/SiC-C_f/SiC and C_f/SiC-metal, such as direct bonding of C_f/SiC-C_f/SiC, indirect bonding of C_f/SiC-C_f/SiC, brazing of C_f/SiC-metal, diffusion bonding of C_f/SiC-metal, online liquid infiltration of C_f/SiC-metal, ultrasonic-assisted joining of C_f/SiC-metal, and electric-assisted joining of C_f/SiC-metal.

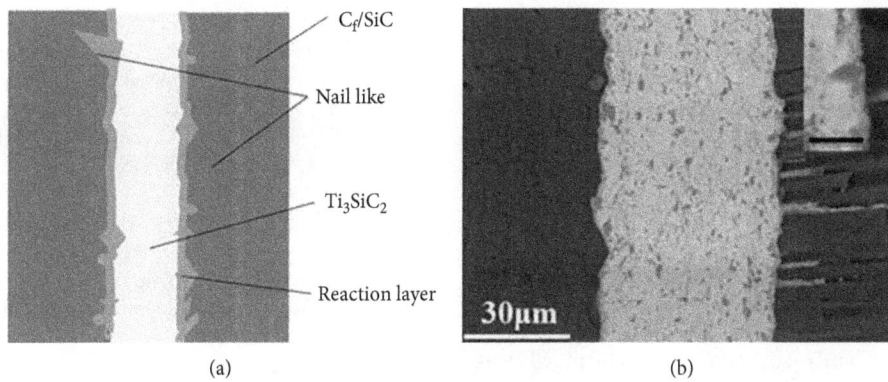

FigURE 16: (a) The schematic diagram of "nail" like and (b) BSE image [101].

To the best knowledge of the authors, however, there has been no systematic summary of the joining of C_f/SiC composites. In this paper, a systematic review of the joining of C_f/SiC composites is conducted, and the aim of this paper is to provide some reference for researchers working on this field.

Conflicts of Interest

The authors declare that they have no conflicts of interest.

Acknowledgments

The authors sincerely thank the Young Elite Scientist Sponsorship (YESS) Program by CAST (No. 2015QNRC001) and the Beijing Institute of Technology Research Fund Program for Young Scholars.

References

[1] A. Luna Ramírez, J. Porcayo-Calderon, Z. Mazur et al., "Microstructural changes during high temperature service of a cobalt-based superalloy first stage nozzle," *Advances in Materials Science and Engineering*, vol. 2016, pp. 1–7, 2016.

[2] A. Baldan, "Progress in Ostwald ripening theories and their applications to nickel-base superalloys. Part I: Ostwald ripening theories," *Journal of Materials Science*, vol. 37, no. 11, pp. 2171–2202, 2002.

[3] M. V. Acharya and G. E. Fuchs, "The effect of long-term thermal exposures on the microstructure and properties of CMSX-10 single crystal Ni-base superalloys," *Materials Science and Engineering: A*, vol. 381, no. 1-2, pp. 143–153, 2004.

[4] B. S. Sung and Y. H. Yun, "SiC conversion coating prepared from silica-graphite reaction," *Advances in Materials Science and Engineering*, vol. 2017, no. 5, pp. 1–8, 2017.

[5] N. S. Jacobson, K. N. Lee, and D. S. Fox, "Reactions of silicon carbide and silicon(IV) oxide at elevated temperatures," *Journal of the American Ceramic Society*, vol. 75, no. 6, pp. 1603–1611, 1992.

[6] S. V. Ushakov and A. Navrotsky, "Experimental approaches to the thermodynamics of ceramics above 1500°C," *Journal of the American Ceramic Society*, vol. 95, no. 5, pp. 1463–1482, 2012.

[7] J. Llorca and R. N. Singh, "Influence of fiber and interfacial properties on fracture behavior of fiber-reinforced ceramic composites," *Journal of the American Ceramic Society*, vol. 74, no. 11, pp. 2882–2890, 1991.

[8] Q. Xie and S. N. Wosu, "Dynamic fracture toughness of TaC/CNTs/SiC CMCs prepared by spark plasma sintering," *Advances in Materials Science and Engineering*, vol. 2015, no. 2, pp. 1–8, 2015.

[9] C. J. Liu, W. F. Ding, T. Y. Yu, and C. Y. Yang, "Materials removal mechanism in high-speed grinding of particulate reinforced titanium matrix composites," *Precision Engineering*, vol. 51, pp. 68–77, 2018.

[10] X. Ma, S. A. Chen, M. Mei et al., "Microstructure and mechanical behaviors of T700 carbon fiber reinforced C/SiC composites via precursor infiltration and pyrolysis," *Materials Science and Engineering: A*, vol. 666, pp. 238–244, 2016.

[11] S. Schmidt, S. Beyer, H. Knabe et al., "Advanced ceramic matrix composite materials for current and future propulsion technology applications," *Acta Astronautica*, vol. 55, no. 3, pp. 409–420, 2004.

[12] L. Wang, F. Hou, X. Wang, J. C. Liu, and A. Guo, "Preparation and mechanical properties of continuous carbon nanotube networks modified C_f/SiC composite," *Advances in Materials Science and Engineering*, vol. 2015, no. 12, pp. 1–7, 2015.

[13] M. Z. Berbon, D. R. Dietrich, and D. B. Marshall, "Transverse thermal conductivity of thin C/SiC composites fabricated by slurry infiltration and pyrolysis," *Journal of the American Ceramic Society*, vol. 84, no. 10, pp. 2229–2234, 2001.

[14] W. Yang, L. Zhang, L. Cheng et al., "Oxidation behavior of C_f/SiC composite with CVD SiC-B_4C coating in a wet oxygen environment," *Applied Composite Materials*, vol. 16, no. 2, pp. 83–92, 2009.

[15] G. Boitier, J. L. Chermant, and J. Vicens, "Bridging at the nanometric scale in 2.5D C_f-SiC composites," *Applied Composite Materials*, vol. 6, no. 5, pp. 279–287, 1999.

[16] P. Tatarko, Z. Chlup, A. Mahajan et al., "High temperature properties of the monolithic CVD-SiC materials joined with a pre-sintered MAX phase Ti_3SiC_2 interlayer via solid-state diffusion bonding," *Journal of the European Ceramic Society*, vol. 37, no. 4, pp. 1205–1216, 2017.

[17] G. B. Zheng, H. Sano, Y. Uchiyama, K. Kobayashi, and H. M. Cheng, "The properties of carbon fibre/SiC composites fabricated through impregnation and pyrolysis of polycarbosilane," *Journal of Materials Science*, vol. 34, no. 4, pp. 827–834, 1999.

[18] K. Jian, Z. H. Chen, Q. S. Ma, and W. W. Zheng, "Effects of pyrolysis processes on the microstructures and mechanical properties of C_f/SiC composites using polycarbosilane,"

Materials Science and Engineering: A, vol. 390, no. 1, pp. 154–158, 2005.

[19] K. Jian, Z. H. Chen, Q. S. Ma, H. F. Hu, and W. W. Zheng, "Processing and properties of 2D-C$_f$/SiC composites incorporating SiC fillers," *Materials Science and Engineering: A*, vol. 408, no. 1, pp. 330–335, 2005.

[20] C. C. Zhou, C. R. Zhang, H. F. Hu et al., "Preparation of 3D-C$_f$/SiC composites at low temperatures," *Materials Science and Engineering: A*, vol. 488, no. 1, pp. 569–572, 2008.

[21] M. Broda, A. Pyzalla, and W. Reimers, "X-ray analysis of residual stresses in C/SiC composites," *Applied Composite Materials*, vol. 6, no. 1, pp. 51–66, 1999.

[22] J.-C. Bae, K.-Y. Cho, D.-H. Yoon et al., "Highly efficient densification of carbon fiber-reinforced SiC-matrix composites by melting infiltration and pyrolysis using polycarbosilane," *Ceramics International*, vol. 39, no. 5, pp. 5623–5629, 2013.

[23] Z. S. Rak, "A process for C$_f$/SiC composites using liquid polymer infiltration," *Journal of the American Ceramic Society*, vol. 84, no. 10, pp. 2235–2239, 2001.

[24] L. B. Li, "Comparison of fatigue life between C/SiC and SiC/SiC ceramic-matrix composites at room and elevated temperatures," *Applied Composite Materials*, vol. 23, no. 5, pp. 913–952, 2016.

[25] G. V. Samsonov and B. A. Kovenskaya, "The nature of the chemical bond in borides," in *Boron and Refractory Borides*, pp. 19–30, Springer-Verlag, Berlin, Germany, 1977.

[26] S. Suyama, T. Kameda, and Y. Itoh, "Development of high-strength reaction-sintered silicon carbide," *Diamond and Related Materials*, vol. 12, no. 3, pp. 1201–1204, 2003.

[27] Y. Z. Zhu, Z. R. Huang, and S. M. Dong, "Manufacturing 2D carbon-fiber-reinforced SiC matrix composites by slurry infiltration and PIP process," *Ceramics International*, vol. 34, no. 5, pp. 1201–1205, 2008.

[28] Q. Zhou, S. M. Dong, X. Y. Zhang, and D. Jiang, "Fabrication of C$_f$/SiC composites by vapor silicon infiltration," *Journal of the American Ceramic Society*, vol. 89, no. 7, pp. 2338–2340, 2006.

[29] M. N. Saleh, Y. Wang, A. Yudhanto et al., "Investigating the potential of using off-axis 3D woven composites in composite joints' applications," *Applied Composite Materials*, vol. 24, no. 2, pp. 377–396, 2017.

[30] D. Y. Fan, J. H. Huang, X. P. Zhao et al., "Joining of C$_f$/SiC composite to Ti-6Al-4V with (Ti-Zr-Cu-Ni)+Ti filler based on in-situ alloying concept," *Ceramics International*, vol. 43, no. 5, pp. 4151–4158, 2017.

[31] Y. Katoh, L. L. Snead, T. Cheng et al., "Radiation-tolerant joining technologies for silicon carbide ceramics and composites," *Journal of Nuclear Materials*, vol. 448, no. 1, pp. 497–511, 2014.

[32] G. D. Li, Y. D. Zhang, C. R. Zhang et al., "Design, preparation and properties of online-joints of C/SiC-C/SiC with pins," *Composites Part B: Engineering*, vol. 48, pp. 134–139, 2013.

[33] Z. B. He, L. T. Zhang, Y. Zhang et al., "Microstructural characterization and failure analysis of 2D C/SiC two-layer beam with pin-bonded hybrid joints," *International Journal of Adhesion and Adhesives*, vol. 57, pp. 70–78, 2015.

[34] J. A. Fernie, *Joining Ceramic Materials*, American Ceramic Society, Westerville, OH, USA, 1997.

[35] X. Y. Wang, Y. Li, S. Z. Wei, and X. D. Ma, "Research progress in connecting techniques of ceramics and metals," *Casting Forging Welding*, vol. 38, no. 13, pp. 144–148, 2009.

[36] P. Tatarko, V. Casalegno, C. F. Hu et al., "Joining of CVD-SiC coated and uncoated fibre reinforced ceramic matrix composites with pre-sintered Ti$_3$SiC$_2$ MAX phase using spark plasma sintering," *Journal of the European Ceramic Society*, vol. 36, no. 16, pp. 3957–3967, 2016.

[37] R. H. Jones, L. Giancarli, A. Hasegawa et al., "Promise and challenges of SiC$_f$/SiC composites for fusion energy applications," *Journal of Nuclear Materials*, vol. 307, pp. 1057–1072, 2002.

[38] G. B. Lin, J. H. Huang, H. Zhang, and H. Y. Liu, "Microstructure and mechanical performance of brazed joints of Cf/SiC composite and Tialloy using Ag-Cu-Ti-W," *Science and Technology of Welding and Joining*, vol. 11, no. 4, pp. 379–383, 2006.

[39] J. K. Li, L. Liu, Y. T. Wu et al., "Microstructure of high temperature Ti-based brazing alloys and wettability on SiC ceramic," *Materials and Design*, vol. 30, no. 2, pp. 275–279, 2009.

[40] G. B. Lin, J. H. Huang, and H. Zhang, "Joints of carbon fiber-reinforced SiC composites to Ti-alloy brazed by Ag-Cu-Ti short carbon fibers," *Journal of Materials Processing Technology*, vol. 189, no. 1, pp. 256–261, 2007.

[41] J. T. Xiong, J. L. Li, F. S. Zhang, and W. Huang, "Joining of 3D C/SiC composites to niobium alloy," *Scripta Materialia*, vol. 55, no. 2, pp. 151–154, 2006.

[42] J. T. Xiong, J. L. Li, F. S. Zhang et al., "Joining of 2D C/SiC composites with niobium alloy," *Journal of Inorganic Materials*, vol. 21, no. 6, pp. 1391–1396, 2006.

[43] Q. Zhang, L. B. Sun, Q. Y. Liu, G. Wang, and Y. Xuan, "Effect of brazing parameters on microstructure and mechanical properties of Cf/SiC and Nb-1Zr joints brazed with Ti-Co-Nb filler alloy," *Journal of the European Ceramic Society*, vol. 37, no. 3, pp. 931–937, 2017.

[44] S. J. Li, Y. Zhou, and H. P. Duan, "Joining of SiC ceramic to Ni-based superalloy with functionally gradient material fillers and a tungsten intermediate layer," *Journal of Materials Science*, vol. 38, no. 19, pp. 4065–4070, 2003.

[45] C. Jiménez, C. Wilhelmi, and T. Speliotis, "Joining of C/SiC Ceramics to Nimonic Alloys," *Journal of Materials Engineering and Performance*, vol. 21, no. 5, pp. 683–689, 2012.

[46] B. Cui, J. H. Huang, C. Cai, S. Chen, and X. Zhao, "Microstructures and mechanical properties of C$_f$/SiC composite and TC$_4$ alloy joints brazed with (Ti-Zr-Cu-Ni)+W composite filler materials," *Composites Science and Technology*, vol. 97, pp. 19–26, 2014.

[47] Y. Liu, Z. R. Huang, and X. J. Liu, "Joining of sintered silicon carbide using ternary Ag-Cu-Ti active brazing alloy," *Ceramics International*, vol. 35, no. 8, pp. 3479–3484, 2009.

[48] S. Rizzo, S. Grasso, M. Salvo et al., "Joining of C/SiC composites by spark plasma sintering technique," *Journal of the European Ceramic Society*, vol. 34, no. 4, pp. 903–913, 2014.

[49] B. Chen, H. P. Xiong, Y. Y. Cheng et al., "Microstructure and strength of Cf/SiC joints with Ag-Cu-Ti brazing fillers," *Journal Materials Engineering*, vol. 329, pp. 27–31, 2010.

[50] D. Y. Fan, J. H. Huang, X. W. Sun et al., "Correlation between microstructure and mechanical properties of active brazed C$_f$/SiC composite joints using Ti-Zr-Be," *Materials Science Engineering: A*, vol. 667, pp. 332–339, 2016.

[51] Q. Y. Tong, L. F. Cheng, and L. T. Zhang, "Microstructure and properties of joints of 2D C/SiC composites," *Materials Engineering*, vol. 11, pp. 14–16, 2002.

[52] Q. Y. Tong, L. F. Cheng, and L. T. Zhang, "On-line joining of 3D fiber reinforced C/SiC composites," *Rare Metal Materials and Engineering*, vol. 33, no. 1, pp. 101–104, 2004.

[53] Q. Y. Tong and L. F. Cheng, "Liquid infiltration joining of 2D C/SiC composite," *Science Engineering Composites Materials*, vol. 13, no. 1, pp. 31–36, 2006.

[54] Y. Katoh, M. Kotani, A. Kohyama et al., "Microstructure and mechanical properties of low-activation glass-ceramic joining and coating for SiC/SiC composites," *Journal of Nuclear Materials*, vol. 283, pp. 1262–1266, 2000.

[55] H. Y. Dong, S. J. Li, Y. Y. Teng, and W. Ma, "Joining of SiC ceramic-based materials with ternary carbide Ti$_3$SiC$_2$," *Materials Science and Engineering: B*, vol. 176, no. 1, pp. 60–64, 2011.

[56] X. B. Zhou, Y. H. Han, X. F. Shen et al., "Fast joining SiC ceramics with Ti$_3$SiC$_2$ tape film by electric field-assisted sintering Technol," *Journal of Nuclear Materials*, vol. 466, pp. 322–327, 2015.

[57] J. Suo, Z. H. Chen, W. M. Han, and W. Zheng, "Joining of ceramic materials by ceramic bonding transformed from silicone resin at high temperature," *Journal of The Chinese Ceramic Society*, vol. 33, no. 3, pp. 386–390, 2005.

[58] P. K. Gianchandani, V. Casalegno, F. Smeacetto, and M. Ferraris, "Pressure-less joining of C/SiC and SiC/SiC by a MoSi$_2$/Si composite," *International Journal of Applied Ceramic Technology*, vol. 14, no. 3, pp. 305–312, 2017.

[59] H. P. Xiong, B. Chen, Y. Pan, H. S. Zhao, and L. Ye, "Joining of C$_f$/SiC composite with a Cu-Au-Pd-V brazing filler and interfacial reactions," *Journal of the European Ceramic Society*, vol. 34, no. 6, pp. 1481–1486, 2014.

[60] H. P. Xiong, B. Chen, W. Mao, and X. H. Li, "Joining of C$_f$/SiC composite with Pd-Co-V brazing filler," *Welding in the World*, vol. 56, no. 1-2, pp. 76–80, 2012.

[61] H. P. Xiong, B. Chen, Y. Pan, W. Mao, and Y. Y. Cheng, "Interfacial reactions and joining characteristics of a Cu-Pd-V system filler alloy with C$_f$/SiC composite," *Ceramics International*, vol. 40, no. 6, pp. 7857–7863, 2014.

[62] E. S. Karakozov, G. V. Konyushkov, and R. A. Musin, "Fundamentals of welding metals to ceramic materials," *Welding International*, vol. 7, no. 12, pp. 991–996, 1993.

[63] M. W. Barsoum, T. El-Raghy, C. J. Rawn et al., "Thermal properties of Ti$_3$SiC$_2$," *Journal of Physics and Chemistry of Solids*, vol. 60, no. 4, pp. 429–439, 1999.

[64] H. L. Liu, C. Y. Tian, and M. Z. Wu, "Technique of joining of Cf/SiC composite via preceramic silicone polysilazane and joining properties," *Chinese Journal of Nonferrous Metals*, vol. 18, no. 2, pp. 278–281, 2008.

[65] B. Zhao, T. Y. Yu, W. F. Ding, and X. Y. Li, "Effects of pore structure and distribution on strength of porous Cu-Sn-Ti alumina composites," *Chinese Journal of Aeronautics*, vol. 30, no. 6, pp. 2004–2015, 2017.

[66] W. F. Ding, C. W. Dai, Y. Tian, J. H. Xu, and Y. C. Fu, "Grinding performance of textured monolayer CBN wheels: undeformed chip thickness nonuniformity modeling and ground surface topography prediction," *International Journal of Machine Tools and Manufacture*, vol. 122, pp. 66–80, 2017.

[67] Y. J. Zhu, W. F. Ding, T. Y. Yu et al., "Investigation on stress distribution and wear behavior of brazed polycrystalline cubic boron nitride superabrasive grains: numerical simulation and experimental study," *Wear*, vol. 376-377, pp. 1234–1244, 2017.

[68] Y. J. Lu, X. Y. Zhang, J. X. Chu, X. Liu, and Z. Fang, "Study on active reactive brazing of C SiC ceramic to Nb alloy," *Chinese Journal of Rare Metals*, vol. 32, no. 5, pp. 636–640, 2008.

[69] C. Y. Liang, Y. G. Du, W. J. Zhang et al., "Joining of C$_f$/SiC composites with Niobium alloy," *Aerospace Materials and Technology*, vol. 39, no. 3, pp. 45–48, 2009.

[70] J. J. Zhang, S. J. Li, H. P. Duan, and Y. Zhang, "Joining of C$_f$/SiC to Ni-based superalloy with Zr/Ta composite interlayers by hot-pressing diffusion welding," *Rare Metal Materials and Engineering*, vol. 31, no. s1, pp. 393–396, 2002.

[71] Z. W. Yang, L. X. Zhang, X. Y. Tian et al., "Interfacial microstructure and mechanical properties of TiAl and C$_f$/SiC joint brazed with TiH$_2$-Ni-B brazing powder," *Materials characterization*, vol. 79, pp. 52–59, 2013.

[72] Z. W. Yang, P. He, and J. C. Feng, "Microstructural evolution and mechanical properties of the joint of TiAl alloys and C/SiC composites vacuum brazed with Ag-Cu filler metal," *Materials Characterization*, vol. 62, no. 9, pp. 825–832, 2011.

[73] Z. H. Zhong, T. Hinoki, H.-C. Jung, Y. H. Park, and A. Kohyama, "Microstructure and mechanical properties of diffusion bonded SiC/steel joint using W/Ni interlayer," *Materials and Design*, vol. 31, no. 3, pp. 1070–1076, 2010.

[74] Y. J. Li, *Selection of Welding Materials*, Chemical Industry Press, Beijing, China, 2004.

[75] B. Riccardi, C. A. Nannetti, T. Petrisor, and M. Sacchetti, "Low activation brazing materials and techniques for SiC$_f$-SiC composites," *Journal of Nuclear Materials*, vol. 307, pp. 1237–1241, 2002.

[76] H. J. Liu, J. C. Feng, and Y. Y. Qian, "Microstructure and strength of the SiC/TiAl joint brazed with Ag-Cu-Ti filler metal," *Journal of Materials Science Letters*, vol. 19, no. 14, pp. 1241-1242, 2000.

[77] M. Singh, R. Asthana, and T. P. Shpargel, "Brazing of ceramic-matrix composites to Ti and Hastealloy using Ni-base metallic glass interlayers," *Materials Science and Engineering: A*, vol. 498, no. 1, pp. 19–30, 2008.

[78] Y. M. He, J. Zhang, X. Wang, and Y. Sun, "Effect of brazing temperature on microstructure and mechanical properties of Si$_3$N$_4$/Si$_3$N$_4$ joints brazed with Ag-Cu-Ti +Mo composite filler," *Journal of Materials Science*, vol. 8, pp. 2796–2804, 2010.

[79] Y. H. Ban, J. H. Huang, H. Zhang et al., "Microstructure of reactive composite brazing joints of C$_f$/SiC composite to Ti-6Al-4V alloy with Cu-Ti-C filler material," *Rare Metal Materials and Engineering*, vol. 38, no. 4, pp. 713–716, 2009.

[80] X. R. Song, H. J. Li, V. Casalegno et al., "Microstructure and mechanical properties of C/C composite/Ti6Al4V joints with a Cu/TiCuZrNi composite brazing alloy," *Ceramics International*, vol. 42, no. 5, pp. 6347–6354, 2016.

[81] Q. Y. Tong, *Microstructure and Properties of the On-Line Liquid Infiltrate Joining of C$_f$/SiC Dissertation for Ph.D. thesis*, Northwestern Polytechnical University, Xi'an, China, 2003.

[82] Q. Q. Ke, L. F. Cheng, Q. Y. Tong et al., "Joining methods for continuous fiber reinforced ceramic matrix composites," *Materials Engineering*, vol. 11, pp. 58–63, 2005.

[83] J. Wang, K. Z. Li, W. Li et al., "The preparation and mechanical properties of carbon/carbon composite joints using Ti-Si-SiC-C filler as interlayer," *Materials Science and Engineering: A*, vol. 574, pp. 37–45, 2013.

[84] Q. Y. Tong, L. F. Cheng, and L. T. Zhang, "Liquid infiltration joining of C/SiC and Nb," *Journal of Aeronautical Materials*, vol. 24, no. 1, pp. 54–56, 2004.

[85] Q. Q. Ke, L. F. Chen, Q. Y. Tong et al., "Microstructure and properties of joints of 2D C/SiC composites by riveting," *Rare Metal Materials and Engineering*, vol. 35, no. 9, p. 1497, 2006.

[86] V. L. Lanin, "Ultrasonic soldering in electronics," *Ultrasonics Sonochemistry*, vol. 8, no. 4, pp. 379–385, 2001.

[87] W. W. Zhao, J. C. Yan, W. Yang, and S. Q. Yang, "Capillary filling process during ultrasonically brazing of aluminium matrix composites," *Science and Technology of Welding and Joining*, vol. 13, no. 1, pp. 66–69, 2008.

[88] M. Naka and M. Maeda, "Application of ultrasound on joining of ceramics to metals," *Engineering Fracture Mechanics*, vol. 40, no. 4-5, pp. 951–956, 1991.

[89] M. H. El-Sayed and M. Naka, "Structure and properties of carbon steel-aluminium dissimilar joints," *Science and Technology of Welding and Joining*, vol. 10, no. 1, pp. 27–31, 2005.

[90] M. H. El-Sayed, K. M. Hafez, and M. Naka, "Interfacial structure and bond strength of ultrasonic brazed Al-304 stainless steel dissimilar joints," *Science and Technology of Welding and Joining*, vol. 9, no. 6, pp. 560–564, 2004.

[91] X. G. Chen, J. C. Yan, S. C. Ren, J Wei, and Q. Wang, "Ultrasonic-assisted brazing of SiC ceramic to Ti-6Al-4V alloy using a novel AlSnSiZnMg filler metal," *Materials Letters*, vol. 105, pp. 120–123, 2013.

[92] X. G. Chen, R. S. Xie, Z. W. Lai et al., "Interfacial structure and formation mechanism of ultrasonic-assisted brazed joint of SiC ceramics with Al 12Si filler metals in air," *Journal of Materials Science and Technology*, vol. 33, no. 5, pp. 492–498, 2017.

[93] M. C. Halbig, R. Asthana, and M. Singh, "Diffusion bonding of SiC fiber-bonded ceramics using Ti/Mo and Ti/Cu interlayers," *Ceramics International*, vol. 41, no. 2, pp. 2140–2149, 2015.

[94] J. K. Liu, J. Cao, X. G. Song, Y. Wang, and J. C. Feng, "Evaluation on diffusion bonded joints of TiAl alloy to Ti_3SiC_2 ceramic with and without Ni interlayer: interfacial microstructure and mechanical properties," *Materials and Design*, vol. 57, pp. 592–597, 2014.

[95] J. Cao, J. K. Liu, X. G. Song, X. Lin, and J. C. Feng, "Diffusion bonding of TiAl intermetallic and Ti_3AlC_2 ceramic: Interfacial microstructure and joining properties," *Materials and Design*, vol. 56, pp. 115–121, 2014.

[96] J. S. Varsanik and J. J. Bernstein, "Voltage-assisted polymer wafer bonding," *Journal of Micromechanics and Microengineering*, vol. 22, no. 2, article 025004, 2012.

[97] C. R. Liu, J. F. Zhao, X. Y. Lu et al., "Field-assisted diffusion bonding and bond characterization of glass to aluminum," *Journal of Materials Science*, vol. 43, no. 15, pp. 5076–5082, 2008.

[98] Q. Wang, Q. H. Li, D. L. Sun, X. Han, and Q. Tian, "Microstructure and mechanical properties of SiC/Ti diffusion bonding joints under electric field," *Rare Metal and Materials*, vol. 45, no. 7, pp. 1749–1754, 2016.

[99] T. Okuni, Y. Miyamoto, H. Abe, and M. Naito, "Joining of silicon carbide and graphite by spark plasma sintering," *Ceramics International*, vol. 40, no. 1, pp. 1359–1363, 2014.

[100] R. Pan, Q. Wang, D. L. Sun, and P. He, "Effects of electric field on interfacial microstructure and shear strength of diffusion bonded α-Al_2O_3/Ti joints," *Journal of the European Ceramic Society*, vol. 35, no. 1, pp. 219–226, 2015.

[101] X. B. Zhou, H. Yang, F. Y. Chen et al., "Joining of carbon fiber reinforced carbon composites with Ti_3SiC_2 tape film by electric field assisted sintering technique," *Carbon*, vol. 102, pp. 106–115, 2016.

[102] T. El-Raghy, M. W. Barsoum, A. Zavaliangos, and S. R. Kalidindi, "Processing and mechanical properties of Ti_3SiC_2: II, effect of grain size and deformation temperature," *Journal of the American Ceramic Society*, vol. 82, no. 10, pp. 2855–2860, 1999.

[103] X. M. Fan, X. W. Yin, Y. Z. Ma, L. Zhang, and L. Cheng, "Oxidation behavior of C/SiC-Ti_3SiC_2 at 800–1300°C in air," *Journal of the European Ceramic Society*, vol. 36, pp. 2427–2433, 2016.

[104] D. Yu, L. Y. Cao, and X. L. Dong, "Effects of pulse field on solution microstructure and solution technique of Al-4% Cu alloy," *Material and Heart Treatment*, vol. 39, no. 2, pp. 144–146, 2010.

23

Protective Ceramic Coatings for Solid Oxide Fuel Cell (SOFC) Balance-of-Plant Components

Raymond L. Winter,[1] **Prabhakar Singh,**[2] **Mark K. King Jr.,**[3] **Manoj K. Mahapatra,**[3] **and Uma Sampathkumaran**[1]

[1]*InnoSense LLC, Torrance 90505, USA*
[2]*Department of Materials Science & Engineering, University of Connecticut, Storrs, CT, USA*
[3]*Department of Materials Science and Engineering, University of Alabama at Birmingham, Birmingham, Alabama 35294, USA*

Correspondence should be addressed to Uma Sampathkumaran; uma.sampathkumaran@innosensellc.com

Academic Editor: Meilin Liu

Solid oxide fuel cells (SOFCs) have the potential to meet the growing need for electrical power generation if the cost per megawatt can be further reduced. Currently, SOFC stacks are replaced too frequently to be cost competitive. SOFC service life can be extended by preventing chromium- (Cr-) bearing species from evaporating from the interior surfaces of balance of plant (BOP) components and poisoning the cathode to increase the lifetime. We have developed yttria-stabilized zirconia (YSZ) and aluminum oxide- (Al_2O_3-) modified sol-gel paints or inks for coating BOP components. 430 stainless steel (430SS) substrates with three surface conditions were coated with the 0.8–1.5 μm thick YSZ and Al_2O_3 paints. The coated 430SS samples were tested for thermal cycling resistance, thermal soak, and Cr evaporation. Thermal soak and thermal cycling test results show promise for the YSZ-coated 430SS substrates. The Cr evaporation test of a coated substrate showed a 51% reduction in Cr generation, when compared with a bare substrate.

1. Introduction

Fossil fuel-based power generation systems are the mainstay of electrical power production in many countries today. These systems, however, contribute to increase in airborne pollutants carbon dioxide (CO_2) emission level leading to potential climate change and global warming. Generally, peak electrical power is met by small- and medium-sized present gas-fired electrical power plants. These plants produce 85% of the air pollution generated by electrical power generation in the USA, along with relatively low fuel consumption efficiency. One of the Department of Energy's (DOE's) goals is to integrate solid oxide fuel cell (SOFC) technologies as an alternative for clean energy systems that can feed into the nation's power grid. Currently, one of the constraints limiting the operational life of a SOFC plant is the poisoning of electrodes from the chromium- (Cr-) bearing volatile materials from stainless steel balance of plant (BOP) components. Various chemical

reactions can take place at a lanthanum strontium manganite (LSM) cathode, due to Cr poisoning. In all cases, the integrity of the LSM is compromised. The result is a change in the chemistry and therefore the electrochemistry of the cathode, rendering it incapable of performing its proper electrochemical function in the SOFC. The extended power plant life is key to making power grid SOFCs a cost-effective reality. The target is a 10-year service life.

Even though perovskite coatings like LSC ($LaScCrO_3$), LSM ($LaSrMnO_3$), and LSCF ($LaSrCoFeO_3$) decrease Cr evaporation [1, 2], they suffer from spallation and difficulty in adhesion of the coatings [3]. Spinel coatings based on cobalt and its oxide combined with other transition metals like MCO ($Mn_{1.5}Co_{1.5}O_4$) have been used [4]. The spinel coatings reduced the oxygen diffusion inward and thereby limited the scale growth under the coatings. Froitzheim and Svensson combined a cerium coating with a cobalt coating to reduce Cr evaporation [5].

To address this need, we developed protective coatings for BOP components. Coatings of refractory ceramics such as yttria-stabilized zirconia (YSZ) and aluminum oxide (Al_2O_3) in the form of a low-cost modified sol-gel formulation can be applied by commercially viable processes such as dip coating or conventional spraying.

The coating not only serves as a thermal barrier but also prevents Cr evaporation. YSZ coatings can be synthesized by using various techniques like electrochemical deposition (EVD) [6], plasma spray, radio frequency (RF) sputtering [7], spray pyrolysis [8], atomic layer deposition (ALD) [9], and sol-gel. A sol-gel dip coating technique was used for coating the 430SS (stainless steel) to form an oxide film. The dip coating is a low-cost coating technique and attractive alternative for large-scale manufacturing applications.

The dipping approach holds promise in specific applications:

(i) Coating small tubes in which solution precursor plasma spray coating is difficult/problematic.

(ii) Where ALD coating is prohibitive for large BOP components.

(iii) A flash fire technique for larger conduit tubes can be applied by either dipping or conventionally spraying followed by infrared (IR) lamp heating inside the tube, for low-cost application.

(iv) Solution precursor plasma spray coats one side of the substrate at a time, so surface oxidation may occur on the opposite side during the initial coating. Dipping can coat both sides simultaneously.

The barrier coating would isolate the BOP component surface from the hot humid gases within the SOFC facility and eliminate or minimize the Cr species evolution at operating temperatures of 700–900°C that can poison SOFC cathodes and degrade electrochemical performance. The engineering goal was to achieve improved stability for >40,000 hours, with an ultimate 88,000-hour or 10-year life goal. Key R&D for advanced energy systems for future coal-based power plants includes developments in protective coatings for SOFC BOP components. SOFC materials must ensure safe, reliable, and continuous operation for 10 years in harsh operating conditions (high temperatures and aggressive cell environments) to fully harness this clean energy source.

2. Materials and Methods

2.1. Materials. Zirconium (Zr) *n*-propoxide, yttrium (Y)-organics, alpha terpineol, ethylene glycol monobutyl ether acetate (BCA), ethyl acetone, diacetone alcohol, ethylcellulose (EC), lactic acid, acetylacetone, and polyethylene glycol (PEG) were procured from commercial vendors. Test samples and metallic substrates consisting of 430SS 1-inch × 1-inch and 4-inch × 4-inch plaques and 1-inch diameter tubes were similarly procured from commercial vendors. The fabricated samples were exposed to the desired atmospheres in an L & L furnace at ~800°C for 10 minutes and cooled to ambient for characterization. An IR heating lamp (Clamp Light Portable Adjustable Pivot Metal Corded Plug-In BK-77937 E225894, with a General Electric 250R40/1 250 W 120 V R40 Med Base Clear Reflector Infrared Heat Lamp) was used to dry the samples. We used a commercial lens dipping holder (BPI Lens Holder II #16150 from Brain Power Inc., Miami, FL) designed to enable coating of the entire substrate in one dip.

2.2. Fabrication Methods. We formulated zirconia-yttria organic sol-gels with various chemical characteristics for dip coating adjusting the carrier solvent concentrations. In fabricating the formulation, the order of addition was critical to success. Initially, the EC was incorporated into a solvent mix of alpha terpineol and BCA. A Zr source is incorporated into a mixture of solvents. Acetylacetone was used to stabilize the zirconium *n*-propoxide, and over a short period of time, it is reported to produce Zr diketonate [10]. This enabled the zirconium material to be readily mixed with water and diacetone alcohol. Viscosity modifiers were added to achieve the appropriate rheological behavior thereby avoiding paint or ink spreading and promoting good surface leveling. This formulation and method were modified over the course of this study; the final process flow method is shown in Figure 1.

By varying the three parameters known to control solvation, the formation of stable nanoparticle sols and precipitates, the performance was optimized. In particular, nanoparticles can be formed in a stable formulation that will subsequently reduce stress during both drying and firing cycles.

From the literature, we know different precursor materials give rise to contrasted nanostructures, based on particle and agglomerate sizes. Xerogels, aerogels, and precipitates can be made by a sol-gel process in the Zr *n*-propoxide-acetylacetone-water-*n*-propanol system. Clear homogeneous sols are made by using a proper amount of acetylacetone and water. The primary control variable for not only producing clear homogeneous sols but also producing sols with some intrinsic nanoparticles in a sol–sol-gel composition is the complexation ratio R as shown in the following equation:

$$R = \frac{[\text{acetylacetone}]}{[\text{Zr}]}, \qquad (1)$$

where R is the main parameter controlling the size of zirconium dioxide (ZrO_2) primary particles. When $R = 0$, precipitates exist. The specific fractal structure of a gel largely determines whether the material becomes a xerogel, aerogel, or precipitate on conventional drying. In all cases, the initial primary units (from the literature) are ~2 nm. So, optimizing R allowed us to formulate a sol-gel with intrinsic particles that act as an in situ filler to avoid cracking during firing. In addition to R (the complexation ratio), W (the hydrolysis ratio) and C (the final metal concentration) also contribute to the stability of the sol-gel, especially with nanoparticles that will later act as stress relief agents during firing [11]. We optimized the performance using statistical design of experiments (SDEs) based on these parameters.

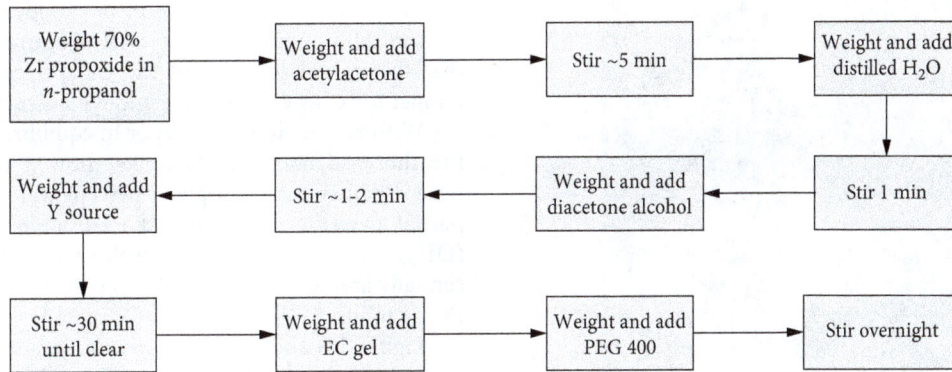

FIGURE 1: Flowchart for fabricating YSZ paint in the laboratory.

We conducted several statistically designed experiments to optimize the rheology of the YSZ and the Al_2O_3 paint for dipping and stability. Stability optimization was targeted to eliminate phase separation in the form of both cloudy multiphase fluids and actual precipitation, which was macroscopically identifiable to the naked eye.

Implicit in the optimization is the generalized Landau–Levich equation used to describe a dip coating process. If sheer withdrawal speed rates keep the system Newtonian (slow, 2.4 in/min to 24 in/min—we ultimately used 4 in/min), then the coating thickness is described by the following Landau–Levich equation as noted in [12–15]:

$$h = 0.945\alpha \times \text{capillary length} \times \text{capillary number}$$
$$= \frac{0.945\,(\eta v)2/3}{[\gamma\,(1/6)\,(\rho g)1/2]}, \tag{2}$$

where h = coating thickness, α = thickening factor, η = viscosity, v = vertical speed, γ = liquid-vapor surface tension, ρ = density, and g = gravity. (The thickness factor accounts for the interaction between the surface tension and the concentration and is approximated.) So, the coating thickness depends upon control variables: viscosity, surface tension, concentration, and density.

2.3. Test Methods. The sol particle sizes were measured by dynamic light scattering (DLS) (University California, Irvine; Biochemistry Dept.; Irvine, CA). A fineness of grind (FOG) gauge (Gardco Fineness of Grind Gauge Model PD-250) was used to determine if there were large particulates, agglomerates, polymer ensembles, or extrinsic particulate impurities in the paint. Viscosity was characterized for various paints during the project. We used a Brookfield DV II+ Pro-LV plate and cone viscometer, although at higher shear rates the Weissenberg effect was a limitation.

The substrate camber was measured using a custom-built manual camber sorter (Figure 2). A gap was produced between the top and bottom plates using shims of known thickness. The gap was then checked using a feeler gauge.

FIGURE 2: Camber sorter. A substrate is slid down the bottom plate through the gap to the end of the two metal plates. The smallest gap distance a substrate passes through is D (Equation (3)).

This measure gap was used for camber measurements. The substrate camber is defined by the following equation:

$$C = \frac{(D - T)}{L}, \tag{3}$$

where the distance setting is D, nominal substrate thickness T, camber value is C, and substrate length is L.

The samples were dip coated using the custom dip coater shown in Figure 3. Continuity of the coating was checked by the Scotch tape test (ASTM D3359-97), optical microscopy, and scanning electron microscopy (SEM) imaging.

The primary tests for the coating qualification included thickness measurements using a FilMetrics F20 Thin Film Measurement System and a Hitachi S4500 SEM (Photo-Metrics Inc.,), elemental analysis by energy dispersive X-ray spectroscopy (EDS), thermal cycling stability at 800°C, and Cr evaporation during exposure to humid air (3% H_2O) at 800°C/500 h.

2.3.1. Thermal Cycling at 800°C. Thermal cycling was performed by exposing the samples to five (5) cycles at 800°C for 10 h followed by removal of the samples from the furnace. The cooling rate was 20°C/min to 600°C and then ambient cooled. This was meant to generate an initial indication of

FIGURE 3: Custom dip coater. It consists of a ring stand, ring stand clip for the samples, a CNC motor Sherline Direct 6559 machine1-axis slide, 13″ table, stepper motor and mount, CNC linear controller held by the ring stand and in a vertical position, and a glass container for the liquid.

thermal expansion and contraction data. Planar SEM micrographs were taken of the surface before and after thermal cycling.

2.3.2. Thermal Soak at 800°C for 100 h. The samples were isothermally aged at 800°C for 100 h in air to look for development of oxide layer growth, spalling, and weight gain. Planar SEM images were taken.

2.3.3. Chromium Evaporation Tests. SOFC cathodes are prone to poisoning and degradation arising from (a) impurities present in the incoming air (intrinsic and extrinsic impurities) and (b) interactions with the electrolyte. Intrinsic gas-phase impurities include water (H_2O), CO_2, and other gas reactants. Extrinsic gas-phase impurities include various phases (some oxygen deficient) of chromium (II) oxide (CrO_x), chromium oxyhydroxide ($CrO(OH)_x$), and other Cr bearing volatile species. Degradation can be caused by solid-gas and solid-solid interactions, exsolution, and compound formation. BOP components and cell interconnections both contribute to Cr evaporation and poisoning of the cathode.

Cr evaporation (transpiration) testing was performed using the test unit shown in Figure 4. The atmosphere was held at 800°C for 500 h, with 3% water (room temperature, RT, humidification) at a flow rate of 300 sccm.

The relative humidity in air at a given temperature is fixed by the vapor pressure. The Cr vapors given off from the substrate surface are recognized to follow the laws of thermodynamics. The resulting theoretical behavior is shown in Figures 5 and 6. Figure 6 shows the dependence of

Cr vapor species concentration on temperature and H_2O concentration expressed as partial pressure. It can be seen that the partial pressure of the Cr species increases with temperature, as expected thermodynamically.

With respect to the Cr vapor in equilibrium with Cr_2O_3, the thermodynamic calculations show a relationship between temperature at a given percent water in air and the partial pressure of Cr plus the partial pressure of Cr_2O $(OH)_2$, in Figure 6. The atmosphere is assumed to be essentially ambient air, with a $P(O_2)$ of 0.209 atm—the normal $P(O_2)$ in air.

Figure 7 is a schematic representation of the typical two-phase boundary and three-phase boundary reactions from Cr-bearing species evaporated from BOP components that can react with an LSM oxide such as $La_{0.7}Sr_{0.3}MnO_3$ or LSM-based cathodes.

In Figure 8, we see the various chemical reactions that can take place at an LSM cathode, due to Cr poisoning. In all the cases, the integrity of the LSM is compromised. The result is a change in the chemistry and therefore the electrochemistry of the cathode, rendering it incapable of performing its proper electrochemical function in the SOFC.

2.3.4. Statistically Designed Experiments. Statistically designed experiments were conducted and analysed with the assistance of Design Expert 8 software (Stat-Ease, Inc.).

3. Results and Discussion

We conducted several statistically designed experiments to optimize the rheology of the YSZ and the Al_2O_3 paint for dipping and stability. Stability optimization was targeted to eliminate phase separation in the form of both cloudy multiphase fluids and actual precipitation, which was macroscopically identifiable to the naked eye. The most significant correlation diagram for YSZ is shown in Figure 9. Note: R = complexation ratio, W = hydrolysis ratio, and C = final metal concentration. Figure 9 shows a statistically significant relationship among separation, R and W. When C is high, a strong dependence exists between R and W. Only when C is high do we find that when R is low, there is phase separation. When W is high, separation increases even more. This correlation has a 93% confidence limit. In looking at the implications, we found phase separation is eliminated if the acetylacetone concentration is high and the water concentration is low.

For alumina formulations, there was severe precipitation. A statistically designed experiment was conducted and the degree of precipitation was indexed from 0 (no precipitation) to 1 (severe precipitation). Precipitation only occurs if the ethyl acetone is high. The concentrations for acetone, aluminum acetylacetonate, and the other constituents had no significant effect on precipitation. The ethyl acetone concentration was lowered to eliminate precipitation.

A FOG gauge was used to determine if there were large particulates, agglomerates, polymer ensembles, or extrinsic particulate impurities in the paint. The results are reported as

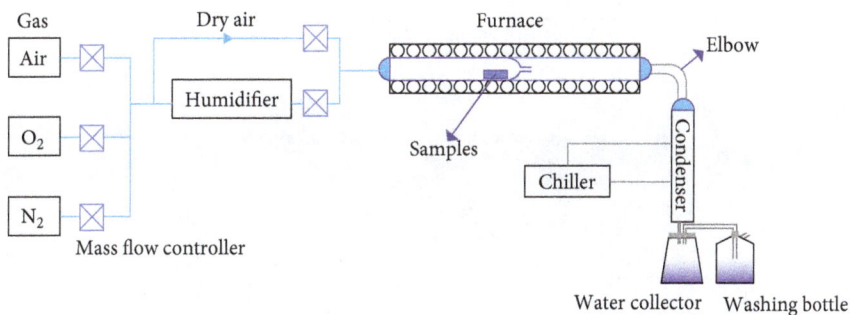

FIGURE 4: Schematic of Cr evaporation test units at University of Connecticut.

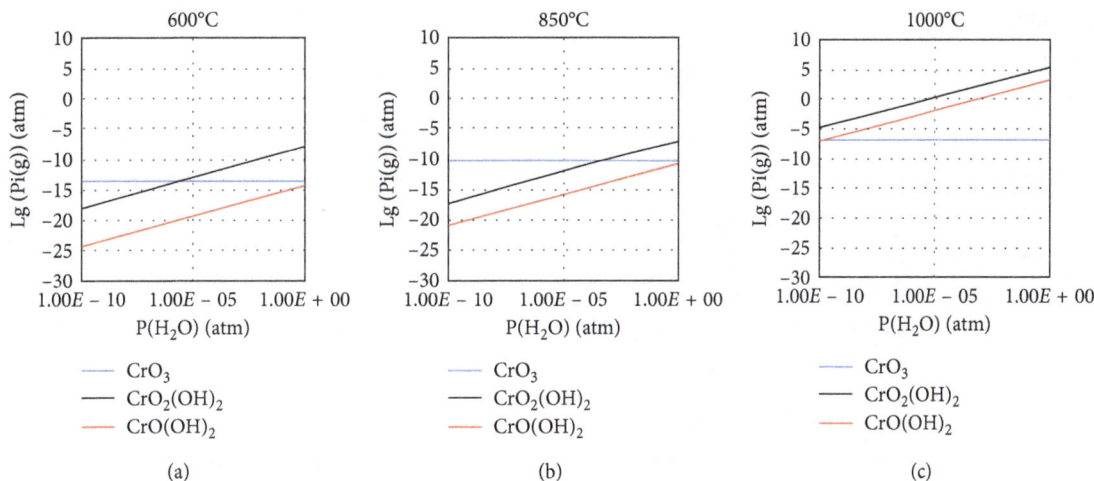

FIGURE 5: Partial pressures of chromium vapor species varying with the partial pressure of H_2O under different temperatures. (a) 600°C; (b) 850°C; (c) 1000°C. Wherein, $P(O_2) = 0.209$ atm (partial pressure of oxygen in air).

1st scratch, 4th scratch, and 50% size, with all measurements at <2 μm. The detection limit is 2 μm. Larger (>2 μm) agglomerates were not observed.

DLS intensity was analysed to determine the particle-size distribution function. For the YSZ formulations, the DLS average particle size was in the range of 1,000–2,000 nm (Table 1). This may be due to, to some extent, an effective particle size for YSZ plus associated layers of EC and water of hydration. The DLS data are consistent with the FOG measurements indicating the particle size for all the YSZ powders is <2 μm. The alumina-based sample shows an initial low DLS particle size of ~4 nm. However, over time the 102615-SZ-3087-Al1 formulation showed precipitation, and this is borne out by the FOG gauge measurements of ~40 μm.

Viscosity of dipping paints or inks were measured. At higher shear rates, the Weissenberg effect was a limitation. After the issue of precipitation was eliminated, the viscosity was reasonably stable, compared with the measurement error. The viscosity was varied over a wide range to develop an appropriate rheology for dipping, which balances a high YSZ or Al_2O_3 content with a low viscosity for good dipping, 2.21–270.7 (at 1.5 RPM) to 115–416.8 (at 9 RPM).

The thicknesses of sample coatings were measured using a Filmetrics Thin-Film Measurement System. The values ranged from 0.20 to 1.66 μm. There is significant variation in

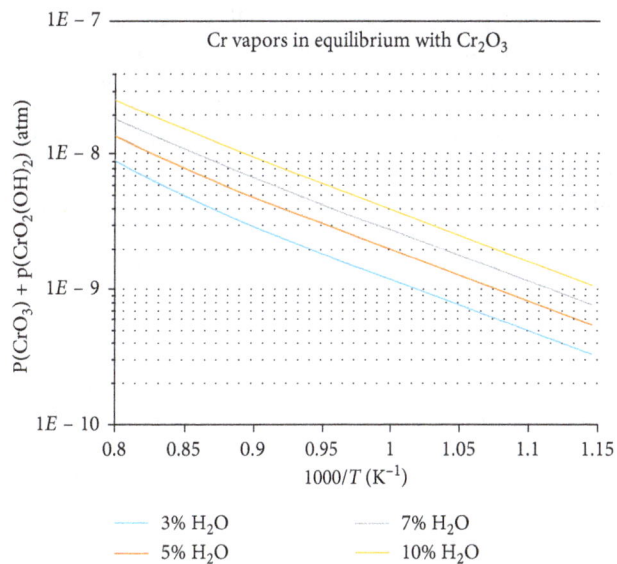

FIGURE 6: Partial pressures of chromium vapor species varying with temperature for different partial pressures of H_2O. Wherein, $P(O_2) = 0.209$ atm (partial pressure of oxygen in air).

thickness, in part dependent upon the uniformity of the substrate. The valleys are filled, and due to surface energy, the asperities are lightly covered. Considering that

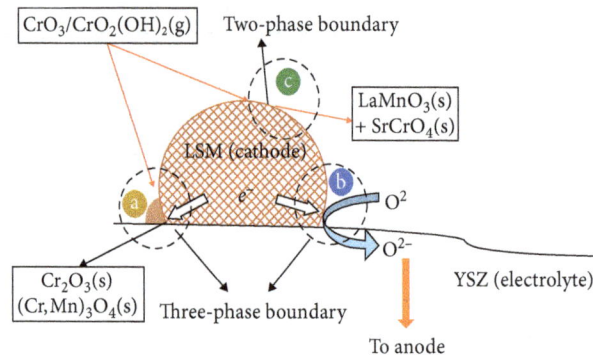

FIGURE 7: A schematic representation of the general types of Cr poisoning reactions that may take place: two-phase boundary reactions and three-phase boundary reactions.

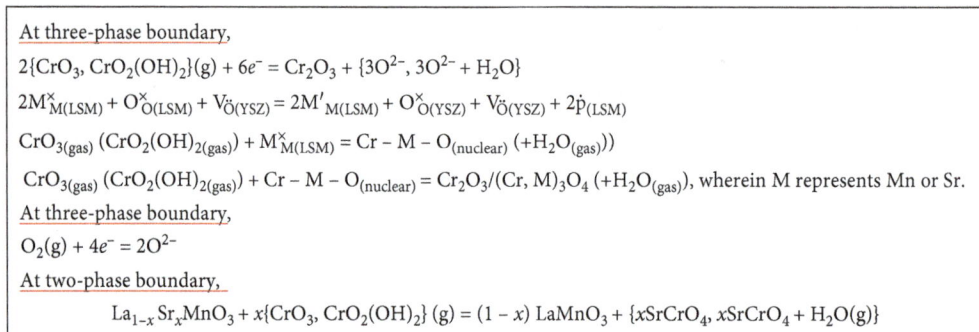

At three-phase boundary,

$2\{CrO_3, CrO_2(OH)_2\}(g) + 6e^- = Cr_2O_3 + \{3O^{2-}, 3O^{2-} + H_2O\}$

$2M^\times_{M(LSM)} + O^\times_{O(LSM)} + V^{\cdot\cdot}_{\ddot{O}(YSZ)} = 2M'_{M(LSM)} + O^\times_{O(YSZ)} + V^{\cdot\cdot}_{\ddot{O}(YSZ)} + 2\dot{P}_{(LSM)}$

$CrO_{3(gas)} (CrO_2(OH)_{2(gas)}) + M^\times_{M(LSM)} = Cr - M - O_{(nuclear)} (+H_2O_{(gas)}))$

$CrO_{3(gas)} (CrO_2(OH)_{2(gas)}) + Cr - M - O_{(nuclear)} = Cr_2O_3/(Cr, M)_3O_4 (+H_2O_{(gas)})$, wherein M represents Mn or Sr.

At three-phase boundary,

$O_2(g) + 4e^- = 2O^{2-}$

At two-phase boundary,

$La_{1-x}Sr_xMnO_3 + x\{CrO_3, CrO_2(OH)_2\} (g) = (1 - x) LaMnO_3 + \{xSrCrO_4, xSrCrO_4 + H_2O(g)\}$

FIGURE 8: Typical reactions that may take place at an LSM cathode from Cr poisoning.

Filmetrics' interferometric technique depended upon having a flat and parallel film structure with respect to the non-transparent substrate, the values are only approximate.

Ten 430SS substrates that had been laser cut into 1-inch × 1-inch substrates were measured for the camber. The average camber was 0.000 ± 0.002 mm/mm (in/in). The maximum camber was 0.003 mm/mm (in/in). This camber can be interpreted as representative of general shape variation. In BOP components, the shape nonuniformity in, for example, ductwork, is expected to be considerably worse.

3.1. Coating 1 in. × 1 in. × 0.025 in. 430SS Substrates. One issue was the need to dip samples twice for complete coverage. We found that, in dipping the entire flat edge of the substrate, there was substantial paint buildup that led to cracking during drying and firing. This effect was minimized by dipping down to one corner (the diamond orientation in Figure 3). To demonstrate a holder design that would enable the coating of the entire substrate in one dip, we obtained a commercial lens dipping holder. To minimize the effects of dust on the uniformity of the coating, a laminar flow booth was used. The dip speed, dip dwell time, and dip depth were regulated by a computer numerical control (CNC) linear controller. The sample with the holder was then transferred to a drying area, where the wet substrate was dried at a fixed distance from an IR heating lamp. After this drying step,

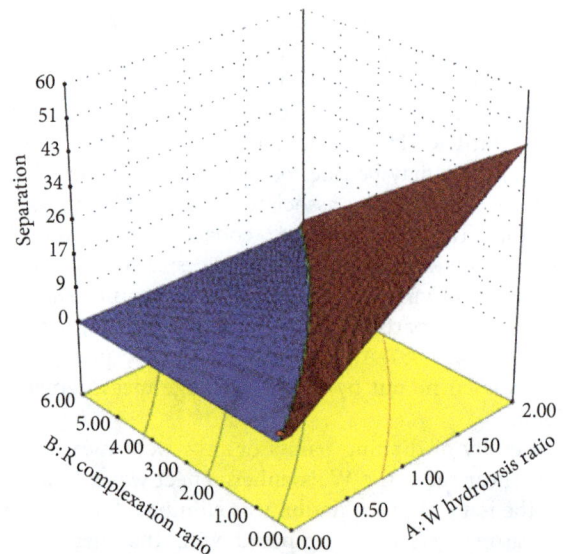

FIGURE 9: Graph showing the correlations between R, W, and separation for high final metal concentration. When C is high, a strong dependence exists between R and W. Only when C is high do we find that when R is low, there is phase separation, and it increases further when W is high (93% confidence limit). We found phase separation is eliminated if acetylacetone is in high concentration and the water concentration is low.

TABLE 1: The average particle size found for several formulation paints from DLS.

Paint formulation	Material	d (nm)	Std. dev. (nm)
090315-SZ-3087-15-NA1	YSZ	1058	345.8
091015-SZ-3087-15-NA1	YSZ	1219	480
101215-SZ-32087-15-NA1	YSZ	1915	1209
102615-SZ-3087-Al1	Al_2O_3	25.62	4.1

some formulations were still not fully cured. These were placed in a small Petri dish with a loose lid to dry. Substrate films were dry before firing.

3.2. Firing Profile through YSZ Nucleation and Growth.
In the literature, nucleation and growth has been reported for the 8 mol% Y-doped ZrO_2 system, with the nucleation range lower than the crystal growth temperature range. The nucleation and growth curves for sol-gel-derived YSZ show nucleation in the range of 380 to $\geq 400°C$ [16] and crystal growth in the range of $400°C–500°C$ for sol-gel-derived YSZ depositions [17]. However, in this work, we found our paint has a typical suspended particle size of $\sim 1\,\mu m$ for YSZ and ~ 25 nm for Al_2O_3. Nucleation and growth was checked by optical and SEM imaging. Little or no such growth was found, with the only identifiable particles being ~ 200 nm.

After exposure to $800°C$ for 10 minutes, the surface showed no evidence of the presence of grain formation. However, after 100 h at $800°C$, substantial sintering neck growth occurred with grain sizes of ~ 200 nm, demonstrating some densification but no grain growth. The grain size is uniform, with no exaggerated grain growth, although some grain growth would be expected for long firing times, >100 h.

3.3. Thermal Cycling and Heat Soak Results.
The samples were exposed to $650°C$ for 10 minutes followed by $800°C$ for 10 minutes to minimize the surface coating spallation during firing as well as removal of organic constituents. The samples are described in Table 2.

The YSZ coating withstands thermal cycling and isothermal aging to 100 h at $800°C$. Al_2O_3 coating sample quality however varied. Sample B137 (Table 2) does not withstand thermal cycling. Sample B108 does withstand thermal cycling; however, the quality is not good.

3.4. Thermal Cycling at $800°C$.
Thermal cycling was performed by exposing the samples for five (5) cycles at $800°C$ for 10 h followed by removal of the samples from the furnace. The cooling rate was $20°C/min$ to $600°C$ and then ambient cooled. This was meant to generate an initial indication of thermal expansion and contraction data. Two samples (YSZ-B131 and Al_2O_3-B137) were tested. Planar SEM micrographs were taken of the surface before and after thermal cycling (Figures 10 and 11).

Above observations clearly indicate that YSZ is superior to Al_2O_3, in that Al_2O_3 spalled off during the thermal cycling, particularly during handling between cycles. The spalling was sufficient to render the weight gain measurements invalid since some materials pulled off and actually decreased the overall weight of the sample.

Evaluation of Multilayer Coated Samples. Characterization of the coating was performed to identify mechanisms needed for uniform coverage. We made multiple coatings on 430SS 1-inch × 1-inch substrates. Each layer was dipped, dried, and fired for 10 minutes at $800°C$. Cr evaporation can be decreased by relatively thin film coatings, which can be classified into two main groups perovskite structure and spinel structure coatings.

From the EDS, with Mn, Cr, and O, with Cr < Mn, the stoichiometry from Figures 12(a) and 12(b) appears to be $Mn_x Cr_3\text{-}xO_4$, with $x \sim 2$. Fe is from the substrate. The tentative conclusion is that we are looking at the 430SS substrate from below the surface and crystals of spinel $Mn_x Cr_3\text{-}xO_4$, with $x \sim 2$.

The effect of using a multiple-layer coating process on the morphology and adhesion of YSZ to 430SS post-annealing treatment was evaluated. Figures 13(a)–13(d) show the surface morphology of the YSZ coating using SEM microscopy. The intensity of the cracks decreased with the increase of the deposition cycles of the YSZ coating. Fewer cracks were observed at the edge of the sample when compared to the center, in contrast to what had been seen with 1-2 layer coatings. However, the edge area also shows fewer cracks, as the number of layers increases.

Figures 13(e)–13(j) show that the addition of the YSZ nanopowder increases the intensity of the cracks for fewer deposition layers (2 and 3), while the intensity of the cracks decreases as the number of layers increases. The YSZ film covered the 430SS substrate. Microcracks observed on all the samples could affect the protection ability of the YSZ film. The round shape of the grain edges observed in Figures 13(g)–13(j) may have formed due to the volatilization of the grain boundary phases.

The energy dispersive X-ray analysis of all the samples showed the presence of Y and Zr in the stoichiometric ratio, within experimental error. The intensity of the cracks on the YSZ coatings decreased with the increase of the deposition cycles both with and without the addition of the nanopowder. The Cr evaporation decreased by 51% for a single YSZ layer coating (B146) sample, with no nanopowder YSZ added. It was then compared with an uncoated sample (B151)—both were tested on a 1 in. × 1 in. 430SS substrate.

3.5. Thermal Soak at $800°C$ for 100 h.
The samples were isothermally aged at $800°C$ for 100 h in air to look for development of oxide layer growth, spalling, and weight gain. Planar SEM images are shown for YSZ-B123 and Al_2O_3-B127 (Figure 14). The content of Zr and Y at the B123 surface is much higher than the B132 sample seen in Figure 15. From EDS, the Cr content at the B123 surface is also less than the

TABLE 2: Characteristics of the substrates used in coating and testing.

Sample	Substrate	Surface	Sol-gel material	Firing
B123	430 1 in. × 1 in. laser cut	2000 grit	111715-3087-SZ-YSZ-Scaleup	650°C-10 min + 800°C-10 min
B131	430 1 in. × 1 in. laser cut	2000 grit	112015-SZ-3087-YSZ-Scaleup-4	650°C-10 min + 800°C-10 min
B132	430 1 in. × 1 in. laser cut	2000 grit	112015-SZ-3087-YSZ-Scaleup-4	650°C-10 min + 800°C-10 min
B127	430 1 in. × 1 in. laser cut	2000 grit	111315-SZ-3087-Al7-4-Scaleup	650°C-10 min + 800°C-10 min
B107	430 1 in. × 1 in. laser cut	2000 grit	111315-SZ-3087-A17-4-Scaleup	650°C-10 min + 800°C-10 min
B108	430 1 in. × 1 in. laser cut	2000 grit	111315-SZ-3087-A17-4-Scaleup	650°C-10 min + 800°C-10 min
B137	430 1 in. × 1 in. laser cut	2000 grit	111315-SZ-3087-Al7-4-Scaleup	650°C-10 min + 800°C-10 min
B138	430 1 in. × 1 in. laser cut	2000 grit	111315-SZ-3087-Al7-4-Scaleup	650°C-10 min + 800°C-10 min
B140	430 1 in. × 1 in. laser cut	Bare	None	None
B141	430 1 in. × 1 in. laser cut	Bare	None	None

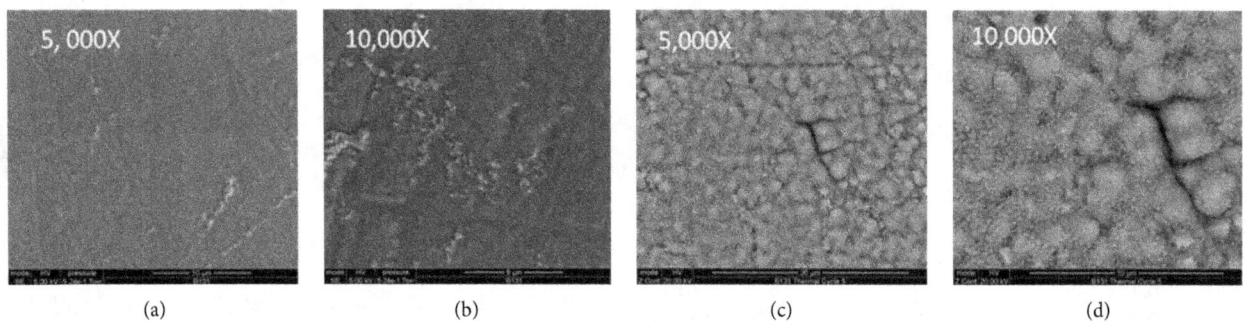

(a)   (b)   (c)   (d)

FIGURE 10: Sample YSZ-B131 before and after thermal cycling. (a) Before thermal cycling, shown at 5,000X. (b) Before thermal cycling, shown at 10,000X. (c) After thermal cycling, shown at 5,000X. (d) After thermal cycling, shown at 10,000X. The nonthermally cycled sample EDX shows only iron (Fe), Y, Zr, and oxygen (O).

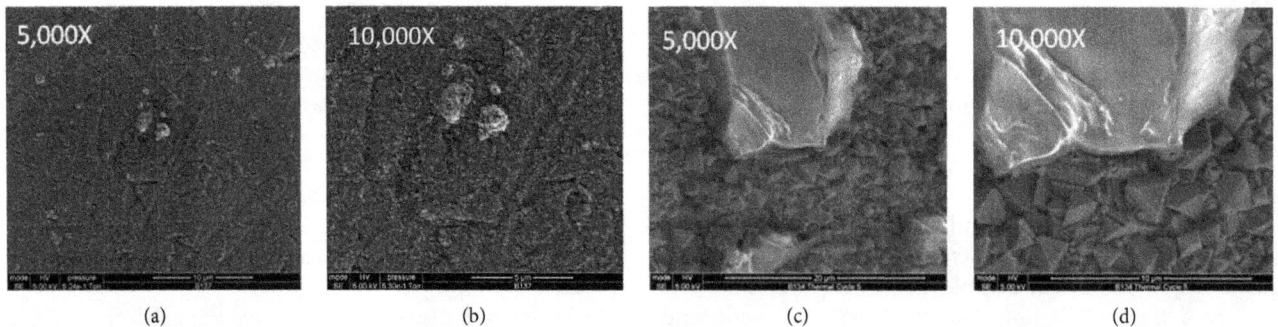

(a)   (b)   (c)   (d)

FIGURE 11: Sample Al2O3-B137 before and after thermal cycling. (a) Before thermal cycling, shown at 5,000X. (b) Before thermal cycling, shown at 10,000X. (c) After thermal cycling, shown at 5,000X. (d) After thermal cycling, shown at 10,000X. The nonthermally cycled sample EDX shows only iron (Fe), aluminum (Al), silica (Si), carbon (C), and oxygen (O). Fe, Si, and C are in the substrate. No Cr or Ni was detected.

B132 sample. The observation is possibly related to the coating thickness and the spatial resolution for EDS analysis. B127 surface morphology is similar to the uncoated sample. However, aluminum (Al) is detected at the surface. It is likely that Al may have reacted with the substrate oxide layer.

EDS was performed on various areas of the YSZ-B123 sample. From the micrograph (Figure 14), the coating appears to be ~1 μm thick. Crystals are seen underneath the coating. It appears that the crystals are probably manganese (Mn)-Cr-iron (Fe)-oxygen (O) spinels, such as $MnFe_2O_4$

and Mn_xCr_3-xO_4. From an EDS, showing Mn, Cr, and O, with Cr < Mn, the surface appears to be Mn_xCr_3-xO_4, with ×~2. Fe is from the substrate. The tentative conclusion is that we are looking at the 430SS substrate from below the surface and crystals of spinel Mn_xCr_3-xO_4, with ×~2.

The SEM image of the cross section of the thermally soaked B-123 (YSZ) sample is shown in Figure 16. Four distinct phases are observed as marked. The EDS spot analyses (Table 3) reveal the composition of these phases. The dark colored phases (spot 1) are enriched of chromium, manganese, and iron while the content of these elements

(a)   (b)

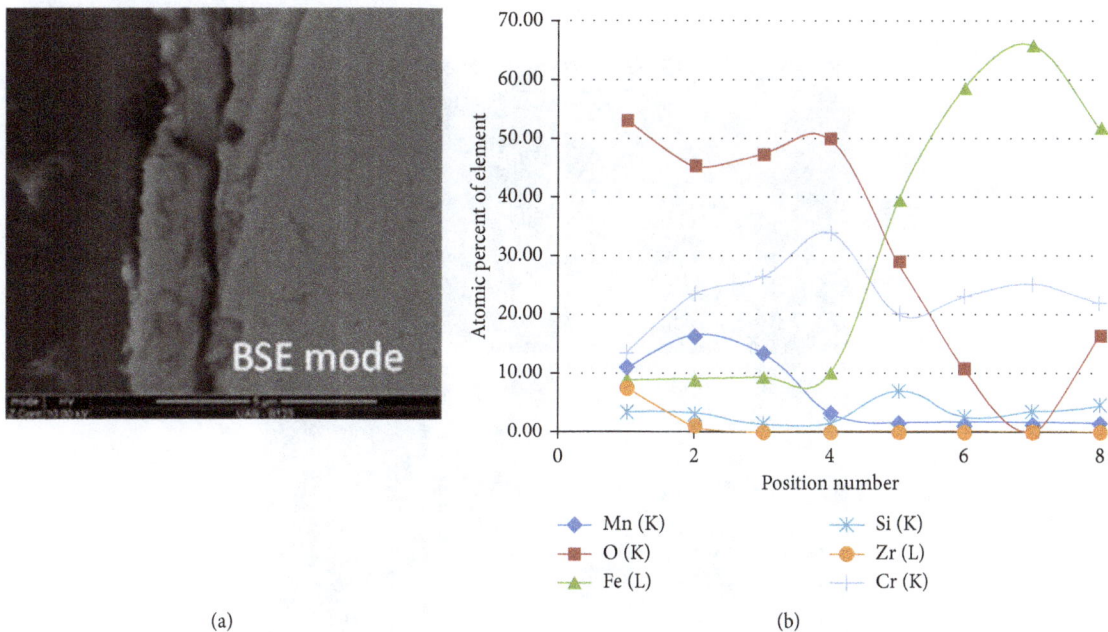

FIGURE 12: YSZ-B123: EDX analyses were run fairly evenly spaced in a line from 1 through 8 for the thermally soaked sample.

decreases for the brighter phases. The oxygen content increases for the brighter phases and is highest for the spots labeled 4. The zirconium content is higher for the spots 1 and 4. Several interferences are made from the observation. Firstly, oxygen and zirconium diffuse inward and other elements diffuse outward. Secondly, silicon diffuses throughout the coating. Thirdly, the oxide layers consist of Cr_2O_3, Mn-Cr-O, and Mn-Fe-O spinels. The content of these phases varies for different layers (spots). Zirconium likely dissolves in these phases in the form of solid-solution.

The EDS line scan is shown in Figure 17. The line scan is consistent with the EDS spot analysis and supports the aforementioned interpretation. However, the cause for the decrease in oxygen content in the spot 1 (Figure 16) region is unknown.

The EDS elemental maps of the B-123 sample are shown in Figure 18. Juxtaposing the elemental images supports the above interpretations.

The SEM image of the cross section of the thermally soaked B-127 (Al_2O_3) sample is shown in Figure 19. Three distinct layers are observed as marked. The EDS spot analyses (Table 4) reveal that the thin dark layer near the interface is enriched with silicon, chromium, and iron. Layer 2 is enriched of chromium while layer 3 is enriched of chromium and manganese. Only a very small quantity of aluminum is detected in layer 3. The layer 1 is likely to be SiO_2, which acts a protective oxide layer. Layer 2 is likely to be Cr_2O_3, and layer 3 is Cr-Mn-O spinel. Al_2O_3 coating may have spalled of the oxidized sample. Therefore, it is not detected.

The line scan in Figure 20 shows the presence of silicon, as seen by the hump in the Si profile followed by humps in chromium and oxygen profile. Lastly, chromium content

decreases and manganese increases as observed by the hump in the manganese profile. These observations support the EDS spot analysis and suggest the presence of the SiO_2 layer followed by Cr_2O_3 and Mn-Cr-O spinel, which is complimented by the EDS elemental maps in Figure 21.

3.6. *Cr Evaporation Tests.* SOFC cathodes are prone to poisoning and degradation arising from (1) impurities present in the incoming air (intrinsic and extrinsic impurities) and (2) interactions with the electrolyte. Intrinsic gas-phase impurities include H_2O, CO_2, and other gas reactants. Extrinsic gas-phase impurities include CrO_x, $CrO(OH)_x$, and other Cr bearing volatile species. Degradation can be caused by solid-gas and solid-solid interactions, exsolution, and compound formation.

A Cr evaporation test was run on two, 1-inch × 1-inch × 0.020-inch, 430SS substrates, one uncoated and one coated. The two samples were run separately. The images of the Cr-bearing evaporated material can be seen visually on the interior of the exit condenser in Figure 22, especially when comparing the two 500 h results. The 500 h image clearly shows the uncoated sample has much more of the yellow-brown condensate than the coated sample. The collaborators at the University of Connecticut extracted the Cr bearing material from the condenser and ran inductively coupled plasma atomic emission spectroscopy (ICP). As can be seen, there is significantly more Cr deposited from the uncoated sample (Figure 22).

The Cr bearing material that was found in the condenser tube at the end of the Cr tester was washed out after the completion of each Cr evaporation test. The resultant solution was taken, and ICP analysis was used to measure the

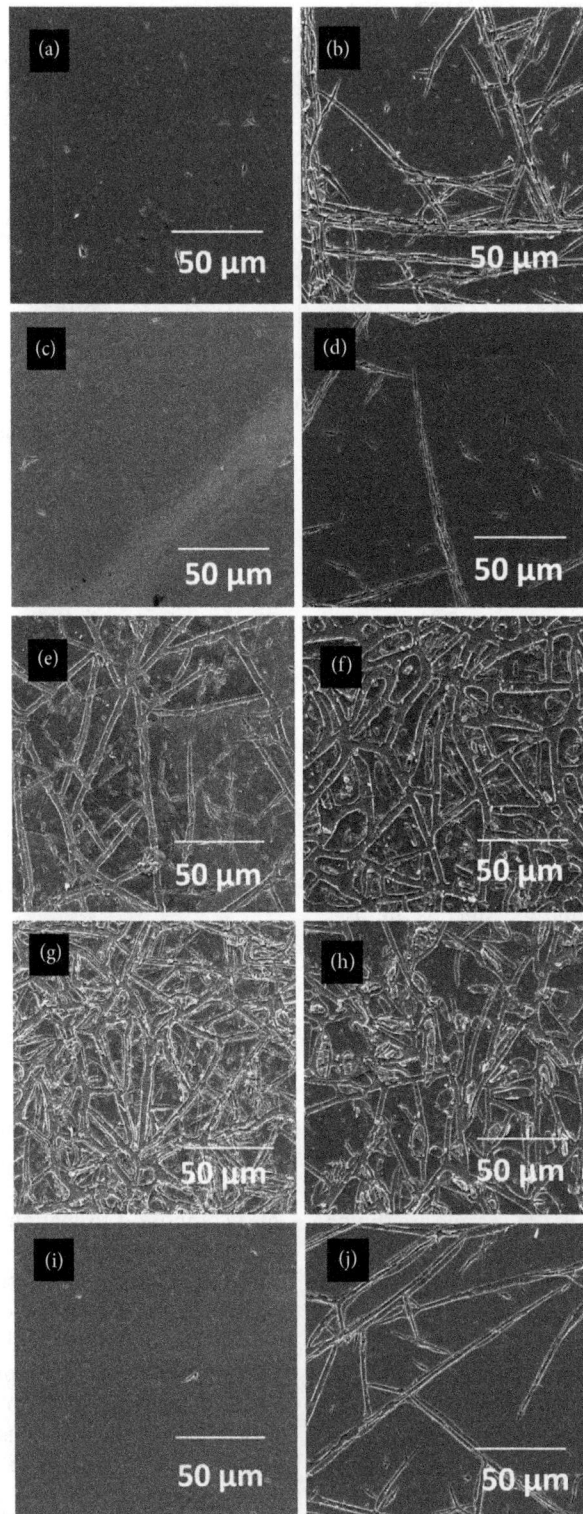

FIGURE 13: YSZ-coated 430 stainless steel. (a) Edge; (b) center of the sample after four (4) deposition cycles using dip coating. (c) Edge; (d) center of the sample after five (5) deposition cycles using dip coating. (e) Edge; (f) center of the sample after two (2) deposition cycles using dip coating. (g) Edge; (h) center of the sample after three (3) deposition cycles using dip coating. (i) Edge; (j) center of the sample after four (4) deposition cycles using dip coating.

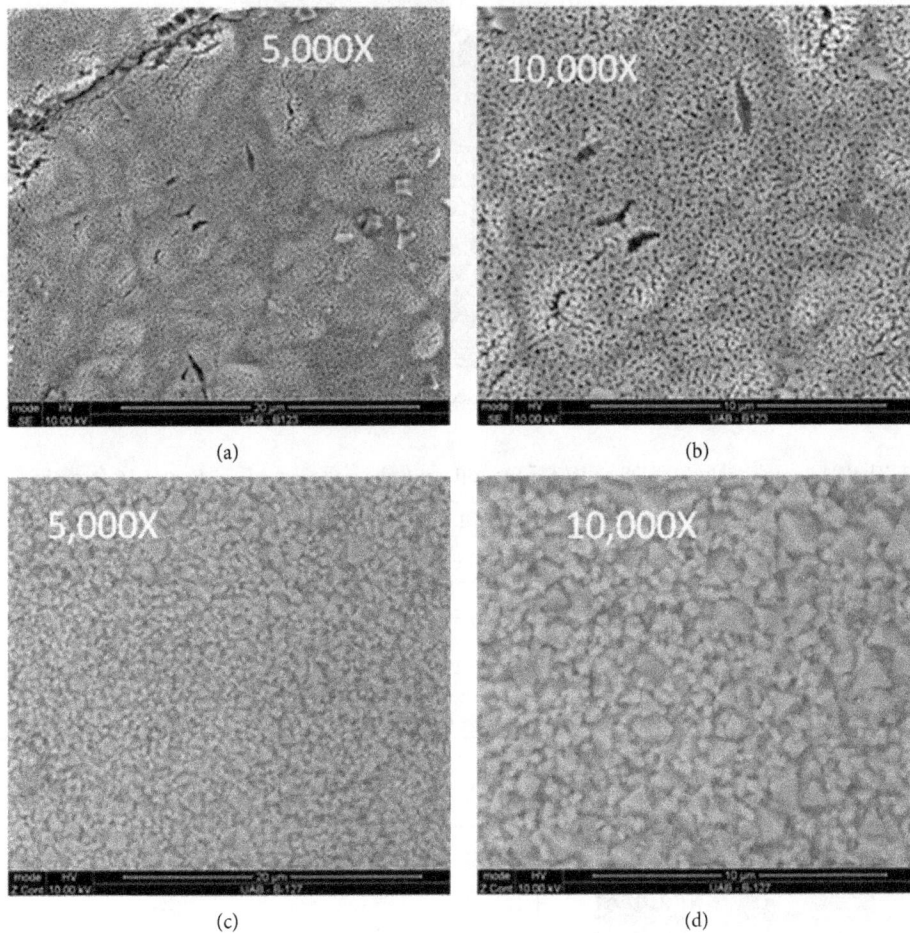

FIGURE 14: Thermal soak samples. (a) Sample YSZ-B123 at 5,000X. (b) Sample YSZ-B123 at 10,000X. (c) Sample Al_2O_3-B127 at 5,000X. (d) Sample Al_2O_3-B127 at 10,000X.

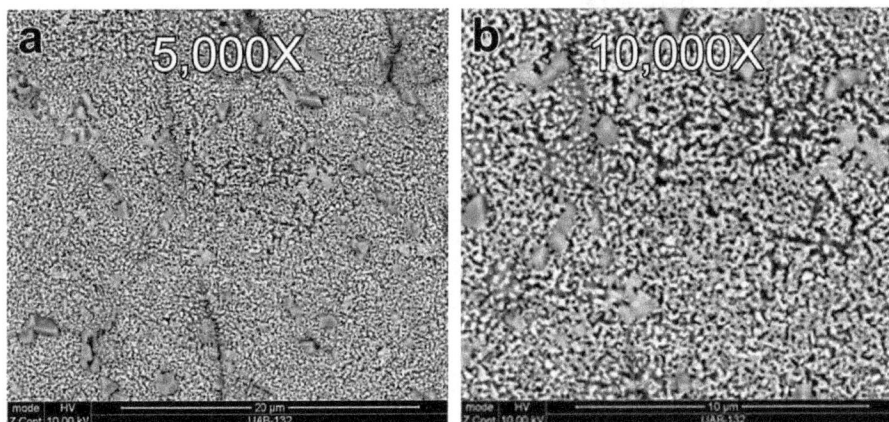

FIGURE 15: Thermal soak samples: YSZ-B132 (a) at 5,000X and (b) at 10,000X.

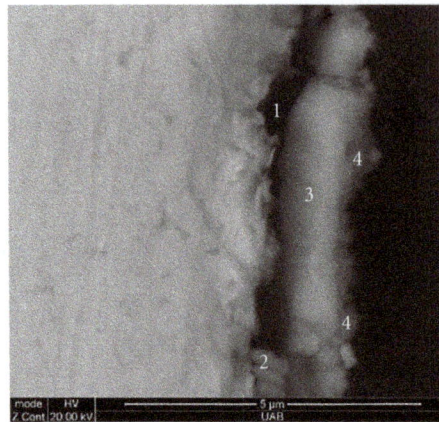

FIGURE 16: SEM image of cross section of the B-123 (YSZ) sample.

TABLE 3: The elemental analysis (EDS spot) for the different phases (marked) in Figure 16 (atomic percent).

Elements	1	2	3	4
O	43.9 ± 4.5	57.5 ± 3.2	63.8 ± 3.7	69.2 ± 2.0
Si	2.3 ± 0.1	1.9 ± 0.3	2.2 ± 0.9	2.6 ± 1.1
Cr	20.5 ± 2.4	19.1 ± 1.1	17.7 ± 1.1	9.7 ± 1.4
Mn	9.7 ± 1.0	6.0 ± 0.6	4.9 ± 0.3	5.2 ± 1.2
Fe	18.3 ± 0.7	13.6 ± 0.9	9.7 ± 2.2	9.5 ± 1.0
Y	0.5 ± 0.1	0.2 ± 0.1	0.2 ± 0.0	0.3 ± 0.0
Zr	4.8 ± 0.5	1.6 ± 0.4	1.4 ± 0.3	3.5 ± 1.3

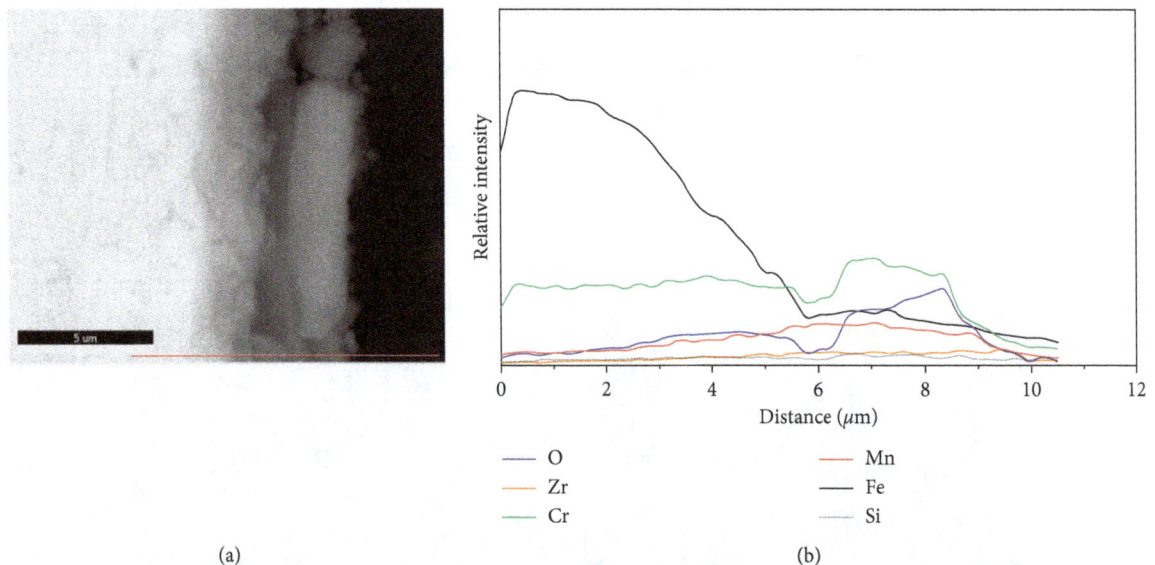

(a)

(b)

FIGURE 17: Line scan analysis of the cross section of the B-123 (YSZ) sample. (a) EDS spot analysis. (b) Line scan.

amount of Cr for the bare and the coated substrates. The data are shown in Table 5. The coating decreased the Cr evaporation by ~51%. This demonstrates that even with some areas not fully covered, there was a reduction in the Cr evaporation due to the coating.

The SEM images of substrates are shown in Figure 23, at different magnifications for 500 h in the 800°C Cr evaporation test. The uncoated 430SS 1-inch × 1-inch × 0.020-inch shows the growth per the thermal growth oxide (TGO) scale. The coated specimen shows the surface with some protective coating, although there are open, uncoated areas that can be seen in some cases for the coated specimen.

We focused on developing and evaluating a coating to reduce or eliminate Cr evaporation for BOP SOFC components. Coating performance was improved and cost minimized by using flash firing at 800°C for 10 minutes.

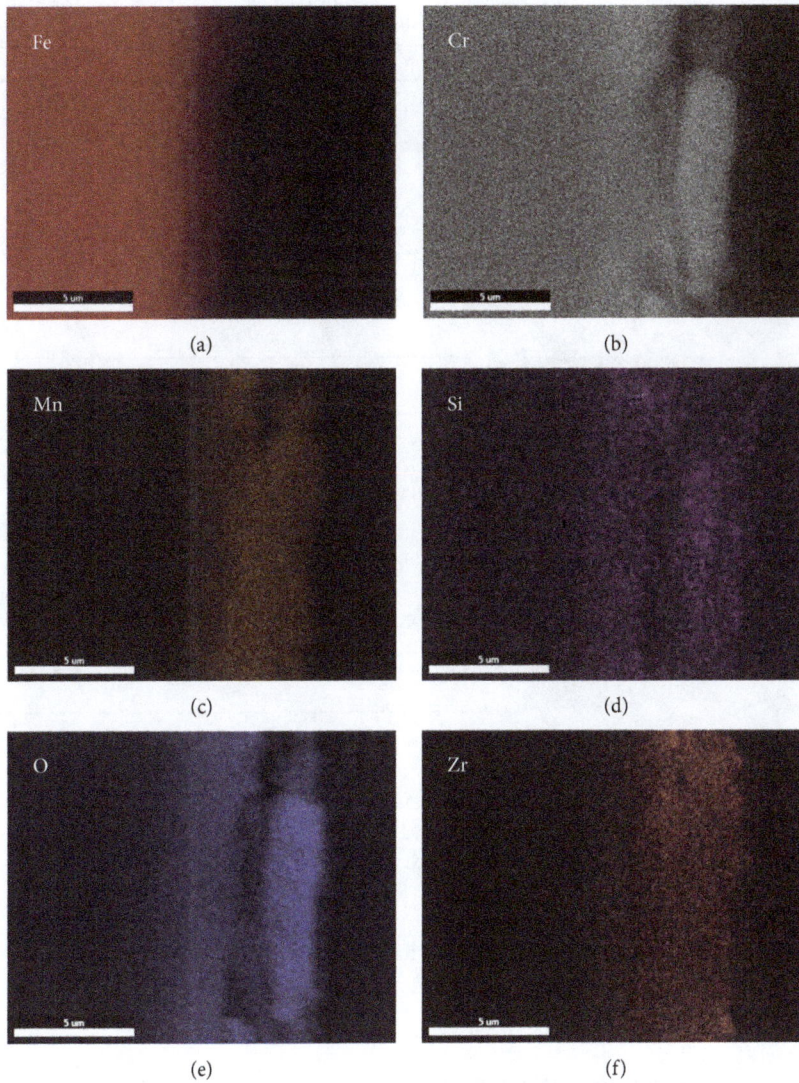

FIGURE 18: Elemental maps of the cross section of the B-123 (YSZ) sample.

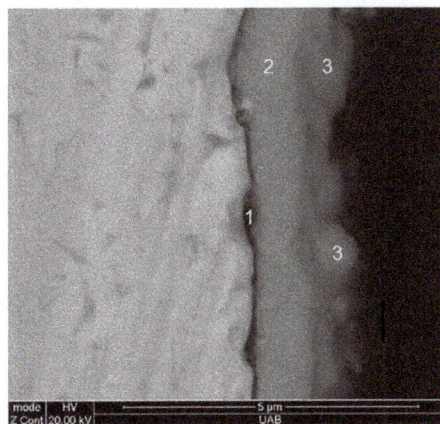

FIGURE 19: SEM image of cross section of the B-127 (Al$_2$O$_3$) sample.

We used a model GS1714 Furnace with Honeywell UDC3300 program control and DC2500 over temperature protection. The samples were vertically mounted in slits cut into a silica ceramic foam monolith ~4 in. × ~4 in. × ~2 in. holder. The holder with the sample was introduced into the hot furnace chamber, held at target test temperature of

TABLE 4: The elemental analysis (EDS spot) for the different phases (marked) in Figure 19 (atomic percent).

Elements	1	2	3
O	33.5 ± 10.1	55.6 ± 0.3	58.6 ± 1.1
Al	0.1 ± 0.2	0.7 ± 0.3	1.7 ± 0.5
Si	4.8 ± 2.5	1.4 ± 0.2	2.2 ± 0.6
Cr	31.3 ± 7.8	28.9 ± 0.2	23.6 ± 1.4
Mn	2.3 ± 0.3	2.3 ± 0.2	9.2 ± 1.0
Fe	27.7 ± 2.6	11.0 ± 0.6	4.7 ± 0.5

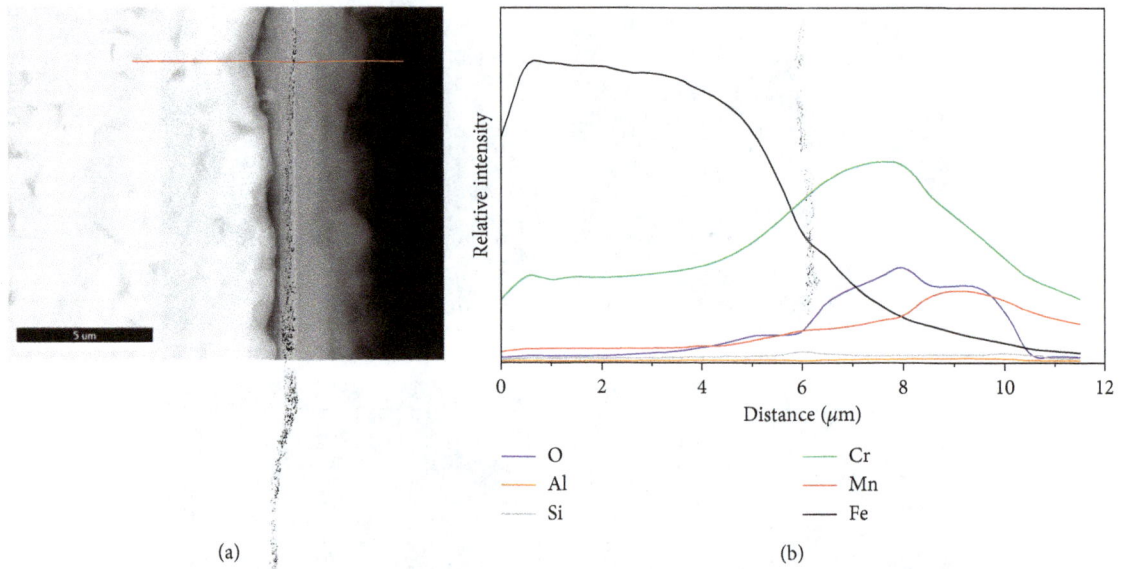

(a)   (b)

FIGURE 20: Line scan analysis of the cross section of the B-127 (YSZ) sample.

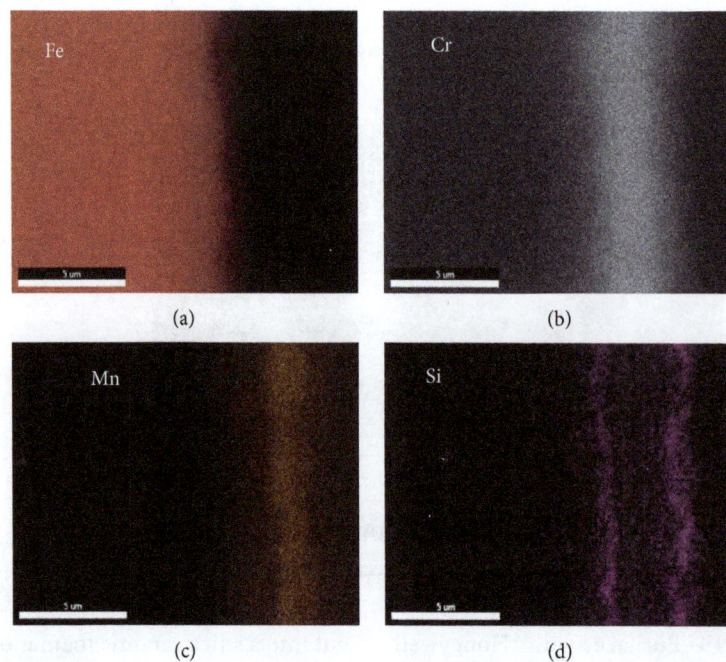

(a)   (b)

(c)   (d)

FIGURE 21: Continued.

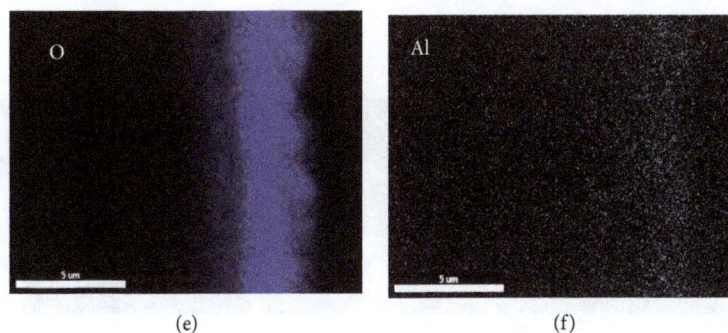

FIGURE 21: Elemental maps of the cross section of the B-123 (YSZ) sample.

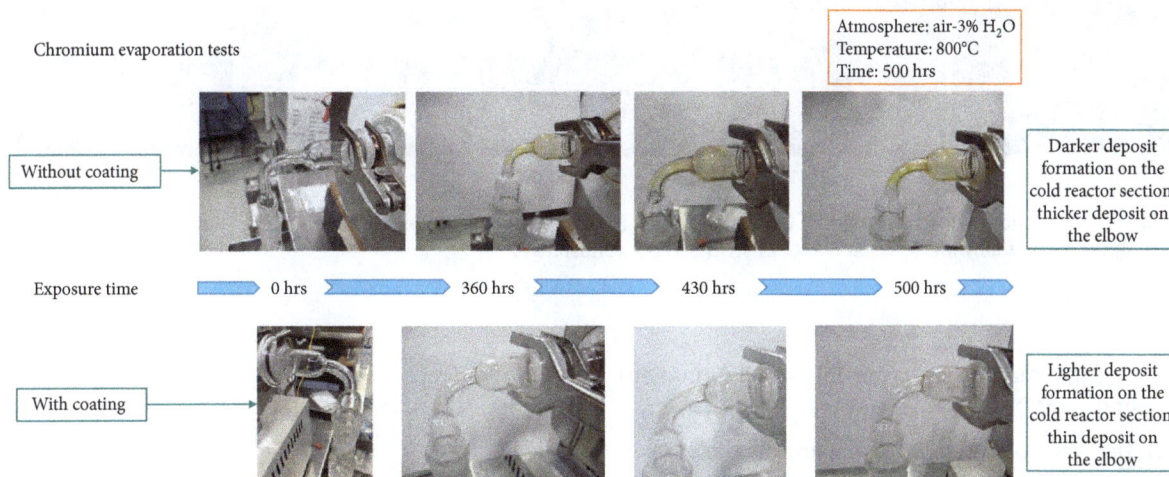

FIGURE 22: Cr evaporation test images of uncoated and YSZ-coated 430SS 1-inch × 1-inch × 0.020-inch samples showing condensate at various intervals for 500 h at 800°C.

TABLE 5: Concentration of Cr found in the materials deposited in the condensers from the Cr evaporation testing at 800°C for 500 h.

Samples	Total Cr mass (g) evaporated during 500 hrs	Surface area of samples (m^2)	Evaporation rate of Cr_2O_3 (kg/m^{2}s)	Partial pressure of Cr vapors in 300 sccm air flux (atm)
B-146 (SS430 coated with YSZ)	$1.89 E - 04$	$1.30 E - 03$	$2.36 E - 10$	$3.56041 E - 08$
B-151 (bare SS430)	$3.87 E - 04$	$1.32 E - 03$	$4.76 E - 10$	$7.27762 E - 08$

800°C and after 10 minutes, the holder with samples was removed and allowed to cool at ambient. After cooling, samples were removed and stored for testing.

Additional coating improvement and potential cost reduction were achieved through improved dipping and handling technology. We identified areas for fine-tuning and more rigorous testing. We had demonstrated the following:

(i) Dipping formulations showed better overall performance for YSZ than Al_2O_3. The clear, nonphase separated modified YSZ sol-gel formulation was easily scaled up from 10 g to ~250 g.

(ii) Initial firing at 800°C resulted in macroscopically continuous coatings, with good adhesion.

(iii) 430SS substrate coatings, after heat thermal soak at 800°C for 100 h, are adherent, ~50% porous on

a nanoscale, and nanograined YSZ. They have a thickness of ~100 nm–1 μm and a pore size of ~ 200 nm.

(iv) Spalling in the form of small flakes could be found under certain process conditions, both before and after firing. This was minimized by optimizing process conditions. We switched to flash firing (as previously described) and redesigned the dip holder design. We also optimized composition.

(v) Although initial flash 800°C firing (as previously described) can result in continuous coatings with good adhesion, subsequent thermal soak at 800°C after 100 h resulted in a nanoporous fine grain (~200 nm) structure and what appears to be

Chromium evaporation tests

Atmosphere: air-3% H_2O
Temperature: 800°C
Time: 500 hrs

(a)

(b)

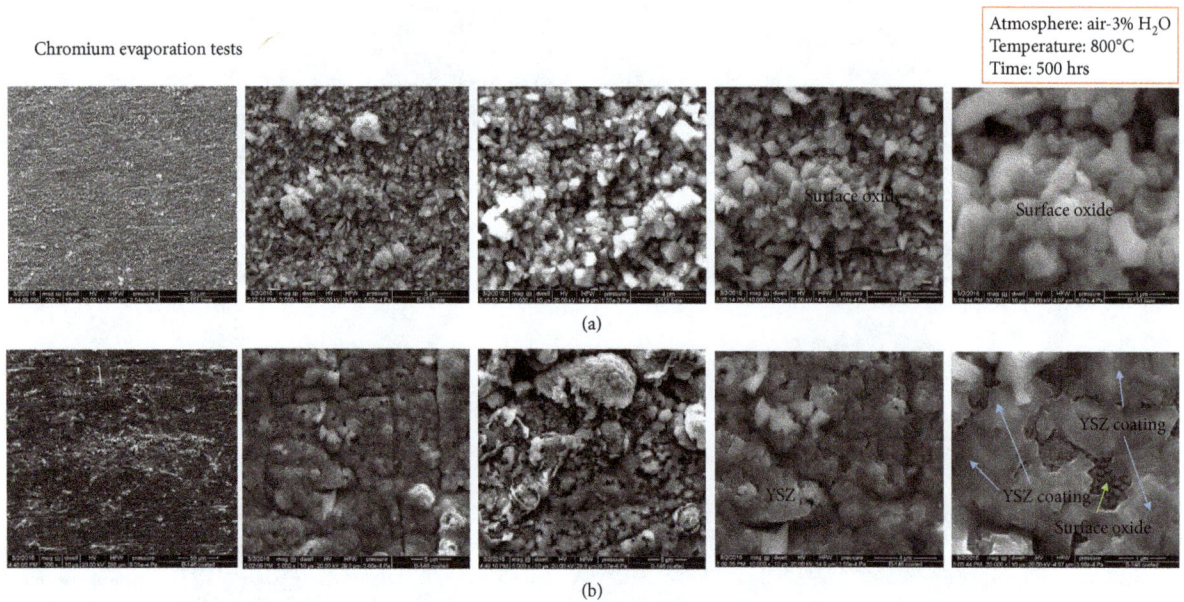

FIGURE 23: Cr evaporation test: SEM images of (a) uncoated and (b) YSZ-coated 430SS substrates (1-inch × 1-inch × 0.020-inch) shown at various magnifications following 500 h at 800°C in air/3% H2O atmosphere.

a surface reaction layer (possible spinel formation) under the YSZ layer.

(vi) The transformation of an apparently dense layer to a nanoporous layer can be explained in several different ways that could be clarified in future work:

   (a) Resolution of SEM for nanopores

   (b) Free volume within the surface state and vitreous state is eliminated by thermal treatment

   (c) Material leaving the YSZ coating, as appears to occur for Y

   (d) A combination of the above

(vii) Literature reports are consistent in recognizing the following:

   (a) Porous structures result from 800°C firings. Diffusion only becomes an effective transport mechanism above 800°C, where time improves densification.

   (b) Coatings containing YSZ are adherent and homogeneous, with cracks that can be reduced by optimizing the alkoxide sol content.

(viii) Some of these issues can be addressed by sequential dip-fire cycles. As the number of layers is increased, the thickness and surface roughness both increase.

(ix) The thickness of the thermally grown oxide (apparently a Mn-Cr-O spinel) is thinner with a YSZ coating (~1.5 μm) than for uncoated samples (~3.5 μm).

(x) Even a relatively thin (100 nm–1 μm) YSZ coating reduced the Cr evolution by ~51% (500 h at 800°C)

compared with an uncoated sample of the same substrate material (430SS). This suggests that with increase in coating thickness and a denser surface coverage, significant reduction to complete elimination of Cr evaporation can be achieved with the YSZ coatings.

(xi) The intensity of the cracks on the YSZ coatings decreased with the increase of the deposition cycles both with and without the addition of the nanopowder.

These initial results have provided insight about where future efforts should be focused.

4. Conclusions

We demonstrated feasibility through dip coating and firing samples of 430SS substrates in various sizes and shapes. Samples were tested by thermal cycling/soak and Cr-species evaporation. A ~51% reduction in Cr-species evaporation was measured for the YSZ-coated substrate compared to an uncoated substrate after 500 h at 800°C. Metallographic cross sections show that the intermetallic layer over the 430SS surface for the BOPSeal-coated surface is about 1/3 the thickness for the non-BOPSeal-coated surface. Results also indicate that the intensity of the cracks on the YSZ coatings decreased with increasing layers, with and without the addition of the nanopowder. We will pursue increasing the number of layers with and without additions of YSZ nanopowder to increase the coating coverage protecting the stainless steel surface. The result is envisioned to substantially decrease/eliminate Cr evaporation from BOP components for SOFCs. Optimized coatings would support protective barrier layers for (1) SOFC

BOP components, (2) bioreactors, and (3) seals in YSZ electrolytes for manifolds used in space applications.

Conflicts of Interest

There are no conflicts of interest. The sol-gel formulations discussed in this paper are proprietary.

Acknowledgments

Mr. Steven R. Markovich, Program Manager, U.S. Department of Energy, National Energy Technology Laboratory, Pittsburgh, is acknowledged for technical discussions during the course of the program. This material is based on the work supported by the U.S. Department of Energy (Office of Science and Office of Fossil Energy (Award Number: DE-SC0013879)).

References

[1] M. Stanislowski, J. Froitzheim, L. Niewolak et al., "Reduction of chromium vaporization from SOFC interconnectors by highly effective coatings," *Journal of Power Sources*, vol. 164, no. 2, pp. 578–589, 2007.

[2] H. Kurokawa, C. P. Jacobson, L. C. DeJonghe, and S. J. Visco, "Chromium vaporization of bare and of coated iron-chromium alloys at 1073 K," *Solid State Ionics*, vol. 178, no. 3-4, pp. 287–296, 2007.

[3] S. Chevalier, C. Valot, G. Bonnet, J. C. Colson, and J. P. Larpin, "The reactive element effect on thermally grown chromia scale residual stress," *Materials Science and Engineering: A*, vol. 343, no. 1-2, pp. 257–264, 2003.

[4] Z. Yang, G.-G. Xia, C.-M. Wang et al., "Investigation of iron–chromium–niobium–titanium ferritic stainless steel for solid oxide fuel cell interconnect applications," *Journal of Power Sources*, vol. 183, no. 2, pp. 660–667, 2008.

[5] J. Froitzheim and J.-E. Svensson, "Multifunctional nanocoatings for SOFC interconnects," *ECS Transactions*, vol. 35, no. 1, pp. 2503–2508, 2011.

[6] U. B. Pal and S. C. Singhal, "Electrochemical vapor deposition of yttria-stabilized zirconia films," *Journal of The Electrochemical Society*, vol. 137, no. 9, pp. 2937–2941, 1990.

[7] F. Smeacetto, M. Salvo, L. C. Ajitdoss et al., "Yttria-stabilized zirconia thin film electrolyte produced by RF sputtering for solid oxide fuel cell applications," *Materials Letters*, vol. 64, no. 22, pp. 2450–2453, 2010.

[8] D. Perednis, O. Wilhelm, S. E. Pratsinis, and L. J. Gauckler, "Morphology and deposition of thin yttria stabilized zirconia films using spray pyrolysis," *Thin Solid Films*, vol. 474, no. 1-2, pp. 84–95, 2005.

[9] J. H. Shim, C. C. Chao, H. Huang, and F. B. Prinz, "Atomic layer deposition of yttria-stabilized zirconia for solid oxide fuel cells," *Chemistry of Materials*, vol. 19, no. 15, pp. 3850–3854, 2007.

[10] Y. Xie, *Solution-based synthesis and processing of nanocrystalline zirconium diborides-based composites*, Ph.D. thesis, Georgia Institute of Technology, Atlanta, GA, USA, 2008.

[11] G. Zhang, Z. Rasheva, J. Karger-Kocsis, and T. Burkhart, "Synergetic role of nanoparticles and micro-scale short carbon fibers on the mechanical profiles of epoxy resin," *Express Polymer Letters*, vol. 5, no. 10, pp. 859–872, 2011.

[12] C. J. Brinker, "Dip coating," in *Chemical Solution Deposition of Functional Oxide Thin Films*, Schneller, Ed., p. 237, Springer-Verla, Wien, Europe, 2013, https://www.unm.edu/~solgel/PublicationsPDF/2013/BrinkerDipCoating2013.pdf.

[13] H. Schmidt and M. Mennig, "Wet coating technologies for glass," in *The Sol-Gel Gateway*, Institut für Neue Materialien, Saarbrücken, Europe, 2000, http://www.solgel.com/articles/nov00/mennig.htm.

[14] I. Strawbridge and P. F. James, "The factors affecting the thickness of sol-gel silica coatings prepared by dipping," *Journal of Non-Crystalline Solids*, vol. 86, no. 3, pp. 381–393, 1986.

[15] H. C. Mayer and R. Krechetnikov, "Landau-Levich flow visualization: revealing the flow topology responsible for the film thickening phenomena," *Physics of Fluids*, vol. 24, no. 5, article 052103, 2012.

[16] T. Barnardo, *Time resolved anomalous small angle X-ray scattering of the sol-gel process*, Ph.D. thesis, Aberystwyth University, Wales, UK, 2010.

[17] S. Heiroth, R. Frison, J. L. M. Rupp et al., "Crystallization and grain growth characteristics of yttria-stabilized thin films grown by pulsed laser deposition," *Solid State Ionics*, vol. 191, no. 1, pp. 12–23, 2011.

Permissions

List of Contributors

Roman Koleňák and Igor Kostolný
Faculty of Materials Science and Technology in Trnava, Slovak University of Technology in Bratislava, Paulínska 16, 917 24 Trnava, Slovakia

Yi Deng
School of Materials Science and Engineering, Sichuan University, Chengdu 610065, China
School of Chemical Engineering, Sichuan University, Chengdu 610065, China
Department of Mechanical Engineering, University of Hong Kong, Hong Kong 999077, China

Yang Zhou, Xiuyuan Shi, Kewei Zhang, Ping Zhang and Weizhong Yang
School of Materials Science and Engineering, Sichuan University, Chengdu 610065, China

Yuanyi Yang
Department of Materials Engineering, Sichuan College of Architectural Technology, Deyang 618000, China

Y. Y. Wang, B. Li, Y. L. Yu and P. S. Tang
Department of Material Chemistry, Huzhou University, Huzhou 313000, China

Catherine Billotte, Edith Roland Fotsing and Edu Ruiz
NSERC-Safran Chair on 3D Composites for Aerospace, Department of Mechanical Engineering, École Polytechnique de Montréal, Centre-Ville Station, Montréal, QC, Canada H3C 3A7

Sylvain Marinel, Nicolas Renaut, Rodolphe Macaigne and Guillaume Riquet
Laboratoire de Cristallographie et Sciences des Matériaux, Normandie Univ, ENSICAEN, UNICAEN, CNRS, CRISMAT, 14000 Caen, France

Etienne Savary
Laboratoire de Cristallographie et Sciences des Matériaux, Normandie Univ, ENSICAEN, UNICAEN, CNRS, CRISMAT, 14000 Caen, France
Université de Valenciennes et du Hainaut-Cambrésis, Boulevard du Général de Gaulle, 59600 Maubeuge, France

Christophe Coureau
SOLCERA, ZI n°1 rue de l'industrie, 27000 Evreux, France

Thibault Gadeyne and David Guillet
SAIREM, 12 Porte du Grand Lyon, 01702 Neyron, France

V. G. Karayannis, A. N. Baklavaridis and A. E. Domopoulou
Department of Environmental Engineering, Technological Education Institute of Western Macedonia, Kila, 50100 Kozani, Greece

A. K. Moutsatsou and E. L. Katsika
School of Chemical Engineering, National Technical University of Athens, Zografou Campus, 15773 Athens, Greece

Ling-yun Han, Yong-chun Shu, Yang Liu, Lu Gao, Qing-tong Wang and Xiao-na Zhang
The Key Laboratory of Weak Light Nonlinear Photonics, Ministry of Education, Nankai University, Tianjin 300457, China

Karolina Beer–Lech, Anna Skic and Krzysztof Gołacki
Department of Mechanical Engineering and Automation, Faculty of Production Engineering, University of Life Sciences in Lublin, Głęboka Street 28, 20-612 Lublin, Poland

Krzysztof Pałka and Barbara Surowska
Department of Materials Engineering, Faculty of Mechanical Engineering, Lublin University of Technology, Nadbystrzycka Street 36, 20-618 Lublin, Poland

Wei Si, Hua-Shen Xu, Ming Sun and Wei-Yi Zhang
School of Materials Science and Engineering, Dalian Jiaotong University, Dalian 116028, China

Chao Ding
Dalian Environmental Monitoring Center, Dalian 116023, China

Mohammed Sabah Ali
Department of Mechanical and Manufacturing Engineering, Faculty of Engineering, Universiti Putra Malaysia, 43400 Serdang, Selangor, Malaysia
Department of Agriculture Machinery & Equipment Engineering Techniques, Technical College, Al-Mussaib, Iraq

M. A. Azmah Hanim
Department of Mechanical and Manufacturing Engineering, Faculty of Engineering, Universiti Putra Malaysia, 43400 Serdang, Selangor, Malaysia
Laboratory of Biocomposite Technology, Institute of Tropical Forestry and Forest Products, Universiti Putra Malaysia, 43400 Serdang, Selangor, Malaysia

S. M. Tahir and C. N. A. Jaafar
Department of Mechanical and Manufacturing Engineering, Faculty of Engineering, Universiti Putra Malaysia, 43400 Serdang, Selangor, Malaysia

Norkhairunnisa Mazlan
Laboratory of Biocomposite Technology, Institute of Tropical Forestry and Forest Products, Universiti Putra Malaysia, 43400 Serdang, Selangor, Malaysia
Department of Aerospace Engineering, Faculty of Engineering, Universiti Putra Malaysia, 43400 Serdang, Selangor, Malaysia

Khamirul Amin Matori
Department of Physics, Faculty of Science, Universiti Putra Malaysia, 43400 Serdang, Selangor, Malaysia

Bronis Baw Psiuk, Anna Gerle and Andrzej Uliwa
Refractory Materials Division, Institute of Ceramics and Building Materials, Toszecka 99, 44-100 Gliwice, Poland

MaBgorzata Osadnik
Institute of Non-Ferrous Metals, Sowińskiego 5, 44-100 Gliwice, Poland

Túlio R. N. Porto, Wanderley F. A. Júnior, Antonio G. B. De Lima, Wanderson M. P. B. De Lima and Hallyson G. G. M. Lima
Department of Mechanical Engineering, Federal University of Campina Grande, Campina Grande 58429-900, Brazil

Yan Jie Guo, Fei Lu, Lei Zhang, He Lei Dong, Qiu Lin Tan and Ji Jun Xiong
Key Laboratory of Instrumentation Science and Dynamic Measurement, Ministry of Education, North University of China, Tai Yuan 030051, China
Science and Technology on Electronic Test and Measurement Laboratory, North University of China, Tai Yuan 030051, China

Lei Zheng, Chen Zhang, Xianglong Dong, Shitian Zhao, Weidong Wu, Zhi Cui and Yaoqing Ren
School of Mechanical Engineering, Yancheng Institute of Technology, No. 1 Middle Road, Xiwang Avenue, Tinghu District, Yancheng 224051, China

Ji-Woon Lee, Changhyun Jin and Soong-Keun Hyun
Department of Materials Science and Engineering, Inha University, Incheon 22212, Republic of Korea

Soon-Jig Hong
Division of Advanced Materials Engineering, Kongju National University, Kongju 32588, Republic of Korea

Chengxi Hu and Wujun Wang
Faculty of Science, Xi'an Aeronautical University, No. 259, West 2nd Ring, Xi'an 710077, China

Yuan Liu
School of Physics and Information Technology, Shaanxi Normal University, Xi'an 710062, China

Bo Yang
Surface and Interface Science Laboratory, RIKEN, 2-1 Hirosawa, Wako-shi, Saitama 351-0198, Japan

Wichita Kayaphan and Pornsuda Bomlai
1Department of Materials Science and Technology, Faculty of Science, Prince of Songkla University, Hat Yai, Songkhla 90112, Thailand
Center of Excellence in Nanotechnology for Energy (CENE), Prince of Songkla University, Hat Yai, Songkhla 90112, Thailand

Xue-Yu Tao
School of Materials Science and Engineering, China University of Mining and Technology, Xuzhou, Jiangsu 221116, China
Jiangsu Province Engineering Laboratory of High Efficient Energy Storage Technology and Equipments, China University of Mining and Technology, Xuzhou, Jiangsu 221116, China
Xuzhou City Key Laboratory of High Efficient Energy Storage Technology and Equipments, China University of Mining and Technology, Xuzhou, Jiangsu 221116, China

Jie Ma, Rui-Lin Hou, Xiang-Zhu Song, Lin Guo, Shi-Xiang Zhou, Li-Tong Guo, Zhang-Sheng Liu, He-Liang Fan and Ya-Bo Zhu
School of Materials Science and Engineering, China University of Mining and Technology, Xuzhou, Jiangsu 221116, China

D. Eliche-Quesada
Department of Chemical, Environmental and Materials Engineering, Advanced Polytechnic School of Jáen, University of Jáen, Campus Las Lagunillas s/n, 23071 Jáen, Spain

M. A. Felipe-Sesé
Department of Chemical, Environmental and Materials Engineering, Advanced Polytechnic School of Jáen, University of Jáen, Campus Las Lagunillas s/n, 23071 Jáen, Spain
International University of La Rioja, Avenida La Paz, No. 137, Logrõno, 26002 La Rioja, Spain

A. Infantes-Molina
Department of Inorganic Chemistry, Crystallography and Mineralogy, Affiliate Unit of the ICP-CSIC, Faculty of Sciences, University of Málaga, Campus de Teatinos, 29071 Málaga, Spain

Chee Sun Lee, Sidek Hj. Ab Aziz and Halimah Mohamed Kamari
Department of Physics, Faculty of Science, Universiti Putra Malaysia (UPM), 43400 Serdang, Selangor, Malaysia

Khamirul Amin Matori and Mohd Hafiz Mohd Zaid
Department of Physics, Faculty of Science, Universiti Putra Malaysia (UPM), 43400 Serdang, Selangor, Malaysia
Materials Synthesis and Characterization Laboratory, Institute of Advanced Technology, Universiti Putra Malaysia (UPM), 43400 Serdang, Selangor, Malaysia

Ismayadi Ismail
Materials Synthesis and Characterization Laboratory, Institute of Advanced Technology, Universiti Putra Malaysia (UPM), 43400 Serdang, Selangor, Malaysia

Keqiang Zhang, Lu Zhang and Rujie He
Institute of Advanced Structure Technology, Beijing Institute of Technology, Beijing 100081, China

Kaiyu Wang and Kai Wei
State Key Laboratory of Advanced Design and Manufacturing for Vehicle Body, Hunan University, Changsha 410082, China

Bing Zhang
Bristol Composites Institute (ACCIS), University of Bristol, Queen's Building, University Walk, Bristol BS8 1TR, UK

Raymond L. Winter and Uma Sampathkumaran
InnoSense LLC, Torrance 90505, USA

Prabhakar Singh
Department of Materials Science & Engineering, University of Connecticut, Storrs, CT, USA

Mark K. King Jr. and Manoj K. Mahapatra
Department of Materials Science and Engineering, University of Alabama at Birmingham, Birmingham, Alabama 35294, USA

Index

www.ingramcontent.com/pod-product-compliance
Lightning Source LLC
Chambersburg PA
CBHW080640200326

41458CB00013B/4694